THE FERNS OF
BRITAIN AND IRELAND

THE FERNS OF BRITAIN AND IRELAND

C. N. PAGE

Royal Botanic Garden, Edinburgh, and
Department of Biological Sciences,
University of Exeter

SECOND EDITION

CAMBRIDGE
UNIVERSITY PRESS

PUBLISHED BY THE PRESS SYNDICATE OF THE UNIVERSITY OF CAMBRIDGE
The Pitt Building, Trumpington Street, Cambridge CB2 1RP,
United Kingdom

CAMBRIDGE UNIVERSITY PRESS
The Edinburgh Building, Cambridge CB2 2RU, United Kingdom
40 West 20th Street, New York, NY 10011-4211, USA
10 Stamford Road, Oakleigh, Melbourne 3166, Australia

First published 1982
Second edition 1997

Printed in the United Kingdom at the University Press, Cambridge

Typeset in 9½/13 pt Times

A catalogue record for this book is available from the British Library

Library of Congress Cataloguing in Publication data

Page, C. N. (Christopher Nigel), 1942–
  The ferns of Britain and Ireland / C. N. Page. – 2nd ed.
    p. cm.
  Includes bibliographical references (p.   ) and index.
  ISBN 0 521 58380 2 (hardback). – ISBN 0 521 58658 5 (pbk.)
    1. Ferns – British Isles.   2. Pteridophyta – British Isles.
I. Title.
QK527.P26   1997
587'.3'0941 – dc20   96–38838   CIP

ISBN 0 521 58380 2 hardback
ISBN 0 521 58658 5 paperback

This book is dedicated to all societies, organisations and individuals in Britain and Ireland, who, by their observations and recordings, help to increase our knowledge of the native wildlife and promote its wellbeing for the future.

Contents

Author's preface to the second edition

Pteridophytes, the ferns and fern-allies, are one of the oldest land-plant groups on earth. Their evolution has been an integral part of the main evolutionary processes that have shaped the environments on this planet. Pteridophytes have existed for an at least three-fold longer period of time than have the flowering plants, and much yet remains to be learned about the specialised ecological strategies that have enabled pteridophytes to survive and produce the diversity of species that we see today.

A mere fifteen years have elapsed since the first edition of this book went to press. But during this time, a number of notable changes and additions to our knowledge of the British and Irish pteridophyte flora have taken place, whilst additional information on many known species has further accumulated. At the time of writing, both versions of the first edition of this book have been out of print for more than five years, and, under the stimulus of Cambridge University Press, and especially the encouragement of Dr Maria Murphy and Mrs Sandi Irvine, to whom I remain grateful, much of this new information, which includes field and biological information as well as taxonomic advances, has been incorporated into this second edition.

Aims of this edition have thus been the amplification of existing descriptions, diagnoses and information at field and distributional level, as well as improving the quality and diversity and accuracy of illustrations. In doing so, I have tried to maintain the style of the text throughout to be easy to read and hopefully understandable to the amateur and beginner, as well as to the serious naturalist and professional – maintaining, perhaps, the tradition of earlier British Victorian fern books such as those of Newman. Today, my publisher calls it 'user-friendly'!

Other pteridophyte books have appeared in this period, notably Jermy & Camus's *Illustrated Field Guide* (1991), Merryweather's *The Fern Guide* (1991), and, on the international front especially Tryon & Tryon's magnificent *Ferns and Allied Plants, with Special Reference to Tropical America* (1982). My own New Naturalist book: *Ferns. Their Habitats in the Landscape of Britain and Ireland* (Collins, 1988) has, I hope, added to our knowledge of the habitats and ecology of native pteridophytes across Britain and Ireland, while other specific aspects have been explored in Dyer & Page's (eds.) *Biology of Pteridophytes* (1985), Camus' (ed.) *The History of British Pteridology* (1991) and Ide, Jermy & Paul's (eds.) *Fern Horticulture: Past, Present and Future Perspectives* (1992).

Particular technical advances that have been made since the first edition of this book was completed include the discovery of spore banks in ferns and our increasing knowledge of gametophyte structure of native species, of pteridophyte life-cycles and of the existence of independent gametophytes in *Trichomanes* in Britain and Ireland. There is a greater knowledge of the genetics, biology, taxonomy and environmental threats posed by Bracken and its control, the discovery of the way in which the megaspores of *Selaginella selaginoides* are ballistically shed, the role that fern spores may play in maintaining whole fern populations, and the use of pteridophytes as indicators of habitat stability as well as of specific landscape change in the British Isles.

We now have a better historic understanding of the fascinating intergeneric hybrid ×*Asplenophyllitis microdon*, as well as of other rare fern hybrids in Guernsey, previously known and now refound. Particular taxonomic advances have been made in the recognition and delimitation of subspecies especially in *Pteridium*, *Dryopteris affinis*, and, in the *Asplenium trichomanes* group, greater understanding (though probably not complete taxonomic resolution) of the *Cystopteris fragilis/dickieana* aggregate, the *Asplenium adiantum-nigrum* group, and amongst the *Diphasiastrum* clubmosses. Considerably greater knowledge has also been gained of the occurrence of natural wild hybrids in *Dryopteris* and in *Equisetum*, and in the latter, several new uniquely British and Irish hybrids have been found and described in combinations that at first seemed so unlikely that they had never been suspected to exist. One entire new fern species, *Pteridium pinetorum*, has also been added to the native flora, and new ploidy levels have been detected at least locally in native material of *Pteridium* and *Isoetes*. Current advances are still being made in other groups, such as in *Athyrium flexile*, gametophyte morphology and taxonomic recognition, the role which soil spore banks might play in restoration ecology, and in recording pteridophyte distributions for a proposed future *Atlas*. Our insular Atlantic pteridophyte flora, unique in Europe as a whole, continues to reveal many new surprises, and no doubt still holds many others!

Additionally, the advent of molecular genetic analysis of ferns is beginning to throw some additional light on the taxonomy and phylogenetic inter-relationships of various pteridophyte taxa, though questions, in my mind, surround the quality of much of this burgeoning data, especially when the data are of a negative type. Further, although such data may give additional leads in overall comparison of phylogenetic affinities (which often seem more reliable at higher than at lower taxonomic levels), molecular evidence alone tells us virtually nothing about what the expression of the genes means in relation to the habitats, ecology, environment or biology of the taxa concerned, and how these relate to its evolutionary success. There are, at last, the beginnings of the realisation that it is vitally important that molecular data be interpreted *not in arrogant isolation from traditional (and especially field) taxonomy and biology, but as an additional and usefully independent unit of it. Such interpretation must always be used in close association with all other available data in drawing sensible and balanced phylogenetic, biological and evolutionary conclusions.* For the morphological structure and behavioural biology of a fern, as a living entity, and careful observations that can be made upon this, remain more sensitive discriminators of its underlying whole genetic make-up than anything yet achieved in a test tube.

In the field, subspecies and hybrids are always ecologically intriguing and part of the process of evolution in our flora, and I have tried to indicate something of this in the field notes, as appropriate. It is also my belief that the ecology of pteridophytes is complex but potentially enormously rewarding to study. For as a result of often unusual requirements and environmental sensitivities, there is high and usually little-appreciated potential in many pteridophyte taxa, in my view, as exacting natural bioindicators of long-term environmental change.

Many of the newer advances in the study of pteridology in these islands have been made through the combination of amateur enthusiasm and professional dedication to research, which the study of ferns and fern-allies has always enjoyed in these islands. Such additions in knowledge are a necessary part of any healthy science. In pteridology in Britain and Ireland, they remain both a thermometer for, and a reflection of, its active research state. The appearance of this edition is thus perhaps a reflection both of the demand for and of the great interest in the significance of this ancient group of plants. Identification of the plants of this group is now seen not merely as an end in itself, but more especially as a beginning of wider awareness of knowledge of the taxonomic status, distribution and ecology of individual species and subspecific groups as a sound basis for establishment of their future conservation.

Last, but not least, it is a pleasure to acknowledge with gratitude the help received from the Directors and staff of the Royal Botanic Garden, Edinburgh, and the Natural History Museum, London, for access to herbarium material in their charge. Permission to reproduce copyright material is also gratefully acknowledged.

C. N. Page
Royal Botanic Garden, Edinburgh,
and Department of Biological Sciences,
University of Exeter

Author's preface and acknowledgements to the first edition

Interest in ferns and allied plants in Britain and Ireland has always been strong, and much fundamental work into the taxonomy of species worldwide has been accomplished here. But, not surprisingly, the native species have also attracted considerable attention. The combination of the relatively small size of the native pteridophyte flora (just over 100 known species and hybrids) and the research attention that has been focused upon it, have made it today, without doubt, the most closely studied fern and fern-ally flora in the world.

The relatively small size of this flora provides, indeed, a strong, positive advantage. It is small enough for the amateur to tackle relatively easily, whilst most parts of the group – representative of a broad cross-section of the Pteridophyta as a whole – can be found in almost all parts of these islands. Even beginners can thus relatively quickly become familiar with the identification of a fairly large part of the native pteridophyte flora, no matter where they are. Encouraged by this, they may well then begin to take an active interest in some of its more specialised aspects, to which they may well be able to contribute useful and original observations in due course. The continuing success of the British Pteridological Society, founded in 1891 following the height of the Victorian 'fern craze', provides ample testimony to the persisting interest, both amateur and professional, horticultural and scientific, in almost all aspects of the native pteridophyte species.

Today there is undoubtedly also a rapidly growing interest in these plants from other quarters: from professionals in other areas of study, from conservationists, planners, amateur botanists, naturalists in many fields and students at all levels. It is essential, therefore, that there be available an account whereby the known species can be identified easily and accurately, summarising existing information from scientific progress into a single comparative account by a professional scientist. This in turn, can help to set a fresh base-line from which further studies can begin.

Such an account, enabling all the known ferns and fern-allies in Britain and Ireland to be critically identified, is now long overdue. This book sets out to try to fill this gap, using new methods of illustration and a new approach to description. The illustrations are prepared directly from authenticated frond material. They attempt to show not only the typical adult form of the plant with a high degree of accuracy, but also its variation, wherever this is sufficiently significant, including the appearance of the juvenile stages. In the descriptions, I have begun each with a brief synopsis of the main features of each plant, and then through the main text have tried to emphasise primarily what each looks like in its native habitat, rather than just as a dried herbarium specimen. There are good reasons for doing this. The most immediate is that ferns and fern-allies are usually more easily identified in the field than when pressed. The second is that, on conservation grounds, I want to encourage the reader to take this book to the plant in the field, and not to assume he or she necessarily has to bring the plant to the book.

Mere identification is not the sole aim of this book. The wealth and depth of research and observation that have been brought to bear by both professional and amateur pteridologists have laid valuable foundations, answering much about the 'what' (taxonomy) and the 'where' (distribution) of native pteridophytes. What I have tried to begin to do here is to take the topic further than is traditional in most field guides, by beginning to ask questions also about 'how' and 'why' in relation to the same fundamental issues – what do we know of the ecology and biology of the plants concerned, and what can we infer from these points about topics for future research and for species conservation?

Today, ferns and their allies still form an important part of our flora. But with modern developments in agriculture, we are in danger of losing much of their wealth before adequate account has yet been taken of it. Everywhere, woodlands and copses are disappearing, along with hedgerows and wetlands of all types. Streamsides and coastlines are becoming changed or polluted; old pastures and marginal land are being ploughed up. Pollution and the widespread use of agricultural herbicides have already had far-reaching effects in lowland Britain. Even in remote upland refuges, ferns are frequently the victims of indiscriminate grazing and moorland burning, and are now threatened with broadcast aerial spraying with pteridocidal chemicals.

Yet ferns are plants with a fascinating ecology – indeed, we are only just beginning to appreciate its intricacies. They are ecological escapists, growing where they do mainly because their habitats are ones where flowering plants do not succeed well. To do so, they have evolved a range of extraordinary ecological strategies. Such an ecology puts many of them in an unusually delicate state of ecological balance, which can be so easily swayed in the direction of their large-scale elimination. The subtleties of their ecology, however, put many also in the position of having considerable, though generally overlooked, potential as natural biological indicators.

In preparing this account, I have thus been particularly conscious of this conservation aspect throughout. But although I have been almost solely responsible for its writing and illustrating, and am thus ultimately responsible for any shortcomings it may have, it would not have been possible without the generous help in many ways of very many colleagues, both professional, amateur and student. Many have helped to clarify my own ideas through criticism of various sections of the text or discussion of particular points in the field, the latter in many parts of these islands and over many years. Much of the consensus of their views and judgements has been combined into the final account. If I have failed anywhere to reflect their views fully, it is either through oversight on my part or because I have had to act as final arbiter on matters of what to include. I hope that such shortcomings will be few, but in such a rapidly developing subject, I shall always welcome any corrections to be made to it.

Amongst the numerous colleagues from discussion with whom the very existence of this book has resulted, I would like to record my thanks to Dr J. Cullen and D. M. Henderson of the Royal Botanic Garden, Edinburgh, for their encouragement to undertake it from the outset. A. C. Jermy and J. A. Crabbe of the British Museum (Natural History) did much also to stress the need for it and stimulate my approach to it, whilst Dr T. G. Walker of the University of Newcastle-upon-Tyne has, for very many years, been a long-forbearing source of inspiration and pteridological guidance. Over many years too, the discussion and encouragement in the field of J. W. Dyce, from an amateur point of view, and of Miss M. A. Barker, from a student point of view, have gone a very long way towards moulding its content into the type of book that has finally emerged.

Of the various specialists with whom I have had the benefit of discussing specific areas, I would like to acknowledge the help of A. Douglas (climatology), Dr R. E. C. Ferreira (geology & soils), Dr Mary Gibby and C. R. Fraser-Jenkins (*Dryopteris*), A. C. Jermy (*Isoetes*, nomenclature and numerous technical points), R. H. Roberts (*Polypodium*), Dr Molly Shivas (Mrs T. G. Walker) (cytology and *Polypodium*), Dr Anne Sleep (*Asplenium*, *×Asplenophyllitis* and *Polystichum*), Dr Rosalind A. H. Smith (conservation) and Dr T. G. Walker (cytology and various technical points).

In addition, the book would not have been possible without the help, at many different times and in many different places, of more than a hundred other local botanists throughout Scotland, England, Wales and Ireland, who have accompanied me on field excursions in their local areas, and from whose observation and local geographic knowledge I have learned very much. To my own university and Field Studies Council course students over many years, I am also grateful for posing many of the questions to which I hope I have now given satisfactory answers, with the wisdom of hindsight, in this book. In addition, Miss Sue Patterson kindly prepared most of the plant habit sketches.

A very important contribution towards the illustrative content was made by my brother, A. C. Page by devising the technique, using a specially set up Rank-Xerox 9400 xerographic copier in prototype form. I would like to record here my gratitude to Rank-Xerox Ltd, Mitcheldean, Gloucestershire, for their kindness in allowing me access to this machine for this unusual purpose, whilst at an engineering development stage. I am indebted, too, to the Kindrogan Field Centre, Perthshire, and the Botany Department, University of Newcastle-upon-Tyne, for access to photographic darkroom facilities in their charge, and to Martin Walters of Cambridge University Press and to Dr Audrey O. Smith for their help in overseeing the publication of the final product.

Lastly, I am grateful to my family not only for putting up with my numerous absences to either the field or the photographic darkroom over many years, but also for tolerating my incessant typing when at home.

C. N. Page
Edinburgh, June 1981

Foreword to the
first edition

Over the last 25 to 30 years botany has been passing through a phase in which it has seemed at times to be in danger of becoming the most neglected of the natural sciences. Yet, throughout this period, popular interest has not only flourished but, if anything, increased, while the oldest branch of the subject – taxonomy – has developed and matured as an astonishingly synthetic activity. This has been achieved through the quiet, yet effective, assimilation and transmutation of data derived from newer, often supposedly more dynamic, aspects of biology as experimental morphogenesis, cytology, genetics, ecology, biochemistry and ecology, biochemistry and biometry, within the existing framework of classical morphology, anatomy and distributional data.

So the best modern taxonomic monographs have to be not merely major synthetic achievements but to be so expressed that their information is readily accessible. Only then are they likely to invite a reader to further observation and study and so bring to amateur or professional alike the deep and satisfying joy of further discoveries, major or miniscule.

It can be said straight away that this new and fascinating monograph on British and Irish Ferns and their allies meets all these criteria. I have had the pleasure and privilege of knowing the author since his student days and, from his first paper almost twenty years ago, I have been aware of, and increasingly impressed by, his enthusiasm and profound knowledge of these plants, attributes which have now matured in this splendid work. Up-to-date in every respect, readable, it is obvious that it is based on intimate personal acquaintance and investigation of ferns whether in the field, experimental garden, laboratory, herbarium or library. Comprehensive yet challenging, it enables the amateur and professional botanist alike to recognise the native and introduced brakes, clubmosses and horsetails and then leads on, implicitly in the sections on 'Variation' and 'Field notes' after each description and, more explicitly, in the concluding chapters to display and illuminate the innumerable problems that still await study by field or laboratory botanists alike.

Ferns have been recognised as an integral part of natural vegetation for over a thousand years in Britain : 'brake' (or bracken) is derived from Old Norse or Middle English and in Victorian times ferns were the object of passionately devoted collectors and horticulturalists. They have been objects of scientific study since Theophrastus. Perhaps it is not wholly inappropriate also to recall that much of the Modern World's material wealth and prosperity has been derived from the use of fossil fuels. Surely the debt we all owe to the Coal Measure ancestors of our modern pteridophytes can best be repaid by ensuring that their modern successors are conserved as part of the natural flora and so bring increasing pleasure and interest to future generations of fern watchers! I believe that the interest which I confidently expect to be engendered in the Ferns and their allies by this book will promote such a case. That, I know, would most please the author.

John Burnett
Principal & Vice-Chancellor
Old College
University of Edinburgh
April 1982

Introduction

What are pteridophytes?

Botanically, the ferns, together with equally ancient plants that share the same special life-cycle, belong to the group Pteridophyta*. All members of Pteridophyta have a life-cycle that includes two distinctive and free-living parts. The first of these is typically large, long lived, and contains an internal water-transporting *vascular system*. This is the *sporophyte* generation – the generation of the plants which is usually seen in the field. As implied by its name, it gives rise to spores (usually in great abundance), and these germinate and grow into a separate generation of the life-cycle, the *gametophyte* generation, which is typically small, short lived and non-vascular. This generation is seldom noticed in the wild. Its plants are typically more or less flat plates of tissue with roots, looking rather like liverworts, and this thalloid form of growth is called the *prothallus*. It is, however, the sexual generation of the life-cycle, and the prothallus gives rise to female *archegonia* (containing the egg cells) and male *antheridia* (containing the male gametes). The male gametes are free-swimming, motile organisms, looking like animal gametes, underlining the very primitive nature of pteridophytes compared with flowering plants. The motile male gametes, of course, need water in which to swim (a thin film will do), and this is one of the factors that confine the pteridophytes mostly to a life in places which are damp at least at certain times of the year. Their primitive affinities, combined with their obvious continuing ecological success in appropriate habitats worldwide, is one of the principal features which make the pteridophytes of considerable botanical and evolutionary interest.

Geographic area covered

Britain and Ireland share the great majority of pteridophyte species in common, and hence are here taken together for botanical purposes as a convenient cohesive unit. The area covered in this book includes the whole of Britain and Ireland and their many outlying islands, from the Channel Islands in the south to the Orkneys and Shetlands in the north.

Significance of the British and Irish pteridophytes in relation to Europe

The insular position of Britain and Ireland gives this region several features of unique interest in western Europe. The prevailing warm, moist, westerly winds bring an ameliorating effect on temperature extremes and a high and frequent precipitation and humidity, which combine to produce an equable oceanic climate. Geologically, rock types are diverse, resulting in an

---

* 'Pteridophyta' (spelt with a capital 'P') is the formal name for the group, which can be more informally referred to as the 'pteridophytes' (small 'p').

1

equally diverse topography of lowlands and uplands. The pteridophyte flora contains a unique phytogeographic assemblage of plants of widely differing taxonomic and geographic affinities and distributions outside the British Isles, but which in these islands may grow unusually close together, giving opportunities for interspecific hybrids to arise in unusual combinations. Further, islands are always of special biological significance through their history of isolation and their often limited floras, whose members have consequently adapted to occupy unusually wide ecological ranges. In combination, these features serve to make the pteridophyte flora of Britain and Ireland one of special botanical significance in Europe as a whole.

Layout of the book

In this book, all known native pteridophyte species and hybrids in Britain and Ireland are described and illustrated, and their ecology discussed.

Preliminary sections include keys (pp. 9–18), discussion of the botanical subdivisions of Britain and Ireland used in recording (pp. 19–24), altitudinal distributions of species (pp. 25–28) and environmental maps (pp. 29–46).

The body of the book (pp. 47–520) deals separately with each species and hybrid. The fern-allies (Clubmosses, Quillworts and Horsetails) complete the species entries at the end. Hybrids are included at the end of the species entries for each genus, in alphabetical order of hybrid name.

This order has been selected to simplify location of particular entries without constant recourse to the Index.

Content of the entries

For ease of comparison, each entry is divided into subheadings as follows.

'Preliminary recognition' attempts to give a brief summary of the principal features of the plant, answering the questions most often asked at the beginning of a field excursion: 'What sort of plant are we looking for?'

Plants are given generalised sizes. These correspond to approximate frond-length measurements as follows: 'very large', over/about 120 cm (*c.* 4 ft); 'large', about 60–120 cm (*c.* 2–4 ft); 'medium', about 30–60 cm (*c.* 1–2 ft); 'small', about 10–30 cm (*c.* 4–12 in); and 'very small', under about 10 cm (*c.* 4 in).

'Occurrence' outlines very generally the geography and habitats of the plant, helping to clarify whether it is a likely one to occur at the site in question.

'Identification' supplements in more detail the 'Preliminary recognition' section, giving details of the main generally useful and taxonomically distinctive features of the plant which can usually be seen in the field, either by the unaided eye or, at the most, by the use of a × 10 hand lens. Detailed description of frond form is kept to a minimum, except where this is not obvious from the illustrations or needs particular emphasis.

'Variation' gives details of how variable a species can be expected to be.

'Possible confusion' groups together other species or hybrids with which confusion seems possible, and differentiates them, with cross-references and comparisons where appropriate.

This should enable the reader to check other possibilities where necessary, and to establish more confidently that the plant in question is, or is not, the one described.

'Technical confirmation' provides a section aimed at the technically minded, with access to modest laboratory facilities, including a good quality microscope. The details given are those which would normally be invoked professionally to resolve the exact taxonomic affinities of the most questionable material. This information is most likely to be called into play in the case of difficult hybrids. The data given usually include the chromosome number of the plant, quoted as '$n =$' for the gametic number and '$2n =$' for the sporophytic number. Details of methods of chromosome counting have been thoroughly described by Professor Irene Manton in her book *Problems of Cytology and Evolution in the Pteridophyta* (Cambridge University Press, 1950), and the majority of those quoted here are counts originally made by her. Spore size is quoted usually only where this is of comparative, and hence diagnostic, value. All measurements given are in micrometres (microns, μm) and refer to the longest diameter of the spore in question, measured across its rounded body, ignoring the depth of surface ornamentation. In all cases, measurements have been established from native material, and where not quoted from other authorities, refer to measurements made by the author using dried spores mounted in a 50% (v/v) glycerine solution.

'Field notes' form a large part of the text for each entry, complementing the information already given under 'Occurrence', and in the sketch-map of species-distribution and calendar of the cycle of annual events, included for most species. It discusses habitats, attempting also, wherever possible, to suggest why that particular plant grows where it generally does, looks like it does, or behaves as it does. There are questions upon which I particularly hope that this book will further stimulate experimental ecological investigation. Finally, I have tried to note any changes in range which may be happening and to inject views on the conservation status of those plants for which this seems particularly significant.

Distribution maps Small inset maps with each species and hybrid show the range of each plant and the approximate density of its occurrence in the area covered by the book, as interpreted from field experience by the author. European distribution maps are included for those species of especially distinctive range on this scale.

Calendar of events The main events in the annual cycle are indicated for most species in the form of a small diagrammatic calendar. This shows:
- the period of expansion of the fronds (first bold dashed line)
- the period from complete expansion to spore maturity (first solid bold narrow line)
- the spore-shedding period (solid bold broad line)
- any period following the end of spore shedding when the fronds remain green (second solid bold narrow line)
- the period during which the fronds die down (second dash bold line)
- any period when fronds are not in evidence (white areas of calendar divided by light vertical dashed lines).

Thus a typical calendar might show:

Jan.	Feb.	Mar.	Apr.	May	Jun.	Jly	Aug.	Sep.	Oct.	Nov.	Dec.

| fronds not in evidence | fronds expanding | fronds expanded | fronds shedding spores | old fronds remaining green | old fronds dying | old fronds dead |

The diagrams are set on a grid of monthly divisions, showing each event to the nearest week. It will be appreciated that these calendars can be averages only. Extremely early or late seasons may well change the patterns by one or two weeks in any season. Species with wide geographic ranges will also vary in season between extreme localities and altitudes and, in such cases, appropriate compensating factors will need to be applied according to the reader's own experience of local conditions.

Silhouette illustrations For most species several silhouettes are given, showing a sample of the normal range of variation which can be expected. All have been made from native plant material. Different species have necessarily had to be reduced by differing amounts to fit them into the book page size, although, on any one page, all are given at the same scale, and the degree of reduction, if any, can be judged from the scale-line given. Although actual plant material has been used to make the illustrations, it should be noted that for all the rarer or more local species, only existing herbarium material has been used.

Photographs are included where they help to illustrate features of the species that are best displayed in living material.

Fern names

In recent years, the scientific names of some ferns have changed many times. Such name changes cause much confusion to the amateur and professional alike, and are the unfortunate result of the application of the rules that govern botanical nomenclature, which were devised in an attempt to produce a stable situation! Those names by which a plant has been formerly known are called 'synonyms'. As a result of name changes, many native pteridophytes have several of these. For simplicity of use, only the most commonly and recently used synonyms are listed beneath the headings in this account. *The remainder are in the index.*

In addition to the names of species, it has become a traditional and generally useful practice to give binomial names to wild hybrid pteridophytes, usually indicating their parentage only at the first mention, or in discussion. Such hybrid binomial names are distinguished from those of species by the insertion of a raised multiplication sign ' × ' before the specific epithet (or before the generic one in the case of an intergeneric hybrid).

Common names are also important and, in ferns, are often more stable and more widely used colloquially than the Latin ones. All entries here are thus also given common names. Those which have been felt to be the most generally useful and least ambiguous have been selected. Hybrids have also been named.

Taxonomic concepts

Most of the taxonomic categories used here follow the latest studies on each group, and the majority of names equate with those used in the *Atlas of Ferns of the British Isles* (Jermy, Arnold, Farrell & Perring, 1978).

I have differed from the above, however, where I have felt this is useful and that there is taxonomic justification for doing so. I have thus continued to maintain *Ceterach* and *Phyllitis* as separate genera from *Asplenium*, on the basis of their distinctive morphology as well as traditional usage. Similarly, for the few hybrids between *Asplenium* and *Phyllitis*, I have maintained the intergeneric notation of [×]*Asplenophyllitis* as a hybrid genus. Two cases where I have maintained former species names instead of more recent ones, because of either taxonomic or nomenclatural doubts about the validity of proposed changes, include retaining the name *Thelypteris palustris* for the Marsh Fern and *Asplenium viride* for the Green Spleenwort.

Purpose of the book

Today, there seems to be a strong awakening of interest in the environment around us, and a growing awareness of the need to conserve carefully the wildlife we still have.

The vegetation of which the pteridophytes are an intimate part forms a valuable resource for research and experiment as well as for teaching and recreation, and a resource which, at least potentially, is naturally recurring. The continued existence of any such resource depends on its effective care and management, founded on detailed knowledge of the plant species concerned. This book is intended to provide the basis of such a guide for the native Pteridophyta, in the hope that it will also stimulate further interest and research into those areas which remain least known – particularly the detailed ecology and biology of the species and hybrids concerned. If this account does anything towards promotion of sound conservation of pteridophytes through stimulation of interest, observation, vigilance and research, as well as providing a means of their identification, it will have achieved its primary aim.

Component parts
of a fern

In the pteridophyte architecture diagram, nomenclature of parts is as follows: *a*, a simple, entire frond; *b*, a once (1 ×) pinnately divided fern frond; *c*, a twice (2 ×) pinnately divided fern frond; *d*, a three (3 ×) times pinnately divided fern frond; *e*, a simplified fern frond showing nomenclature of main parts; *f*, close-up details of a single sorus; *g*, a simplified fern plant; *h*, a simplified clubmoss; *i*, a simplified horsetail.

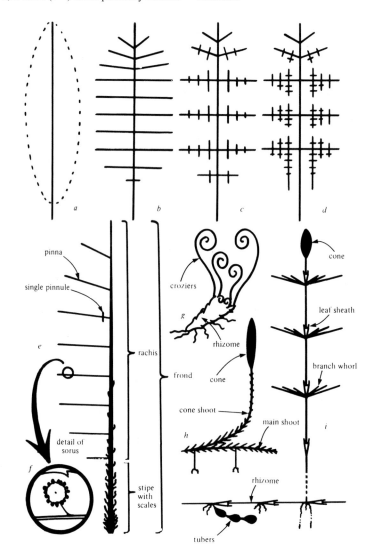

Glossary of terms

acidic	a habitat poor in basic minerals, giving a low pH reading
adnate	attached by the whole width of the base
annulus	the row of specialised cells with thickened walls surrounding each sporangium in ferns
antherozoid	the motile male gamete of the fern
apiculum	pl. **apicula**, the pointed apex of the cones of the evergreen horsetails
apogamous	a peculiar breeding system in which prothalli give rise directly to sporophyte plants without fertilisation
auricles	rounded, 'ear-like', lobes forming part of the frond
basic	a habitat rich in basic minerals, giving high pH readings
blade	the leafy part of the frond
bulbils	buds which break away from the adult sporophyte to form new, whole plants, as in *Huperzia selago*
chromosome	the thread-like bodies contained within every cell which carry the genetic make-up of the individual. The number of them is constant within species (or subspecies)
circinate	coiled in a flat spiral like a watch-spring, resembling the ornamentation at the head of a violin
cone	the specialised portion of the shoot of fern-allies which bears the sporangia, from which the spores are released
deltoid	shaped like a gothic arch, the sides more curved than straight
denticulate	with small teeth, usually with reference to the margin of the frond
dimorphic	having fronds or shoots of two distinct types (one of which is normally vegetative, the other fertile)
echinate	with long, hedgehog-like spines, as in the spores of *Cystopteris fragilis*
eutrophic	a habitat type (usually aquatic) rich in minerals
farinose	covered with whitish glands, giving the surface a mealy appearance, as the surface of the fronds of *Dryopteris submontana*
fetch	the distance on a straight line over which wind movement can be operative in inducing wave-action on a water surface, such as that of a large lake or the sea. Small expanses of water have insufficient fetches for large waves to be created, even by strong winds
flushed (ground)	sloping ground through which there is a steady percolation of water bringing mineral enrichment, especially of chemical bases
flushing (of fronds)	the sudden resumption of vigorous growth, with return of favourable growing conditions, usually in spring
frond	the aerial axes of the fern sporophyte
gametophyte	the minor generation of the fern life cycle which is non-vascular and liverwort-like, i.e. the prothallus
genome	a single, usually identifiable, set of chromosomes

habitat	the place where a plant occurs and grows, and which is often characteristic of each species
hyaline	translucent and whitish, as in the margins of the teeth of *Equisetum palustre*
indusium	pl. **indusia**, a protective flap or other device protecting the sorus, of a characteristic shape and structure in each genus
lamina	pl. **laminae**, the leafy tissue of the blade of a frond
lanceolate	lance shaped – having a narrow curved outline, broadest at or just below the middle and tapering increasingly towards both ends
morphology	the structure of a plant, its architecture
oligotrophic	a habitat type (usually aquatic) poor in minerals
orchreolae	sing. **orchreola**, the basal sheaths out of which the whorled branches arise in *Equisetum*
paraphyses	sterile hairs, usually amongst the sporangia such as in *Polypodium cambricum*
pellucid	translucent but coloured, as in the green quills of *Isoetes*
perispore	the outer coat of a spore, which is often ornamented
pinna	pl. **pinnae**, one of the first major divisions of a fern frond
pinnate	divided in a basically herringbone-like fashion, or like a feather
pinnule	one of the divisions of a pinna, i.e. a secondary division of the frond
prothallus	pl. **prothalli**, the liverwort-like flattened structure which is the gametophyte generation of the pteridophyte life-cycle, growing from the spore and bearing male (antheridia) and female (archegonia) organs
pubescent	finely and evenly downy, with short, soft hairs, as the surface of the stipe of *Phegopteris connectilis*
rachis	pl. **rachides**, the central 'midrib' of the frond
rugose-verrucose	with a wrinkled, warty appearance of ridges and low projections, as in the spore surface of *Cystopteris dickieana*
sorus	pl. **sori**, a group of sporangia together having a shape and position characteristic of a genus
spinose	bearing spines (usually few and widely spaced in ferns)
spinulose	bearing small spines (usually numerous and close together in ferns)
sporangia	sing. **sporangium**, structures containing the spores
sporophyte	the dominant generation of the fern life-cycle which is vascular and leafy, i.e. the fern plant proper
stipe	the 'stalk' of the frond
submarginal	set a short way in from the margin
triangular	in the shape of a triangle
tripartite	having three major parts, as in the frond of *Gymnocarpium dryopteris*
ultimate segment	an ultimate unit of division of a fern frond
undulate	with a wavy edge or surface
vallecular canals	the large canals situated beneath the furrows in the stem and rhizome of *Equisetum*

Chart keys
to the main
pteridophyte
groups

The following charts (1–6, pp. 10–14), used in order, should enable the reader to locate fairly quickly the main taxonomic group or groups to which any unknown native pteridophyte probably belongs.

Further identification should then be made directly by diagnostic comparison in the main text between the members of the group(s) indicated.

If the plant in question has some very distinctive features, the multi-access key which follows the charts (pp. 15–18) might prove of additional preliminary value.

This approach, it is hoped, will lead more rapidly to a diagnosis involving comparison of possible variation, as well as likely geography, seasonal state and ecology, than could be obtained within the mechanical limitations of a long, purely dichotomous, written key.

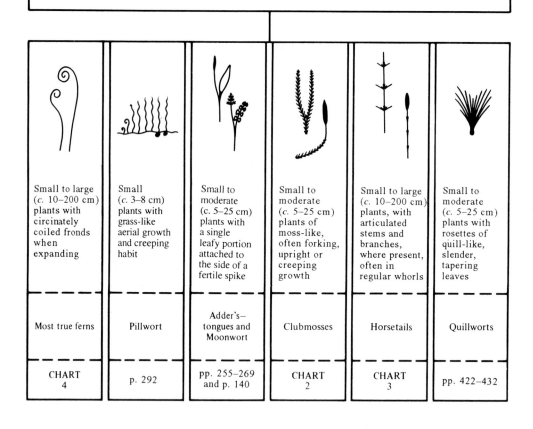

CHART 1

PTERIDOPHYTA

(Plants with a green, vascular generation which
dominates the life-cycle and produces spores)

Small to large (c. 10–200 cm) plants with circinately coiled fronds when expanding	Small (c. 3–8 cm) plants with grass-like aerial growth and creeping habit	Small to moderate (c. 5–25 cm) plants with a single leafy portion attached to the side of a fertile spike	Small to moderate (c. 5–25 cm) plants of moss-like, often forking, upright or creeping growth	Small to large (c. 10–200 cm) plants, with articulated stems and branches, where present, often in regular whorls	Small to moderate (c. 5–25 cm) plants with rosettes of quill-like, slender, tapering leaves
Most true ferns	Pillwort	Adder's—tongues and Moonwort	Clubmosses	Horsetails	Quillworts
CHART 4	p. 292	pp. 255–269 and p. 140	CHART 2	CHART 3	pp. 422–432

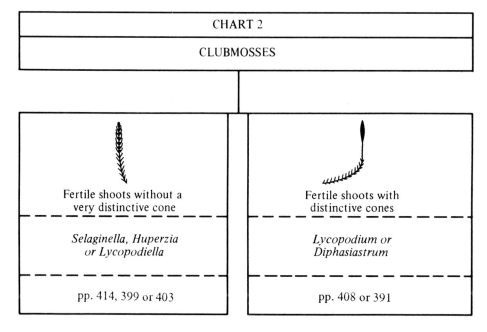

CHART 2

CLUBMOSSES

Fertile shoots without a very distinctive cone	Fertile shoots with distinctive cones
Selaginella, Huperzia or Lycopodiella	*Lycopodium or Diphasiastrum*
pp. 414, 399 or 403	pp. 408 or 391

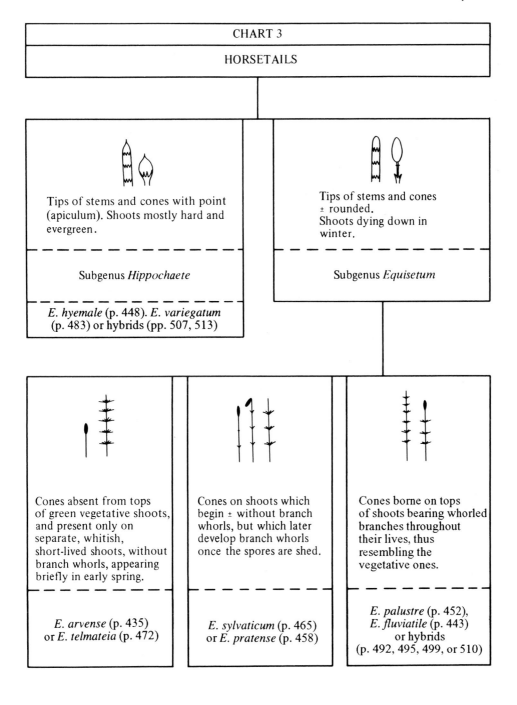

CHART 3

HORSETAILS

Tips of stems and cones with point (apiculum). Shoots mostly hard and evergreen.

Subgenus *Hippochaete*

E. hyemale (p. 448). *E. variegatum* (p. 483) or hybrids (pp. 507, 513)

Tips of stems and cones ± rounded.
Shoots dying down in winter.

Subgenus *Equisetum*

Cones absent from tops of green vegetative shoots, and present only on separate, whitish, short-lived shoots, without branch whorls, appearing briefly in early spring.

E. arvense (p. 435) or *E. telmateia* (p. 472)

Cones on shoots which begin ± without branch whorls, but which later develop branch whorls once the spores are shed.

E. sylvaticum (p. 465) or *E. pratense* (p. 458)

Cones borne on tops of shoots bearing whorled branches throughout their lives, thus resembling the vegetative ones.

E. palustre (p. 452), *E. fluviatile* (p. 443) or hybrids (p. 492, 495, 499, or 510)

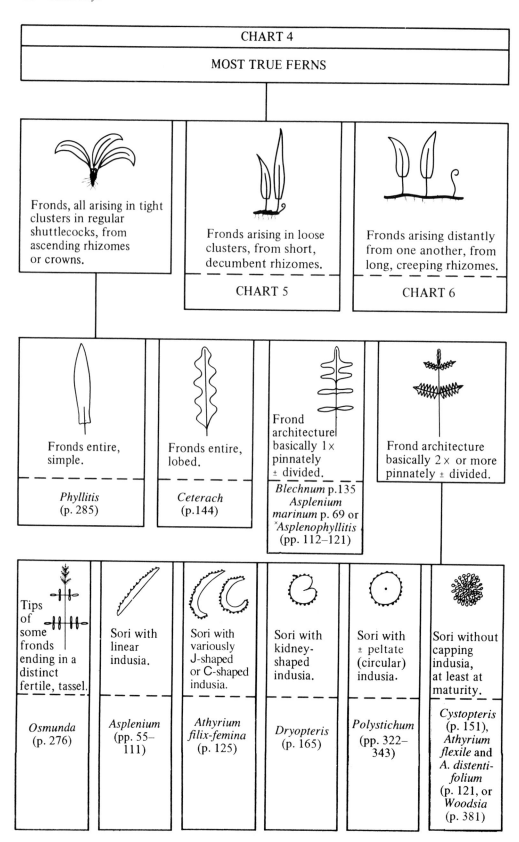

CHART 4

MOST TRUE FERNS

Fronds, all arising in tight clusters in regular shuttlecocks, from ascending rhizomes or crowns.

Fronds arising in loose clusters, from short, decumbent rhizomes.

CHART 5

Fronds arising distantly from one another, from long, creeping rhizomes.

CHART 6

Fronds entire, simple.

Phyllitis (p. 285)

Fronds entire, lobed.

Ceterach (p.144)

Frond architecture basically 1× pinnately ± divided.

Blechnum p.135 *Asplenium marinum* p. 69 or ×*Asplenophyllitis* (pp. 112–121)

Frond architecture basically 2× or more pinnately ± divided.

Tips of some fronds ending in a distinct fertile, tassel.

Osmunda (p. 276)

Sori with linear indusia.

Asplenium (pp. 55–111)

Sori with variously J-shaped or C-shaped indusia.

Athyrium filix-femina (p. 125)

Sori with kidney-shaped indusia.

Dryopteris (p. 165)

Sori with ± peltate (circular) indusia.

Polystichum (pp. 322–343)

Sori without capping indusia, at least at maturity.

Cystopteris (p. 151), *Athyrium flexile* and *A. distenti-folium* (p. 121, or *Woodsia* (p. 381)

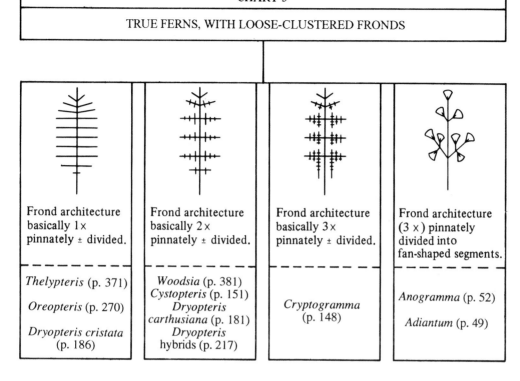

CHART 5

TRUE FERNS, WITH LOOSE-CLUSTERED FRONDS

Frond architecture basically 1× pinnately ± divided.	Frond architecture basically 2× pinnately ± divided.	Frond architecture basically 3× pinnately ± divided.	Frond architecture (3×) pinnately divided into fan-shaped segments.
Thelypteris (p. 371) *Oreopteris* (p. 270) *Dryopteris cristata* (p. 186)	*Woodsia* (p. 381) *Cystopteris* (p. 151) *Dryopteris carthusiana* (p. 181) *Dryopteris* hybrids (p. 217)	*Cryptogramma* (p. 148)	*Anogramma* (p. 52) *Adiantum* (p. 49)

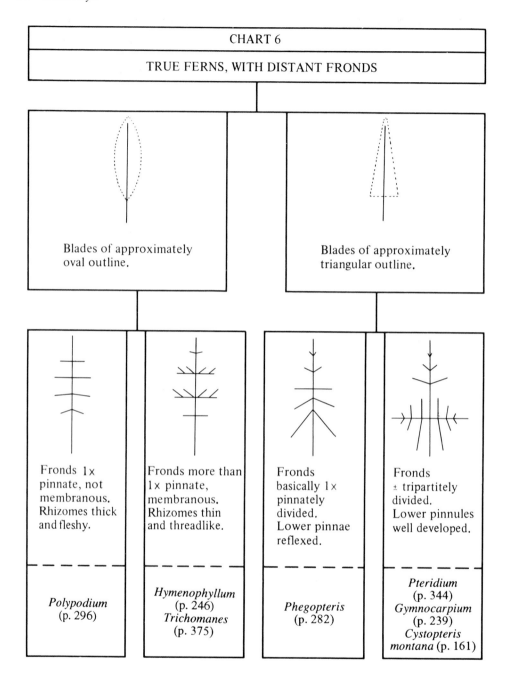

CHART 6

TRUE FERNS, WITH DISTANT FRONDS

Blades of approximately oval outline.

Blades of approximately triangular outline.

Fronds 1× pinnate, not membranous. Rhizomes thick and fleshy.

Polypodium (p. 296)

Fronds more than 1× pinnate, membranous. Rhizomes thin and threadlike.

Hymenophyllum (p. 246)
Trichomanes (p. 375)

Fronds basically 1× pinnately divided. Lower pinnae reflexed.

Phegopteris (p. 282)

Fronds ± tripartitely divided. Lower pinnules well developed.

Pteridium (p. 344)
Gymnocarpium (p. 239)
Cystopteris montana (p. 161)

Multi-access key to main pteridophyte groups

This list groups together those pteridophytes which share various conspicuous or distinctive features, often enabling them to be quickly singled out from the great majority of others. Inspection of the following groupings may thus help to reduce a larger number of possible identities to a smaller number of likely candidates.

Clubmosses with cones very distinct from the vegetative parts	*Diphasiastrum alpinum* *Diphasiastrum* ×*issleri* *Lycopodium annotinum* *Lycopodium clavatum*
Clubmosses with cones only semi-distinct from the vegetative parts	*Huperzia selago* *Lycopodiella inundata* *Selaginella selaginoides*
High montane clubmosses	*Diphasiastrum alpinum* *Huperzia selago* *Lycopodium annotinum* *Lycopodium clavatum* *Selaginella selaginoides*
Quillworts in lakes and tarns	*Isoetes echinospora* *Isoetes lacustris*
Quillworts in pools drying in summer	*Isoetes histrix*
Horsetails with evergreen aerial shoots	*Equisetum hyemale* *Equisetum variegatum* *Equisetum* ×*trachyodon*
Horsetails with winter-deciduous aerial shoots	*Equisetum arvense* *Equisetum fluviatile* *Equisetum palustre* *Equisetum pratense* *Equisetum sylvaticum* *Equisetum telmateia* *Equisetum* ×*bowmanii* *Equisetum* ×*dycei* *Equisetum* ×*font-queri* *Equisetum* ×*litorale* *Equisetum* ×*mildeanum* *Equisetum* ×*moorei* *Equisetum* ×*rothmaleri* *Equisetum* ×*wilmotii*

Horsetails with unbranched or very sparsely branched shoots	*Equisetum fluviatile* *Equisetum hyemale* *Equisetum palustre* *Equisetum variegatum* *Equisetum* ×*dycei* *Equisetum* ×*trachyodon*
Horsetails with very densely and regularly whorled branches	*Equisetum arvense* *Equisetum pratense* *Equisetum sylvaticum* *Equisetum telmateia*
Horsetails with regularly branched branches	*Equisetum sylvaticum* *Equisetum* ×*bowmanii*
Plants with a single fertile spike (bearing whorled leaves or a single leafy branch)	*Botrychium lunaria* *Equisetum arvense* (spring cone shoot) *Equisetum telmateia* (spring cone shoot) *Ophioglossum azoricum* *Ophioglossum lusitanicum* *Ophioglossum vulgatum*
Plants with ± fleshy aerial parts (shoots or fronds)	*Asplenium marinum* *Botrychium lunaria* *Equisetum arvense* (spring cone shoot) *Equisetum telmateia* (spring cone shoot) *Ophioglossum azoricum* *Ophioglossum lusitanicum* *Ophioglossum vulgatum*
Plants with a grass-like structure, but coiled when emerging	*Pilularia globulifera*
Ferns with a membranous lamina	*Hymenophyllum tunbrigense* *Hymenophyllum wilsonii* *Trichomanes speciosum*
Ferns with fronds ending in a fertile tassel	*Osmunda regalis*
Ferns with entire fronds	*Ophioglossum azoricum* *Ophioglossum lusitanicum* *Ophioglossum vulgatum* *Phyllitis scolopendrium*
Ferns with simply lobed or herringbone-like fronds	*Asplenium trichomanes* *Ceterach officinarum* *Blechnum spicant* *Polypodium australe* *Polypodium interjectum* *Polypodium vulgare* *Polypodium* ×*font-queri* *Polypodium* ×*mantoniae* *Polypodium* ×*shivasiae*
Ferns with rounded or fan-shaped segments	*Adiantum capillus-veneris* *Anogramma leptophylla* *Asplenium ruta-muraria*

		Asplenium trichomanes
		Asplenium viride
Ferns with forking fronds		*Asplenium septentrionale*

Ferns with a distinctive smell when frond surfaces are lightly bruised or are dried
{
new-mown hay: *Dryopteris aemula*
balsam: *Dryopteris submontana*
apple: *Gymnocarpium dryopteris*
lemon: *Oreopteris limbosperma*

Ferns with very dense, pale brown scales over back of frond — *Ceterach officinarum*

Ferns with very dense, pale brown scales among much of stipe
- *Dryopteris affinis*
- *Dryopteris submontana*
- *Dryopteris* ˣ*complexa*
- *Dryopteris* ˣ*remota*

Ferns with sparse, white scales on stipe and rachis, conspicuous at the crozier stage
- *Oreopteris limbosperma*
- *Phyllitis scolopendrium*
- all *Polystichum*
- ˣ*Asplenophyllitis microdon*

Ferns with edges of segments turning distinctly downwards (convex above) (especially in plants in exposed conditions)
- *Athyrium distentifolium*
- *Athyrium filix-femina*
- *Cystopteris fragilis*
- *Dryopteris dilatata*
- *Dryopteris filix-mas*
- *Gymnocarpium dryopteris*
- *Gymnocarpium robertianum*
- *Oreopteris limbosperma*
- *Phegopteris connectilis*
- *Pteridium aquilinum*
- *Pteridium pinetorum*

Ferns with edges of segments turning distinctly upwards
- *Dryopteris aemula*
- *Dryopteris oreades*
- *Dryopteris* ˣ*pseudoabbreviata*

Ferns with fronds usually less than 30 cm tall
- *Adiantum capillus-veneris*
- *Anogramma leptophylla*
- most *Asplenium*
- *Athyrium flexile*
- *Ceterach*
- *Cryptogramma crispa*
- all *Cystopteris*
- *Gymnocarpium dryopteris*
- most *Gymnocarpium robertianum*
- *Hymenophyllum tunbrigense*
- *Hymenophyllum wilsonii*
- *Phegopteris connectilis*
- most *Polypodium*
- most *Polystichum lonchitis*
- *Trichomanes speciosum*
- all *Woodsia*
- most ˣ*Asplenophyllitis*

Ferns with fronds often about 1 m tall, or more   *Athyrium filix-femina*
Dryopteris affinis
Dryopteris dilatata
Dryopteris filix-mas
Dryopteris ˣ*complexa*
Oreopteris limbosperma
Polystichum setiferum
Pteridium aquilinum

Botanical subdivisions of Britain and Ireland

For marshalling distributional data and for biological discussion of particular geographic areas, it is convenient to subdivide the area of Britain and Ireland into smaller, readily defined and widely recognised parts. Two systems are in use, each of which has its particular advantages.

The vice-county system

This is the older of the two systems, introduced by H. C. Watson in 1852. It has thus been in continual use for over 100 years and, until recently, most botanical distributional data were based upon it.

In this system, Britain and Ireland are divided into numbered vice-counties (usually abbreviated to 'v.c.'s'), of which there are 113 for England, Wales, Scotland and the Channel Islands, and 40 for Ireland. The vice-counties are based on present or former county divisions of Britain and Ireland. The smaller counties are left intact and larger ones subdivided, to provide units of approximately equal area, each of which has a name and number (see map, p. 24). The vice-county divisions are stable, widely recognised, and do not change with political reorganisations.

This system thus provides (1) conveniently sized units for collection and collation of local biological data, and (2) a system for mapping geographic distributions. Although much of the original value of the system for distributional mapping has now been superseded by the grid-square system (see below), the vice-county system still provides some advantages. It is a convenient basis for well-defined similar-sized areas on which to base local field studies, such as the production of local floras and local distribution maps. One practical result of this has been the establishment of a series of vice-county recorders, each responsible for the collation of distributional data within their vice-county areas.

The vice-county system also provides an overall geographic framework to which it is often convenient to refer when discussing and comparing ranges, in an easily understood and visualised form, which cannot be easily conveyed by continually referring to rather abstract grid co-ordinates.

The grid-square system

Britain, Ireland and the Channel Islands area are covered, for mapping purposes, by kilometre-based grid-square systems, to which any point can be referred by quoting its exact co-ordinates along horizontal and vertical axes. These grids are universally recognised for mapping purposes, enabling political, physical, geological, climatic, biological and other maps to be exactly compared.

Biologically, the grid-square system provides a more detailed basis for mapping the distribution of species than does the vice-county system, which it has almost completely replaced for this purpose. A 10-km square basis is conventionally taken as the base-unit of most biological maps, and the presence or absence of a species in any particular 10-km square generates a spot which is then computer plotted for that square. The whole of the ferns and their allies have already been critically mapped on this basis for Britain and Ireland (Jermy *et al.*, 1978), and in some areas, schemes for mapping on a more detailed basis locally have been considered. Such maps give an objective picture of the relative ranges of native species, and can be updated readily. They can also include different symbols to represent different data, such as past and present ranges, thus helping to show how these might be changing. The data on which these maps are based are collected by contributors and specialists, including very many amateur naturalists, all over Britain and Ireland.

The 100-km squares are each identified by paired letters of classification, though for computer recording purposes, their numerical equivalents (see map, p. 23) are preferred. The Irish grid is similar to that of England, Wales and Scotland, but has a different point of origin and is aligned a few degrees differently to the latter. The Irish grid uses a single letter notation for each of the 100-km squares. The Channel Islands are on a third, UTM (Universal Transverse Mercator), grid. There are no numerical equivalents of their grid letters, and the alphabetic notation has thus to be used in recording in these areas.

The grid squares for England, Wales and Scotland appear in large scale on Ordnance Survey maps of the 1 : 50 000 and other series. Those of the Irish Grid appear on the Ordnance Survey maps of Ireland of various scales, and those for the Channel Islands on Ordnance Survey maps of available scales for these areas. All three grids appear as 100-km squares on maps showing detailed climatological data in the *Climatological Memorandum* series issued by the English Meteorological Office, and on most geological maps published by the Geological Survey.

The Vice-county numbers and their corresponding Vice-counties

England and Wales

1. West Cornwall (with Scilly Isles)	21. Middlesex
2. East Cornwall	22. Berkshire
3. South Devon	23. Oxfordshire
4. North Devon	24. Buckinghamshire
5. South Somerset	25. East Suffolk
6. North Somerset	26. West Suffolk
7. North Wiltshire	27. East Norfolk
8. South Wiltshire	28. West Norfolk
9. Dorset	29. Cambridgeshire
10. Isle of Wight	30. Bedfordshire
11. South Hampshire	31. Huntingdonshire
12. North Hampshire	32. Northamptonshire
13. West Sussex	33. East Gloucestershire
14. East Sussex	34. West Gloucestershire
15. East Kent	35. Monmouthshire
16. West Kent	36. Herefordshire
17. Surrey	37. Worcestershire
18. South Essex	38. Warwickshire
19. North Essex	39. Staffordshire
20. Hertfordshire	40. Shropshire (Salop)

41. Glamorgan
42. Breconshire
43. Radnorshire
44. Carmarthenshire
45. Pembrokeshire
46. Cardiganshire
47. Montgomeryshire
48. Merionethshire
49. Caernarvonshire
50. Denbighshire
51. Flintshire
52. Anglesey
53. South Lincolnshire
54. North Lincolnshire
55. Leicestershire (with Rutland)
56. Nottinghamshire

57. Derbyshire
58. Cheshire
59. South Lancashire
60. West Lancashire
61. South-east Yorkshire
62. North-east Yorkshire
63. South-west Yorkshire
64. Mid-west Yorkshire
65. North-west Yorkshire
66. Durham
67. South Northumberland
68. North Northumberland (Cheviot)
69. Westmorland with N. Lancashire
70. Cumberland
71. Isle of Man

Scotland
72. Dumfriesshire
73. Kirkcudbrightshire
74. Wigtownshire
75. Ayrshire
76. Renfrewshire
77. Lanarkshire
78. Peeblesshire
79. Selkirkshire
80. Roxburghshire
81. Berwickshire
82. East Lothian (Haddington)
83. Midlothian (Edinburgh)
84. West Lothian (Linlithgow)
85. Fifeshire (with Kinross)
86. Stirlingshire
87. West Perthshire (with Clackmannan)
88. Mid Perthshire
89. East Perthshire
90. Angus (Forfar)
91. Kincardineshire
92. South Aberdeenshire

93. North Aberdeenshire
94. Banffshire
95. Moray (Elgin)
96. East Inverness-shire (with Nairn)
97. West Inverness-shire
98. Argyll Main
99. Dunbartonshire
100. Clyde Isles
101. Kintyre
102. South Ebudes
103. Mid Ebudes
104. North Ebudes
105. West Ross
106. East Ross
107. East Sutherland
108. West Sutherland
109. Caithness
110. Outer Hebrides
111. Orkney Islands
112. Shetland Islands (Zetland)

Channel Islands
113. Channel Islands

Ireland

H 1.	South Kerry	(113)	H 9.	Clare	(121)
H 2.	North Kerry	(114)	H 10.	North Tipperary	(122)
H 3.	West Cork	(115)	H 11.	Kilkenny	(123)
H 4.	Mid Cork	(116)	H 12.	Wexford	(124)
H 5.	East Cork	(117)	H 13.	Carlow	(125)
H 6.	Waterford	(118)	H 14.	Leix (Queen's County)	(126)
H 7.	South Tipperary	(119)	H 15.	South-east Galway	(127)
H 8.	Limerick	(120)	H 16.	West Galway	(128)

H 17. North-east Galway	(129)	H 29. Leitrim	(141)
H 18. Offaly (King's County)	(130)	H 30. Cavan	(142)
H 19. Kildare	(131)	H 31. Louth	(143)
H 20. Wicklow	(132)	H 32. Monaghan	(144)
H 21. Dublin	(133)	H 33. Fermanagh	(145)
H 22. Meath	(134)	H 34. East Donegal	(146)
H 23. West Meath	(135)	H 35. West Donegal	(147)
H 24. Longford	(136)	H 36. Tyrone	(148)
H 25. Roscommon	(137)	H 37. Armagh	(149)
H 26. East Mayo	(138)	H 38. Down	(150)
H 27. West Mayo	(139)	H 39. Antrim	(151)
H 28. Sligo	(140)	H 40. Londonderry	(152)

For Ireland, vice-county numbers are prefixed with an 'H' (for *Hibernia*), to distinguish them from the corresponding British numbers. (The numbers ('113' onwards) shown after in parentheses, are the equivalent Irish numbers which have been employed in fossil plant records, and follow Godwin (1956: 70), but '113' repeats that used for the Channel Islands).

Captions to Maps 1, 2 and 3 on pages 23, 24 and 25

Map 1. (p. 23) The 100-km squares of the British and Irish National Grids showing the conventional 100-km square reference letters and their numerical equivalents used for botanical recordings. (Reproduced by permission of the Ordnance Survey.)

Map 2. (p. 24) Vice-county boundaries in Britain and Ireland. (Redrawn from Perring & Walters (1976).)

Map 3. (p. 28) Simplified topographical map of Britain and Ireland, showing areas below sea level, lowlands 0–600 ft (*c.* 183 m), hill areas between 600 ft and 3000 ft (*c.* 914 m), and mountains above 3000 ft.

Map 1 23

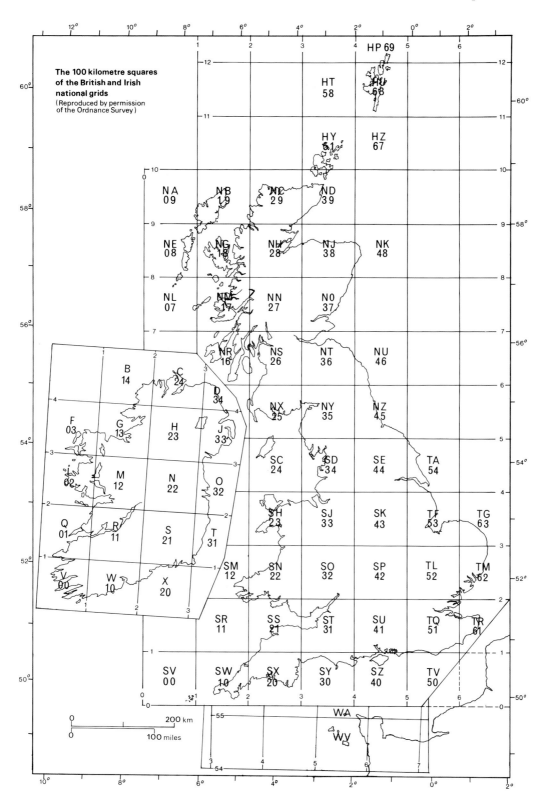

The 100 kilometre squares
of the British and Irish
national grids
(Reproduced by permission
of the Ordnance Survey)

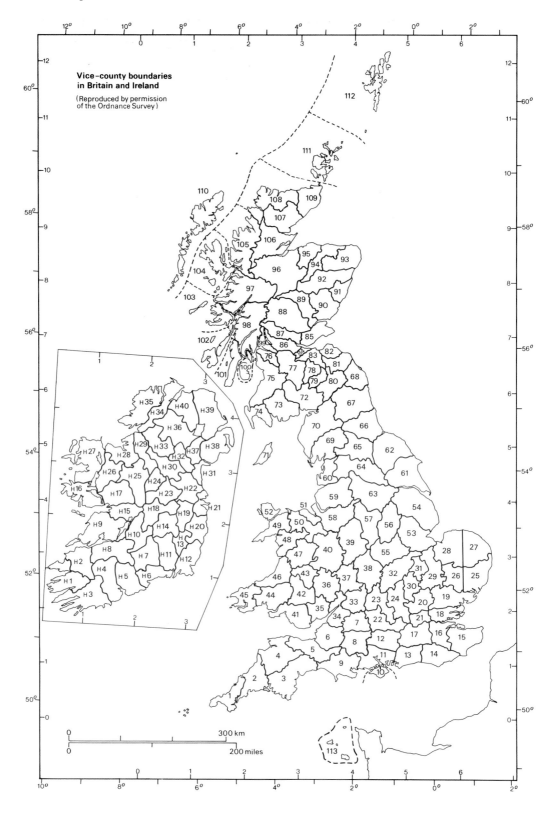

Vice-county boundaries
in Britain and Ireland
(Reproduced by permission
of the Ordnance Survey)

300 km

200 miles

Altitudinal distribution
of native pteridophytes

Although the highest altitudes in Britain and Ireland (just over *c*. 1340 m (4400 ft) on Ben Nevis in Scotland) are only about one third of the height of the highest Alpine peaks of mainland Europe, they are nevertheless sufficiently high for there to be an appreciable zonation of species with altitude.

The normal lapse rate (the rate at which the temperature falls with increase in altitude) produces a temperature reduction of about 1.1 deg.C for every 200 m rise of altitude (about 1 deg.F for every 300 ft). This ensures that maximum temperatures at higher altitudes are not as high as those at lower elevations, whilst minimum temperatures are also appreciably lower.

In Ireland, areas of land over *c*. 305 m (1000 ft) are not extensive, and lie mainly around the periphery of the country, whilst only scattered summits, and not large areas, exceed *c*. 760 m (2500 ft). In Britain, there is a general increase in altitude northward and westward, with most land over *c*. 305 m lying in Devon, Wales, through the Pennine chain of northern England, the English Lake District and much of Scotland. Appreciable amounts of land over *c*. 760 m are confined mainly to North Wales, the English Lake District and Scottish Highlands, and in the latter there is more land over this altitude than in the rest of Britain and Ireland together (Map 3, p. 28).

The distribution of the main mountain masses strongly influences the climate of these areas, and hence the distribution of habitats appropriate for more alpine species. In addition to the tendency for such species to descend to lower altitudes at more northerly latitudes, there is also a tendency for alpine species to descend to lower altitudes in a westerly direction, towards the more oceanic coasts. This is a result of the peculiarly 'oceanic' climate of these areas (see p. 41), where the lack of hot summer conditions enables alpines to survive in some places nearly at sea-level. Because such oceanic conditions influence, to a greater or lesser extent, the climate of the whole of Britain and Ireland, species more typical of much higher altitudes in continental Europe occur at generally lower elevations here. There is thus a considerable 'telescoping' of altitudinal ranges, with an unusually great range of overlap between high-altitude and low-altitude species.

The following chart (pp. 26–7) summarises the altitudinal ranges of most pteridophytes in Britain and Ireland, compiled in part from Salter (1928) and Wilson (1956) and in part from the author's own observations. Solid bold lines mark the main-range band of each plant, broken lines the more occasional occurrences and extremes.

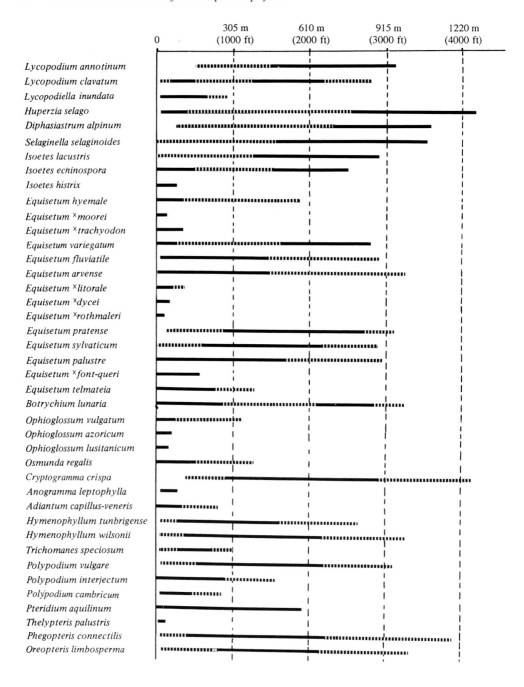

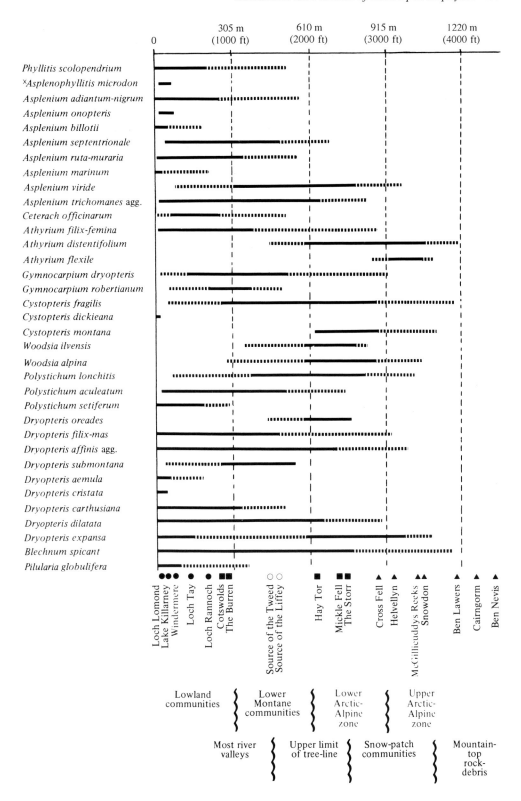

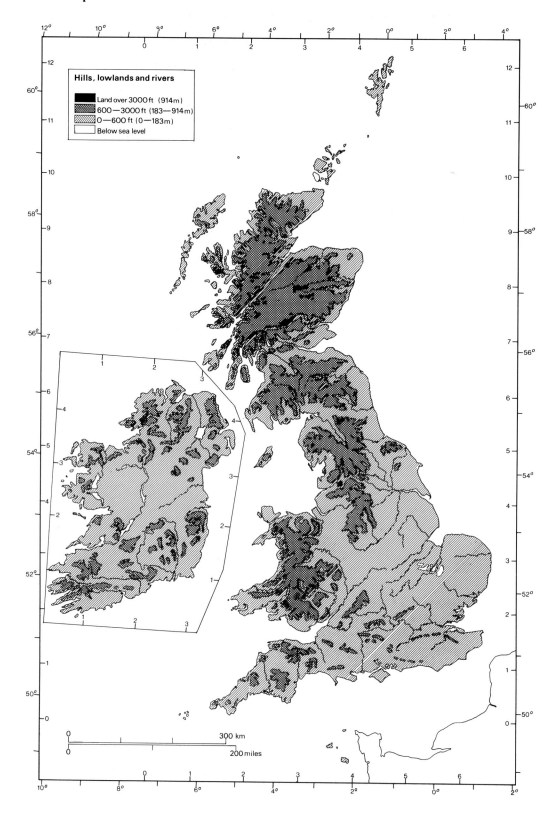

Hills, lowlands and rivers

Land over 3000 ft (914 m)
600—3000 ft (183—914 m)
0—600 ft (0—183 m)
Below sea level

300 km
200 miles

Environmental factors influencing native pteridophyte ranges

The overall ranges of native ferns in Britain and Ireland are determined by a complex interplay of many different factors, of which the most obvious are historic, topographic, edaphic (relating to rock and soil types) and climatic ones. It is very probable that each of these factors can be broken down into many separate components, and these components combine in different ways to influence the ranges of different species.

Although pteridophytes are probably influenced by many of the same general environmental factors that influence flowering plants, there are probably components of these factors to which ferns and fern-allies are especially susceptible. The exact factors responsible for delimitation of the native species ranges in every case await precise ecological field and laboratory experimentation. Some impression, however, of the significance of components can be gained by comparison of the ranges and the density of the occurrence of each species (see text entries) with maps of the distribution of environmental factors and some of their separate components.

A series of maps is thus included here showing the distribution of the main environmental components of historic, topographic, edaphic and climatic factors, that seem to the author to be the ones most likely to be of greatest significance in influencing native pteridophyte ranges.

The base-map used retains all parts of Britain and Ireland covered by this book (including Ireland, Orkney, Shetland and Channel Islands) in their correct relative geographic positions.

Map 4 (p. 31) Simplified geological map of Britain and Ireland, showing the distribution of principal outcrops of rock of differing periods and origins.
Map 5 (p. 32) The distribution of the extent of maximum southern ice limits in Britain and Ireland of the last (Weichselian) and previous pre-Weichselian (Gipping and Lowestoftian) glacial advances (the maximum Pleistocene glacial advances), their centres of ice accumulation and principal directions of ice flow, and the approximate position of the shore-line during the glacial maximum.
Map 6 (p. 33) The distribution of February mean daily minimum temperatures (°C) in Britain and Ireland, for the period 1941–70 for Britain and 1931–60 for Ireland. Readings are based on shade (or *screen*) temperatures, recorded from approximately 1.25 m (4 ft) above the ground, in a

Stevenson screen, over short grass turf. In order to prevent the map becoming as complex as a topographic one, individual station temperatures are reduced to those which they would be if they were all situated at mean sea-level. In practice, for any one station, there would be a temperature *decrease* of approximately 0.5 deg.C for each further 100 m above sea-level, colder than indicated on the map. Note the geographic extent of relatively high values over south-west Ireland, and their northward range along western coasts.
Map 7 (p. 34) The distribution of July mean daily maximum temperatures (°C) in Britain and Ireland, for the period 1941–70 for Britain and 1931–60 for Ireland. Readings are based on screen temperatures, as with those of February minima (q.v.), and are corrected to sea-level. In practice, there would be a temperature *decrease* below

those indicated on the map of about 0.7 deg.C for each 100 m rise in altitude above sea-level, increasing to a lapse rate of about 0.9 deg.C for every 100 m above 400 m on exposed hilltop sites. Note the geographic extent of relatively low values in western Scotland, and their southward range along western coasts.

Map 8 (p. 35) The distribution of mean annual accumulated temperatures in Britain and Ireland, in day-degrees F above 42.8°F (6°C), for the period 1881–1915.

Map 9 (p. 36) The distribution of winter (December–March) accumulated temperatures in day degrees F above 42.8°F (6°C), for the period 1881–1915.

Map 10 (p. 37) The distribution of average annual rainfall in Britain and Ireland for the period 1941–70.

Map 11 (p. 38) The distribution of the mean annual number of wet days in Britain and Ireland for the period 1951–60, based on the meteorological category of a 'wet day' as a period of 24 h in which 0.04 in. (1 mm) of rain is recorded.

Map 12 (p. 39) The distribution of areas of varying average summer potential water deficit in Britain and Ireland.

Map 13 (p. 40) The distribution in Britain of percentage area of standing water, recorded on the basis of hygrometric areas.

Map 14 (p. 41) The distribution of the range of average monthly temperature in Britain and Ireland for the period 1901–30. The isotherms link areas of comparable annual temperature ranges, with the thermically most-oceanic areas shaded darkest and the most-continental lightest. Such differences between mean monthly temperatures of

the warmest and coldest months are thought to provide a relatively good thermic indicator of oceanic versus continental conditions and appear most relevant to vegetation with a low canopy.

Map 15 (p. 42) The distribution of average annual means of relative humidity in Britain and Ireland at 1300 hours (1 p.m.) GMT each day, for the period 1921–35. Note that the isopleths are compiled from readings taken duing the (generally) least humid part of each day, and this is at a different point in daily cycle to those of temperature, with which they should thus not be directly compared. Compared with maximum and minimum temperature data the humidity readings for this diagram are recorded at relatively few (about 125) stations, and are thus subject to considerable interpolation.

Map 16 (p. 43) The distribution in Britain and Ireland of the average annual means of percentage daytime for which bright sunshine is obscured. The map shows the relative values of frequency of obscured bright sunshine, giving a qualitative assessment of the incidence of cloud-effect upon vegetation.

Map 17 (p. 44) The distribution of the mean annual number of days in Britain and Ireland with a minimum temperature of less than 0°C, for the period 1956–70.

Map 18 (p. 45) The distribution of the mean number of days in Britain and Ireland with snow lying at 0900 hours GMT, for the period 1941–70. It can be seen how closely the isopleths follow the contour lines.

Map 19 (p. 46) The distribution of annual arithmetic mean sulphur dioxide concentrations over Britain.

Acknowledgements are gratefully made for permission to use data on which maps are based to the Botanical Society of the British Isles, British Pteridological Society, Ordnance Survey, Institute of Geological Sciences. Meteorological Office, and Irish Meteorological Service. Crown copyright material is used with permission of the Controller, Her Majesty's Stationery Office.

Map 4 31

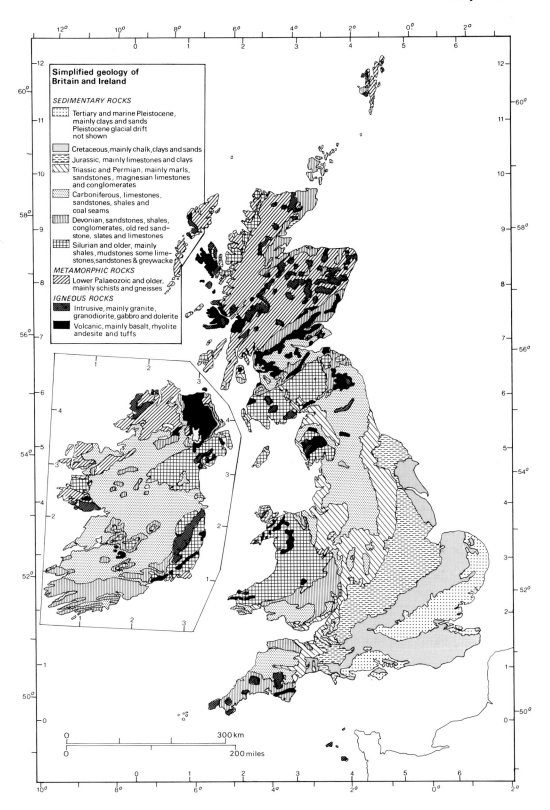

Simplified geology of
Britain and Ireland

SEDIMENTARY ROCKS

Tertiary and marine Pleistocene,
mainly clays and sands
Pleistocene glacial drift
not shown

Cretaceous, mainly chalk, clays and sands

Jurassic, mainly limestones and clays

Triassic and Permian, mainly marls,
sandstones, magnesian limestones
and conglomerates

Carboniferous, limestones,
sandstones, shales and
coal seams

Devonian, sandstones, shales,
conglomerates, old red sandstone, slates and limestones

Silurian and older, mainly
shales, mudstones some limestones, sandstones & greywacke

METAMORPHIC ROCKS

Lower Palaeozoic and older,
mainly schists and gneisses

IGNEOUS ROCKS

Intrusive, mainly granite,
granodiorite, gabbro and dolerite

Volcanic, mainly basalt, rhyolite
andesite and tuffs

300 km

200 miles

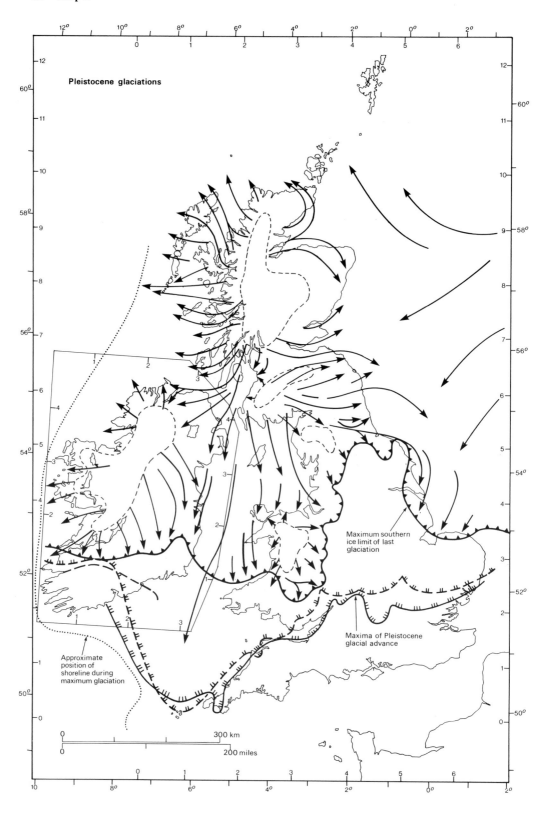

Pleistocene glaciations

Maximum southern
ice limit of last
glaciation

Maxima of Pleistocene
glacial advance

Approximate
position of
shoreline during
maximum glaciation

300 km

200 miles

Map 6 33

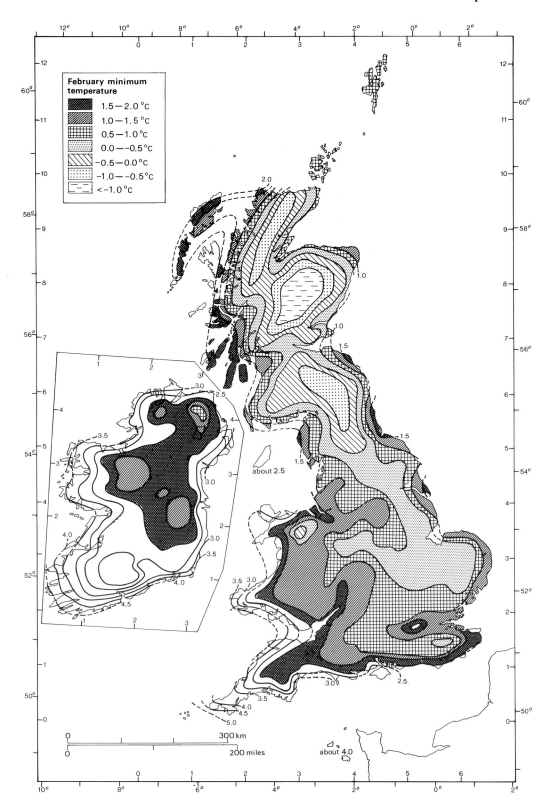

February minimum
temperature

1.5 — 2.0 °C
1.0 — 1.5 °C
0.5 — 1.0 °C
0.0 — −0.5 °C
−0.5 — 0.0 °C
−1.0 — −0.5 °C
< −1.0 °C

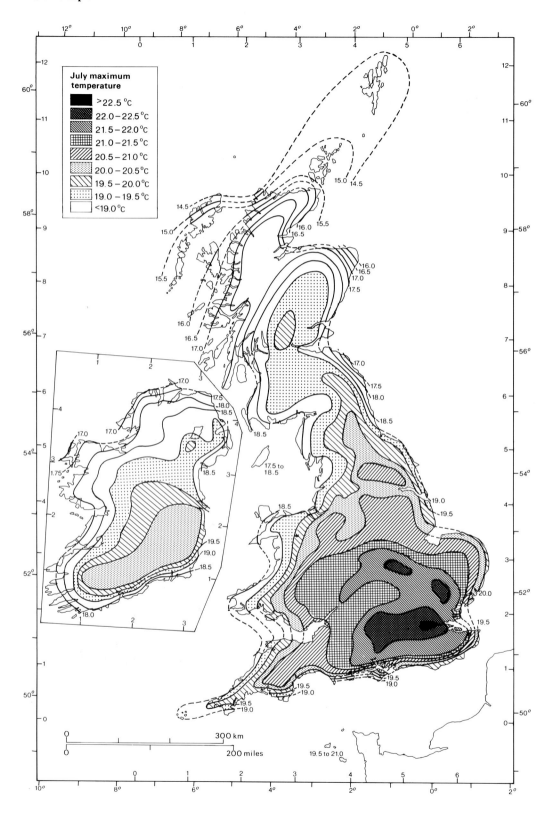

July maximum
temperature
>22.5 °C
22.0 – 22.5 °C
21.5 – 22.0 °C
21.0 – 21.5 °C
20.5 – 21.0 °C
20.0 – 20.5 °C
19.5 – 20.0 °C
19.0 – 19.5 °C
<19.0 °C

300 km
200 miles

Map 8   35

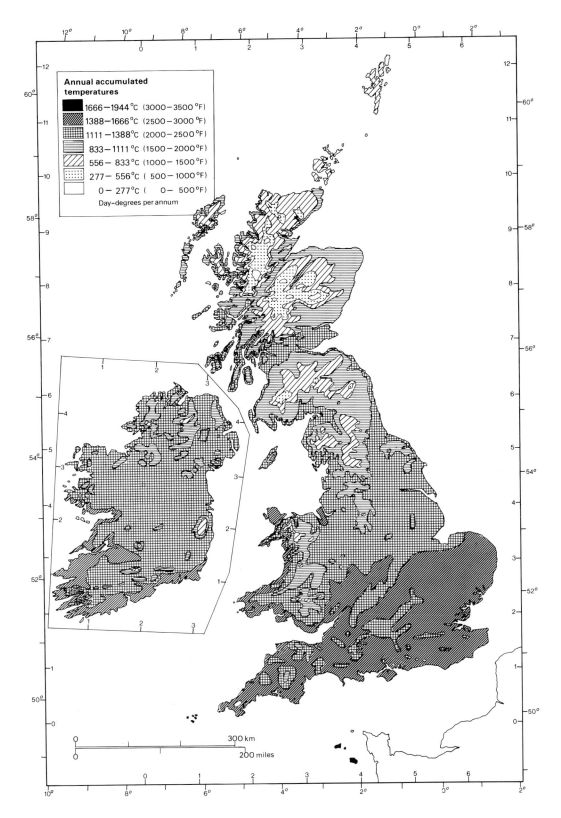

Annual accumulated
temperatures

■ 1666–1944 °C (3000–3500 °F)

1388–1666 °C (2500–3000 °F)

1111–1388 °C (2000–2500 °F)

833–1111 °C (1500–2000 °F)

556– 833 °C (1000–1500 °F)

277– 556 °C (500–1000 °F)

0– 277 °C (0– 500 °F)

Day-degrees per annum

300 km

200 miles

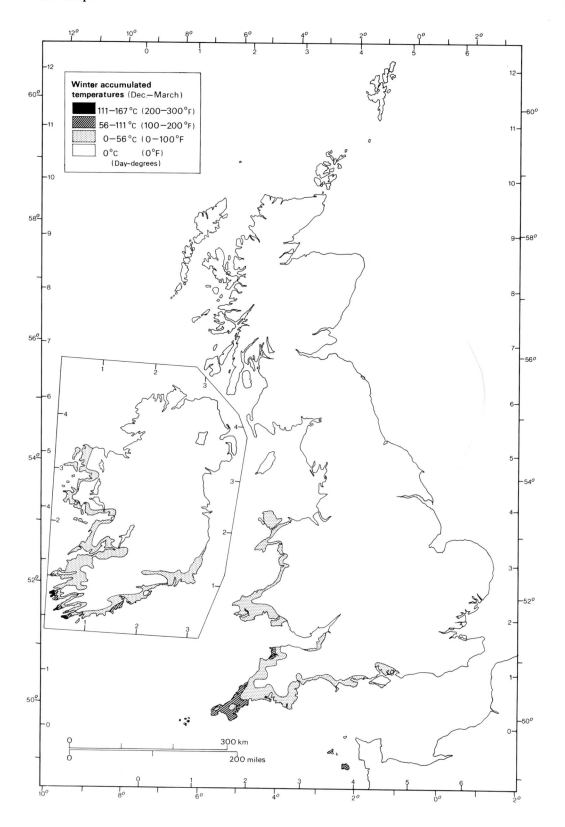

Winter accumulated
temperatures (Dec.—March)

111–167 °C (200–300 °F)
56–111 °C (100–200 °F)
0–56 °C (0–100 °F)
0 °C   (0 °F)
(Day-degrees)

300 km
200 miles

Map 10   37

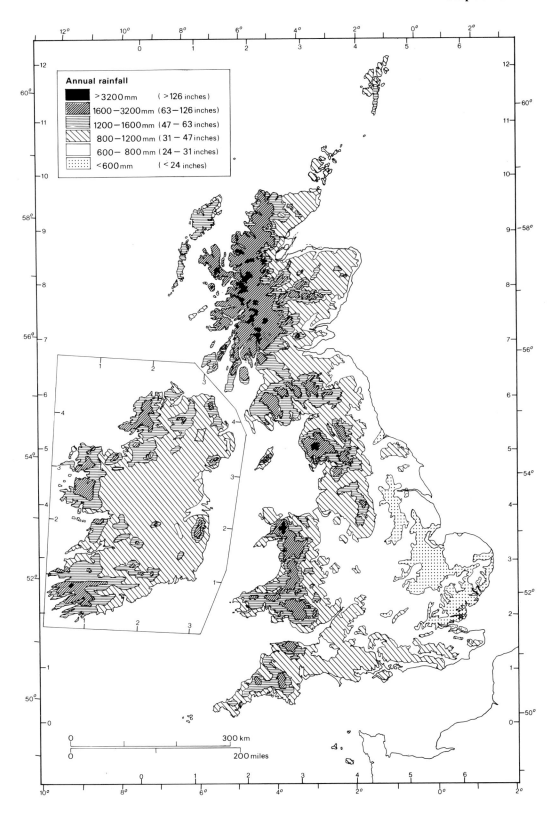

Annual rainfall

>3200 mm   (>126 inches)
1600−3200 mm (63−126 inches)
1200−1600 mm (47 − 63 inches)
800−1200 mm (31 − 47 inches)
600− 800 mm (24 − 31 inches)
<600 mm   (< 24 inches)

300 km
200 miles

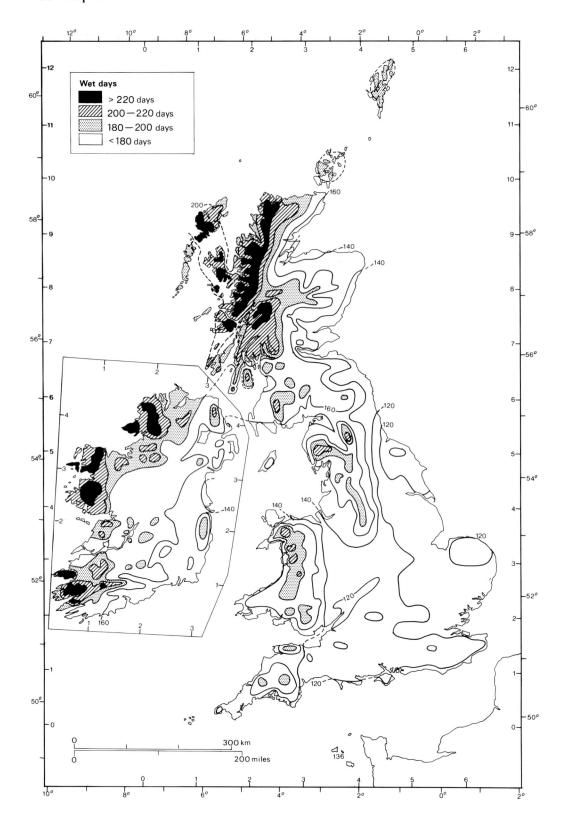

Map 12 39

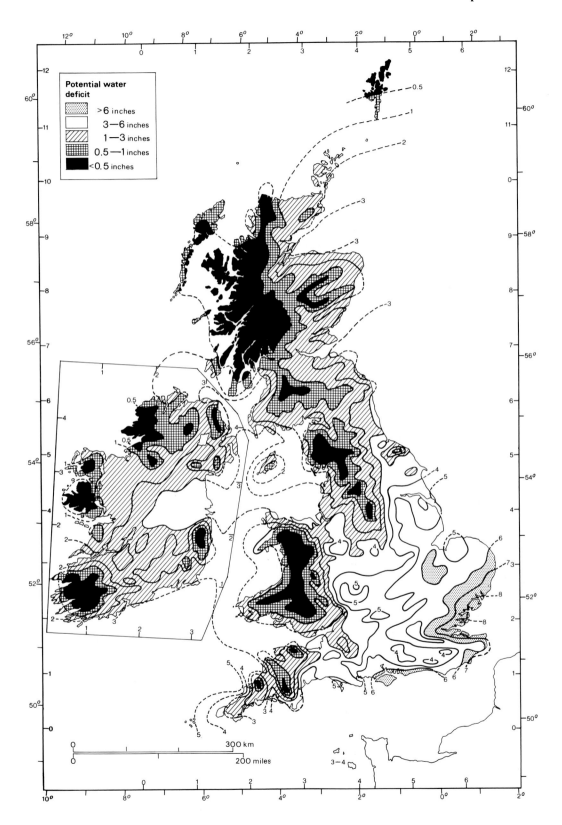

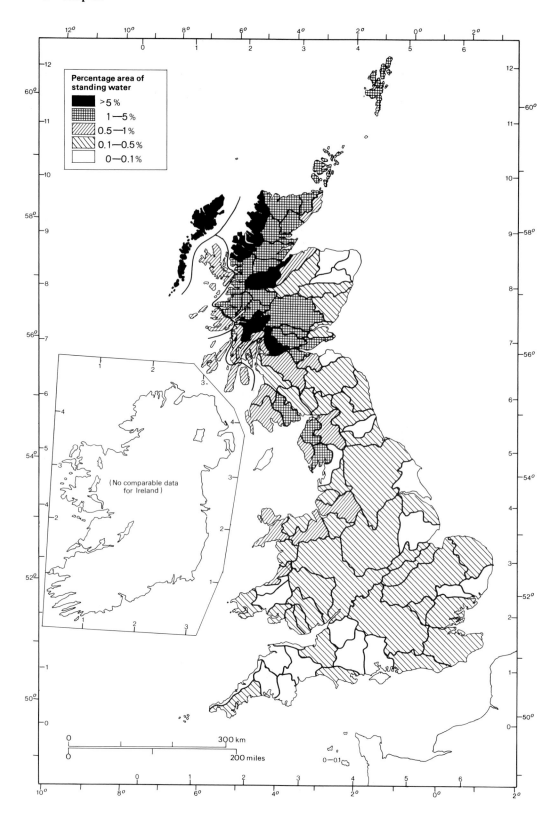

Percentage area of
standing water

■	>5 %
▦	1—5 %
▨	0.5—1 %
▧	0.1—0.5 %
□	0—0.1 %

(No comparable data
for Ireland)

300 km

200 miles

0—0.1

Map 14 41

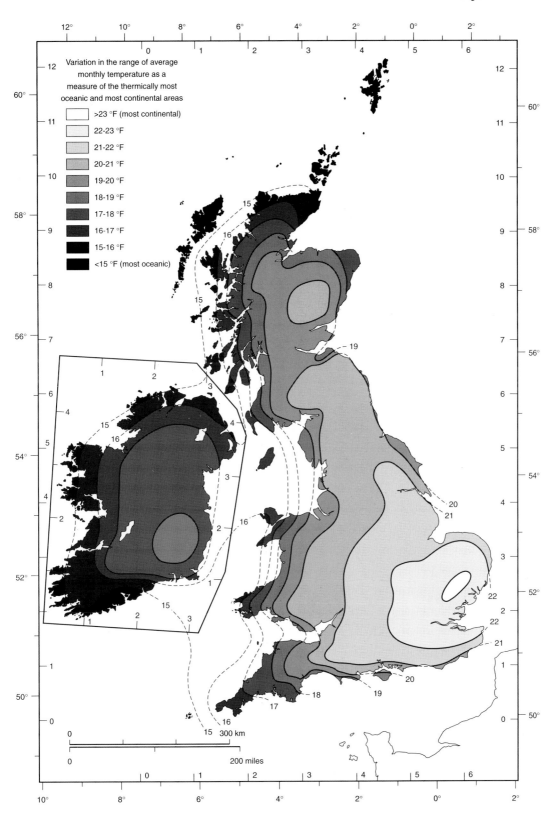

Variation in the range of average
monthly temperature as a
measure of the thermally most
oceanic and most continental areas

>23 °F (most continental)
22-23 °F
21-22 °F
20-21 °F
19-20 °F
18-19 °F
17-18 °F
16-17 °F
15-16 °F
<15 °F (most oceanic)

300 km

200 miles

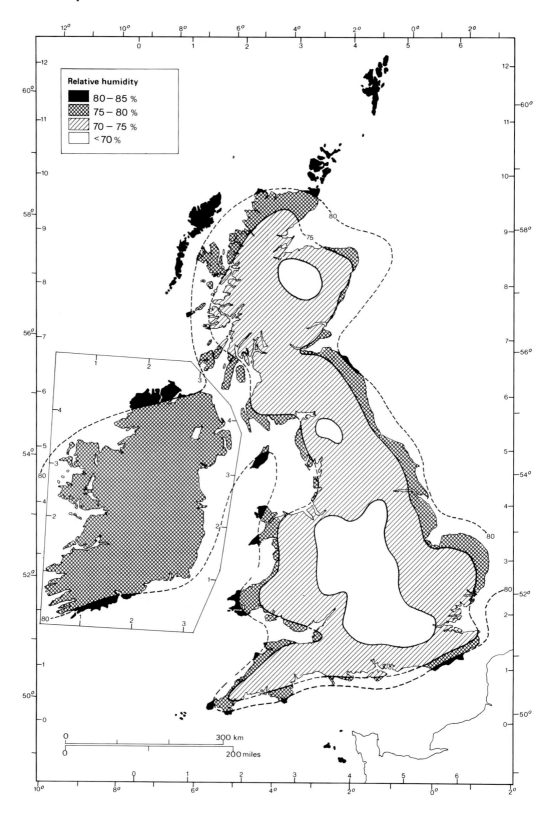

Relative humidity

80 – 85 %
75 – 80 %
70 – 75 %
< 70 %

80

75

80

80

80

80

300 km

200 miles

Map 16 43

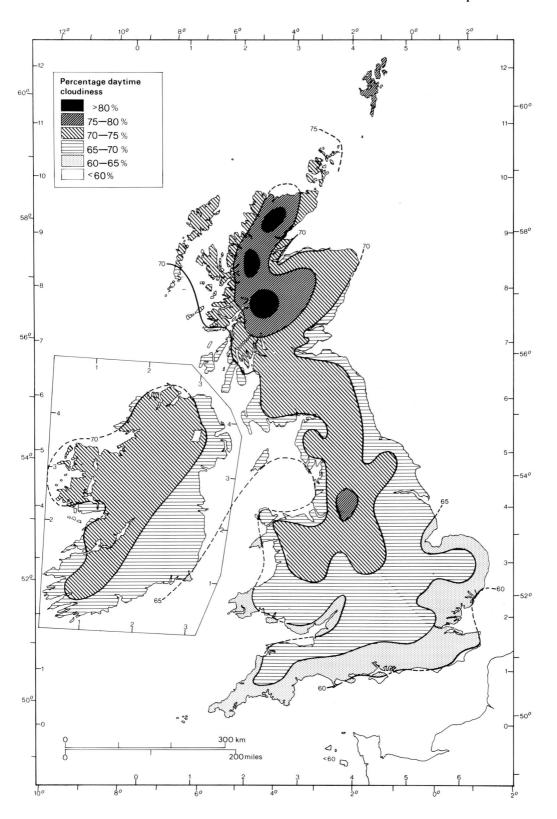

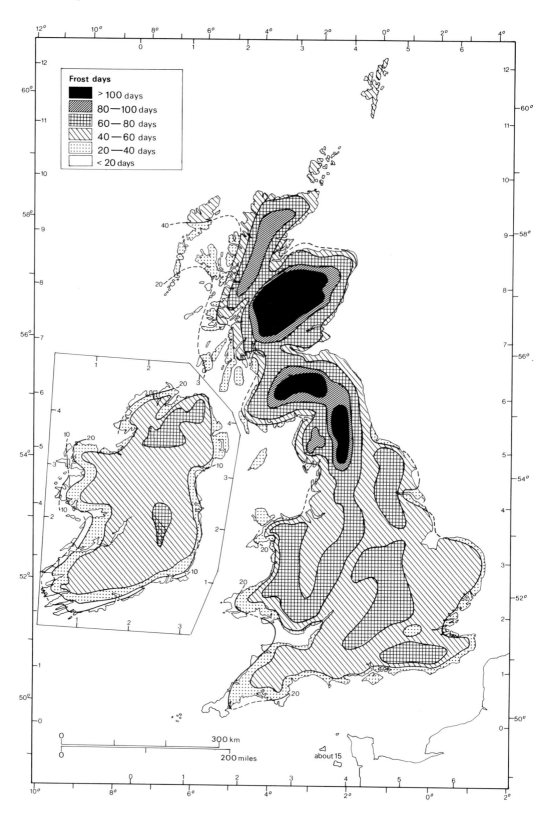

Frost days

> 100 days
80—100 days
60—80 days
40—60 days
20—40 days
< 20 days

Map 18 45

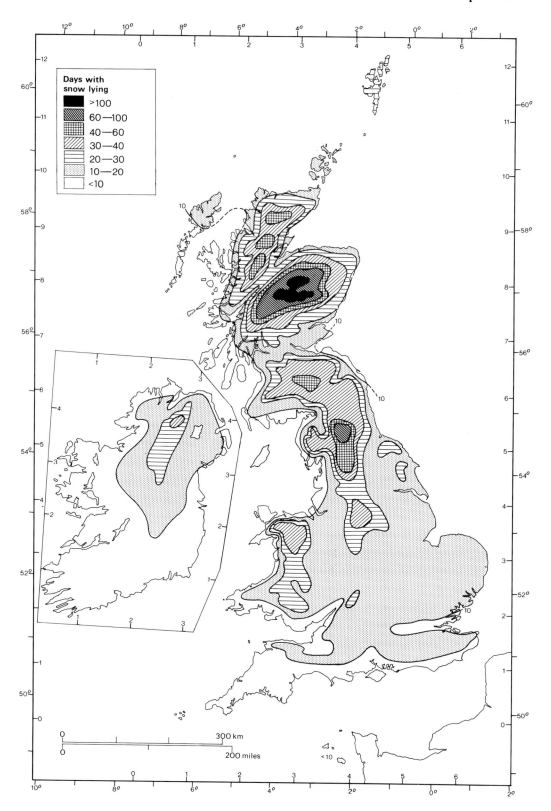

Days with
snow lying

>100
60—100
40—60
30—40
20—30
10—20
<10

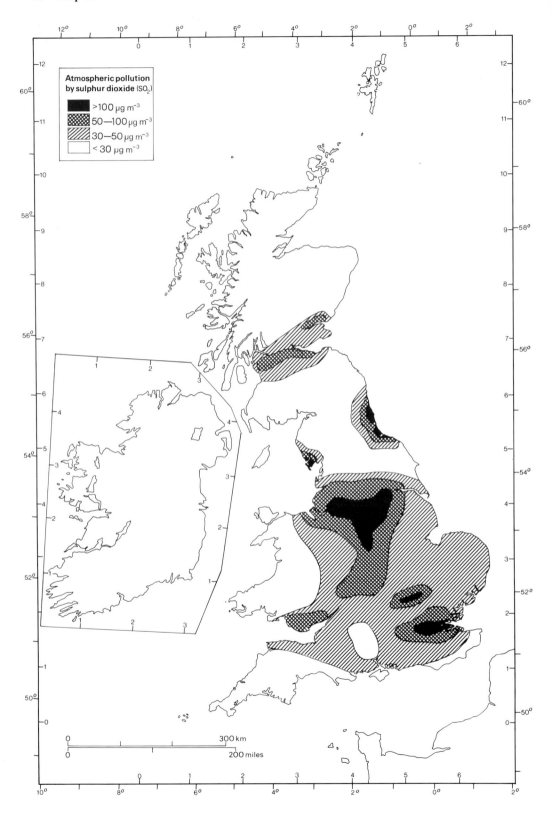

Atmospheric pollution
by sulphur dioxide (SO₂)

>100 µg m⁻³

50—100 µg m⁻³

30—50 µg m⁻³

< 30 µg m⁻³

Ferns

Adiantum capillus-veneris L.   Maidenhair Fern

Preliminary recognition A small to moderate-sized, delicate-looking fern, divided into numerous, small fan-shaped segments, borne on very slender, wiry black stipes.

Occurrence Local and rare in a few stations, chiefly in coastal areas of south-west England and western Ireland, mainly on damp, calcareous sea cliffs. Plants occasionally establish in damp mortar elsewhere, or freely in old greenhouses.

Identification Plants vary greatly in size, from having all fronds about 5 cm in length to over 40 cm in exceptional localities. The fronds stand more or less erect, or hang pendulously from steep surfaces. They are of a triangular-ovate or rhomboidal overall outline and the degree of division of the frond varies with its size, with smaller specimens simple, pinnate or bipinnate, whilst larger ones become bipinnate or tripinnate. Whatever the frond size, *the ultimate segments are all generally similar, about 0.6–3.0 cm diameter, and distinctly kidney shaped, wedge shaped or fan shaped, with straight sides and curved outer margins, which are often deeply and irregularly notched* (incised). The lamina of each segment is very delicate, thin, membranous, and a slightly translucent vivid pale green, sometimes emerald when young, but becoming duller and more leathery with age. The surface has a slight waxy bloom, which prevents wetting of the surface by repelling moisture in small droplets. Each segment has numerous small, forked veins, but no midrib. Besides being notched, the margins of sterile segments have numerous fine teeth. Those of fertile segments have *lobes of the margin turned down onto the underside of the frond to form kidney-shaped or crescent-shaped transverse marginal flaps, which act as indusia and protect the sori beneath.* These flap-like indusia are green or colourless at first but become brown at maturity.

All the segments are borne on extremely thin, hair-like, but wiry, polished minor branches, arising from a *slender, wiry, polished black or deep purple-brown stipe*, which occupies about half the length of the frond. The stipe is smooth, apart from a tuft of narrowly linear pointed dark scales near its base. The fronds arise from slender, creeping rhizomes, which are blackish

Jan.	Feb.	Mar.	Apr.	May	Jun.	Jly	Aug.	Sep.	Oct.	Nov.	Dec.

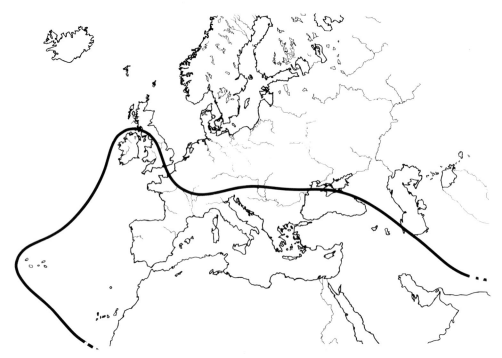

Mediterranean-Atlantic range of *Adiantum capillus veneris* in Europe.

in colour, paler only at the crowns, and have the older parts covered with the persistent bases of previous year's fronds. The rhizomes branch freely, and can eventually build up large patches. Fronds flush very early in spring (March and early April), and in sheltered places, are sub-evergreen, lasting into the following winter. Emerging fronds in early spring look at first like delicate, small, pink spheres. In quantity, dense growths of fully expanded fronds look like masses of curly hair.

Variation Plants are widely variable in size and consequent degree of frond dissection, but also vary in the size of the ultimate segments, their degree of margin serration and depth of notching. Many of the more extreme forms have been selected and brought into cultivation as horticultural varieties or 'cultivars'.

Possible confusion Despite their variation, the delicate, distinctly fan-shaped segments, readily distinguish *Adiantum* from any other British or Irish fern. Only the leaves of Meadow Rue (*Thalictrum* spp.) have a superficial resemblance.

Technical confirmation Plants are diploids, with $n = 30$, $2n = 60$ chromosomes.

Field notes Maidenhair Fern is a mainly Mediterranean species which is a rare and local, highly frost-tender, calcicolous fern in Britain and Ireland. It is confined mainly, in natural situations, to sheltered faces of sea cliffs, which are screened from mid-day sun, and where there is abundant seepage of lime-rich water building up calcareous tufa deposits. In the more exposed of these sites, the fronds remain stunted, and rhizome growth through the tufaceous surface layers as well as local spore recolonisation can build up extensive cushion-like patches, with large drifts of cascading tiny fronds extending over many square metres. In more sheltered crevices, as well as around the

Adiantum capillus-veneris: a, Aran Isles, Co. Galway; *b,* Co. Clare; *c–d,* Cheshire; *e,* Isle of Man; *f–h,* West Cornwall.

mouths of small caves, more scattered plants usually hang and develop much more fully.

Most English and Welsh habitats are restricted to within a few metres of the sea. In its western Irish stations, however, such as in the limestone rock of the Burren and in the Aran Islands in Galway Bay,

plants occur near to the sea typically within the recessed shelter of deep rock fissures and miniature caves. Here they can be especially luxuriant in the vicinity of small seepage lines, with even the tips of the fronds scarcely emerging into full daylight. Plants also extend further inland into the deeper grykes of limestone pavements, where they have been recorded several kilometres inland and up to an altitude of about 240 m (800 ft). Further away from coasts, in all areas, it seems to be ecologically replaced in damp calcareous habitats by the much more hardy Brittle Bladder-fern (*Cystopteris fragilis*).

Maidenhair Fern is a very rapidly growing and usually gregarious species, often able to pioneer damp calcareous surfaces in areas of suitable micro-climate. It has been recorded from a number of artificial calcareous situations, including damp quarry spoil of black marble in Co. Galway, damp crevices of roadside walls in Co. Donegal, damp mortar of heated greenhouses in various parts of England (as well as sometimes extensively establishing as an understanding plant within them), and in brick mortar of an abandoned railway platform in the Wirral Peninsula of Cheshire. In the latter locality, plants occur in considerable numbers, and are believed to have originated as escapes through spore establishment from plants formerly grown in hanging baskets to decorate the railway station in its more active days.

Native plants undoubtedly suffered from collection through horticultural over-popularity in the nineteenth century. But, through its rapid rate of growth, Maidenhair Fern currently maintains itself well in some of its remaining, totally inaccessible, cliff stations. Their particularly early flushing in spring enables them to make use of the vernal aspect of these habitats before becoming partially shaded by other, taller summer herbage. Thereafter, the fronds persist well into the following winter, presumably gaining considerable benefit, not only from the lack of frosts, but also from the long length of growing-season available.

Anogramma leptophylla (L.) Link   Jersey Fern

Preliminary recognition A small fern, with erect fronds of delicate appearance arising annually and hence lacking a discernible woody rhizome.

Occurrence A Mediterranean species which reaches the Channel Islands (Guernsey and Jersey), where it grows on a few damp, exposed roadside banks, regenerating annually.

Identification Plants each have minute rhizomes, from which arise annually rather few (usually about 4–5), quite delicate, *bright golden-green* fronds, the tallest usually to about 7 cm in height, occasionally to 10 cm, arising in a tufted fashion from the apices of minute, scarcely discernible, short-lived rhizomes. Whole plants arise anew annually, *with successive fronds gradually increasing in size and degree of division during the year*. The earliest ones are purely vegetative, of more or less spreading habit and have a usually simple trifid lamina. The later ones stand more strongly erect, are fertile and become progressively more compound, with bipinnate or tripinnately divided blades, *with broadly triangular pinnae* divided each into *a few wedge-shaped slightly stalked pinnules, with rounded tips*. The stipes are about half the length of the fronds, or more, are very slender, and of greenish-straw colour, becoming darker towards their bases. The tallest fronds usually have proportionately the longest stipes, and their

Mediterranean-Atlantic range of *Anogramma leptophylla* in Europe.

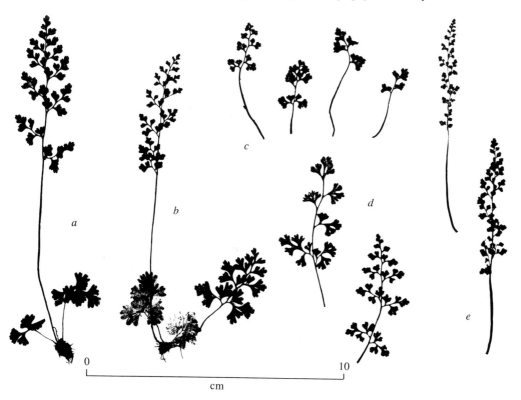

Anogramma leptophylla: a–d, Jersey, Channel Islands; *e*, Guernsey, Channel Islands.

pinnules bear *a few small, linear sori radiating on each pinnule along the veins, each without an indusium* and all becoming more or less confluent on each pinna at maturity.

An excellent colour illustration of the plant *in situ* is given in Le Sueur (1984). (It is also illustrated in colour on a 3p postage stamp of 1972 issued by the Jersey Postal Authorities.)

Variation Not a very variable species, except in the heteromorphy of the successive fronds borne on each plant.

Possible confusion Their small plant size, delicate fronds and relatively long, slender stipes, make these plants unlikely to be confused with any other fern.

Technical confirmation Plants are diploids with probably $n = 29$, $2n = 58$ chromosomes.

Field notes *Anogramma leptophylla* is a species with a curiously disjunct and widely fragmented range, which includes especially the Mediterranean region and Macaronesia, but also discontinuous areas through east Africa south to Madagascar and the Cape, in India and Australasia, and western South America. Such a range seems highly relictual, and is perhaps typical of a once wider one of probably Miocene age (Page 1973*b*, 1977).

In these islands, it occurs only in the Channel Islands, where it has a single station on Guernsey and several on Jersey. Here, plants are mainly winter and early spring growing, shedding their spores as early as April – a habit especially characteristic of a species whose main centre of distribution is in submaritime habitats of southern Europe and especially around the shores of the Mediterranean. The plant is unusual amongst ferns in regenerating each year as an annual.

It grows mostly on steep, moist, sunny, south to west-facing laneside banks, built of granite stone and clay, mostly in otherwise largely uncolonised patches, where there is some surface erosion helping to minimise severe competition.

In Jersey, *Anogramma* was first found in 1852, and by 1853 was known to be widely distributed on the banks of exposed lanes. Today, Le Sueur (1984) notes that the plant occurs locally in sheltered pockets of stony hedgebanks of varying aspect that are damp in winter but fairly dry in summer, and that it is typically associated with the liverwort *Lunularia cruciata* (L.) Dum.

On Guernsey, *Anogramma* was first discovered by G. Derrick in the parish of St Saviour's in 1877. It was reported to grow 'in every favourable spot throughout 200 feet of a hedgebank' (Marquand, 1901). It was suggested to be native to the island and perhaps be the last remnant of a once more common plant. McClintock (1975) records that, by 1957, the lanebanks in the supposed original locality had been rebuilt and lacked even a hedge, and that the fern was not there, but was refound 270–365 m to the north in the parish of Catel. In this locality it still survives, at an altitude of *c.* 180 m. Here it grows on a somewhat exposed, eroding steep slope, with few nearby associates except scattered Gorse (*Ulex europaeus*), Wood Sage (*Teucrium scorodonia*), Wall Pennywort (*Umbilicus rupestris*), Celandine (*Ranunculus ficaria*) and Lanceolate Spleenwort (*Asplenium billotii*).

As indicated by Marquand (1901), such a species might have once occurred more widely in these islands in more natural, surface-eroding habitats, before the construction of the present fields and lanes. The hedgebank habitats themselves may well be many centuries in age, and have probably been in an ideal position to conserve some of the original native flora which was present at the ancient time of the hedgebank construction (Page 1988*b*: 134).

Clearly, even today, the plant remains very local in the Channel Islands, but occurs sometimes in large numbers, with populations varying in size from year to year. McClintock (1975) rightly states that today the sole Guernsey site is in a position vulnerable to modern traffic and chemicals, and this may apply to its Jersey sites too. Le Sueur (1984) notes that the Jersey populations thrive best after years which have had a hot summer and a mild, wet winter. Such seasons would, clearly, appear to approximate most closely to Mediterranean growing conditions, and it seems likely that annual weather patterns must substantially and continually closely influence the year-to-year success of annual re-establishment of the unusual life-cycle of this species in such sites on the very northern margin of its European range.

Asplenium adiantum-nigrum L.

Black Spleenwort

Preliminary recognition A small to medium-sized wintergreen fern, with fairly finely cut distinctly triangular glossy-green blades borne on shining, deep red-brown or black stipes.

Occurrence Present throughout much of Britain and Ireland, becoming frequent in the south and west, especially in coastal districts, in rocky places, cliffs, lanebanks and occasionally on mortared walls.

All records

Identification On all plants, fronds are rather few and arise from stocky, short, creeping dark-brown rhizomes. Plants commonly have fronds 30–40 cm or more in length when in sheltered situations, although they are usually much smaller than this when in exposed sites. The fronds are bipinnately to tripinnately cut, and *conspicuously triangular-outlined*, with *glossy green blades borne on shining, red-brown or nearly black stipes*. The dark stipe colour gives way at about the level of the lowest pinnae on both surfaces of the frond to a green rachis. In most large specimens, the stipes are strongly ascending, and much of the blade is held in an ascending direction, arching slightly to become more nearly horizontal only towards its tip. On such specimens the pinnae, which are usually held perpendicularly to the rachis or swept slightly upward, are also tilted to become more nearly horizontal. On smaller specimens, the fronds are usually less strongly ascending, and the stipes much shorter.

The undersurface of the blade bears sparsely scattered, small, narrow dark scales (visible under a hand lens), especially along the pinna midribs. There are also *short, linear sori over much of the undersurface of most fronds*, often close enough to appear confluent on each segment at maturity, when they readily yield quantities of sooty black spores.

Overall frond outline, cutting of the blade and shape of the ultimate segments are all highly variable between specimens and especially between habitats and localities, and differ still more significantly between the two subspecies (q.v.).

Subsp. *adiantum-nigrum*

Distribution: that of the species.

This is the widespread type of *A. adiantum-nigrum*, which is itself highly variable in size and frond form. Specimens have stipes usually as long or slightly longer than the blade, and blade

Jan.	Feb.	Mar.	Apr.	May	Jun.	Jly	Aug.	Sep.	Oct.	Nov.	Dec.

outlines which typically form a moderately narrow triangle, the width of the blade typically about half that of its length. The pinnae sweep somewhat forwards at an angle of about 60° to the rachis, and are themselves of moderately narrowly tapering outline, with the pinnules somewhat crowded with acutely toothed segments, the pinnae fairly strongly inclined and tapering to a moderately acute, though usually not highly attenuate, apex, and of an overall deep mid-green colour when fully expanded.

This widespread plant occurs throughout the whole of Central and southern Europe, reaching its most northerly points in the Faroes and south-western Norway, becoming less extensive in the countries fringing the Mediterranean, where it is widely sympatric with *A. onopteris* (q.v.). Outside Europe, it has a wide essentially Tethyan range, stretching from the Azores and Madeira eastwards to the Himalayas, with scattered stations in eastern central and southern Africa, the Mascarenes, Australia, Polynesia, and sparsely in the southern Rocky Mountains.

Subsp. *corrunense* Christ

Distribution: Confined to scattered outcrops of ultrabasic (especially serpentine) rock, occurring highly locally from Cornwall (Lizard) to a number of widely scattered sites in Scotland as far west as the Isle of Rhum and north to the Shetland Islands.

This much more localised variant of *A. adiantum-nigrum* is itself moderately variable in size and frond form. Specimens have stipes as long or usually shorter than the blade, and blade outlines which typically form a wide triangle, the width of the blade typically over half (often nearly 3/4) that of its length. The pinnae are angled only slightly forwards (*c.* 70° to 80° to the rachis) and are thus more nearly perpendicular to it, and are themselves of fairly broadly and obtusely tapering outline with the pinnules more distant and composed of more obtusely fan-shaped segments, the pinnae usually relatively widely spaced and spreading, tapering rather suddenly to a bluntly obtuse apex, and of an overall rich grass-green colour when fully expanded.

This plant of limited range was formerly thought to be conspecific with the European *A. cuneifolium* Viv., which it much resembles, but is distinguished from it by its tetraploid status, showing its affinity with other *A. adiantum-nigrum* (cf. diploid in *A. cuneifolium*). Outside Britain and Ireland, it has stations in Spain, France (Alsace) and south-west Norway (J. Vogel, personal communication). Within Britain and Ireland, it is widely sympatric with subsp. *adiantum-nigrum*, replacing it on ultrabasic rocks. It is unknown elsewhere, and its principal stations are thus within these islands.

Variation In each subspecies wide variation in the degree of cutting of the frond occurs, especially between small and large specimens, and plants respond in size profoundly to differences in situation. In addition, there is considerable inherent variation in frond form. Some populations are more variable than others, and extreme individuals can occur.

Much variation in all specimens in stipe length and habit of the frond is probably largely environmentally induced. Small specimens occurring on the most bare rocky faces have small mature fronds of short stipe length, widest blade shape and flat and spreading habit, growing close to the rock surface. Individuals in luxuriant deep rock crevices, however, usually have more nearly erect mature fronds borne most often on proportionately long stipes, their narrower blades with pinnae nearly horizontally rotated.

0
10
cm

Asplenium adiantum-nigrum, typical lanebank forms: *all* West Cornwall.

Possible confusion Although *A. adiantum-nigrum* is much more common and widespread in Britain and Ireland than its near-ally *A. onopteris*, technical confirmation may be needed to separate closely similar forms of each. All plants of *A. adiantum-nigrum* can be separated from *A. onopteris* by their tetraploid chromosome number (cf. diploid in *A. onopteris*), spores which are distinctly larger and darker with a more irregular wing, and presence of the narrow, dark scales on the underside of the blade (absent or extremely sparse in *A. onopteris*).

The presence of uniformly good spores separated *A. adiantum-nigrum* from *A.* ×*ticinense* and *A.* ×*contrei* (q.v.), its hybrids with *A. onopteris* and *A. septentrionale* respectively.

Technical confirmation Plants are tetraploids, with $n = 72$, $2n = 144$ chromosomes, with spores mostly about 38–52 μm in length. Prothalli grow closely appressed to the substrate, bear numerous short rhizoids, and have elongate trichomes present around the thallus margins. Samples of *A. adiantum-nigrum* from throughout its range contain C-glycosylxanthones, which are also present in European diploid *A. onopteris* but absent from the *A. cuneifolium* parent (Richardson & Lorenz-Libeurnau, 1982).

Field notes Black Spleenwort is a widely ranging and fairly vigorous fern in Britain and Ireland, seldom occurring above about 600 m altitude and thriving best near the sea, where conditions of good daylight combine with mild winter temperatures, high humidity and reasonable shelter. It is intolerant of dense herbaceous competition, tree overgrowth and grazing, and in exposed situations becomes very stunted. It occurs on a wide variety of well-drained, fairly basic, light, sandy soils or clay banks, and on fissured, slightly permeable but well-drained exposures of mainly non-acidic rocks such as schistose grits, basalts, limestones and calciferous sandstones, but sometimes also on shales and slates, especially on loose coastal and disused quarry screes*.

In most of its habitats, *A. adiantum-nigrum* occurs as scattered individuals, though sometimes quite numerous populations are present locally where sufficiently extensive rocky habitats are more uniformly present, such as in coastal cliff habitats widely scattered around mainly western coasts. Such coastal populations typically reach their greatest development where hard-rock cliffs are interbedded with mixtures of occasional tougher shales or softer slates, giving rise to local screes of clattering pieces of plate-like rock lying steeply in an unconsolidated jumble, the whole scree loosely perched over hard-rock shoulders. In such sites, especially where also semi-exposed but south facing, warm and sunny, some of the largest and oldest clumps of *A. adiantum-nigrum* can occur, each marking an island of least mobility within the scree surface's mosaic of mobility, with bare unstable portions sliding all around it. Such microcosms of

pioneering vegetation are probably mostly initiated by the fern, and thereafter only gradually accumulate a humus content largely from mosses and the fern's own frond decay, bound only within the zone of the fibrous roots of the fern. Into these eventually also typically establish a few other wind-pruned vascular associates, typically including scattered miniature shrubs of Gorse (*Ulex europaeus*), wind-blasted Wood Sage (*Teucrium scorodonia*) and tightly leaved shoots of low-profiled English Stonecrop (*Sedum anglicum*) or compact and clinging Ivy (*Hedera helix*).

In south-west England, specimens of *A. adiantum-nigrum* of sometimes highly varying size and form are particularly abundant on earth-and-rock lanebanks of clay soils, where light surface erosion opens new soil exposures, verge trimming reduces competition, and there is good local shelter. In such habitats, it is often associated with *Phyllitis scolopendrium* and *Polystichum setiferum*. Further north, it is extensively replaced by larger ferns such as *Dryopteris filix-mas* and *D. dilatata* in increasingly cool and damp lanebank habitats, and in such areas it becomes more confined to low altitude riverine cliffs and screes. Plants also sometimes extensively colonise old quarry workings, and in coastal districts, are also frequent in mortar of old walls. Further inland, where such mortar habitats are more prone to prolonged drying out, *A. adiantum-nigrum* is largely replaced by *A. trichomanes* subsp. *quadrivalens* and *A. ruta-muraria*,

* CAUTION – *loose, sunny, scree slopes where* A. adiantum-nigrum *grows in quantity are often also habitats of abundant adders.*

Asplenium adiantum-nigrum, young frond stages (after Orth, 1938).

while in northern districts, *A. adiantum-nigrum* sometimes associates with *Cryptogramma crispa* in disused slate quarry workings, to be replaced entirely by this fern species (q.v.) in natural scree slopes at generally higher altitude, where increased leaching further ensures the enhanced acidity of the environment.

Plants of subsp. *corrunense* are very much more local in occurrence, but are sometimes fairly frequent in their isolated habitats with very great distances between many of these (e.g. between the Cornish Lizard populations and those of western Ireland and western and northern Scotland). Those I have seen (Lizard, Perthshire and Isle of Rhum populations) grow in a variety of rocky habitats, including bare rock crevices and the interstices of scree slopes. In the Scottish sites, as seems to be the case with most other populations, plants grow mostly on inland cliff flanks in relatively exposed situations. In such habitats, the majority of plants are usually of small stature and have few vegetation associates. On Rhum, plants also occur on nearby dry-stone walls, where the loose rock and rubble from which these are made is that of local peridotite lava rock.

In the Cornish Lizard site, plants of subsp. *corrunense* grow mostly in semi-sheltered wind-sheltered niches between the major boulders of south-west-facing large-boulder scree, sloping steeply to the sea. The rock is pure serpentine, mostly of hard texture, but has also a rich melange of numerous colourful rock veinlets of varying physical texture and chemical composition included within it, and individual plants form clumps that may be of appreciable age.

In each of its sites, plants of subsp. *corrunense* associate with very few other ferns and with only a limited, but interesting range of flowering plants, which, in Scotland, can include Northern Rock-cress (*Cardaminopsis petraea*) and Vernal Sandwort (*Minuartia verna*) and, in Cornwall, Sea Asparagus (*Asparagus officinalis* subsp. *prostratus*).

Other species variously noted (mainly in the northern Scottish sites) include *Antennaria dioica*, *Campanula rotundifolia*, *Cochlearia officinalis*, *Hypericum pulchrum*, *Leontodon autumnale*, *Linum catharticum*, *Lotus corniculatus*, *Minuartia verna*, *Polygala serpyllifolia*, *Solidago virgaurea*, *Stellaria holostea*, *Luzula pilosa*, *Festuca ovina*, *Melampyrum pratense* and *Asplenium viride* (Dr Mary Gibby, personal communication). The fairly high frequency of individuals of subsp. *corrunense* in most sites probably reflects the overall marked lack of com-

0   10
cm

Asplenium adiantum-nigrum, all Lizard serpentine examples: West Cornwall.

petition for such habitats, directly in a rock type yielding several toxic (poisonous) heavy metals, including chromium, cobalt and nickel.

In each of their sites, plants of subsp. *corrunense* grow with their roots usually in close and constant contact with bare mineral rock surfaces of ultrabasic character, to the metal-yielding excesses of which individual populations have clearly become closely adapted. In this respect, it is significant that plants of this *A. cuneifolium*-like form occur *only* on ultrabasic rocks (for which they are thus extremely good indicators), and are replaced in nearby mesic rock habitats entirely by plants with the 'normal' frond form of the species. Each is thus absent from

the habitats of the other, despite their occurrence each within abundant spore exchange distance. Experimental cultivation at Edinburgh of spore-reared material (ex Lizard serpentine) shows that intergametophytically selfed progeny breed 100% true for the frond morphology of their wild populations, as do some of the other more morphologically extreme forms of this species from other localities. Further, reciprocal transplants of spore-reared material using serpentine and non-serpentine soils each display significantly better sporophytic growth success when reared on their native soils in contrast to growth of reciprocal transplants. The occurrence of like morphological

Asplenium adiantum-nigrum, serpentine populations, mainland Scottish examples: *a–c*, Sutherland; *d*, Inverness-shire; *e*, Ayrshire.

0   10

cm

Asplenium adiantum-nigrum, serpentine populations: *all* Isle of Rhum peridotite examples, Inner Hebrides, Western Scotland.

forms on distant like rock types, wherever these may occur, seems to suggest independent origins, in the absence of competing true *A. cuneifolium* in these islands, of similarly adapted and similarly structured populations from variable former ones of the 'parent' *A. adiantum-nigrum* subspecies at many different times and in many different places, wherever such ultrabasic rocks have occurred, probably each achieving some degree of local genetic isolation across prominent edaphic boundaries.

Elsewhere around these islands, the widely scattered nature of coastal, as well as some inland populations of the overall populations of *A. adiantum-nigrum* probably ensures that in many of these too there is often considerable genetic isolation resulting in significantly inbreeding populations, and that selection of the progeny to the often exacting climatic and edaphic extremes of many sites is perhaps a population norm. In some often smaller sites, occasional plants occur whose frond

Asplenium adiantum-nigrum, Irish specimens with a morphology closely approaching that of Macaronesian *A. onopteris: a*, Wexford; *b, c*, North Kerry.

size and structure approaches the more finely dissected and narrow, acutely tapering pinna form of *A. onopteris* (q.v.), the continental and especially Macaronesian (but not Irish) forms of which they resemble. A chromosome count is clearly needed to distinguish such material (all *A. adiantum-nigrum* is tetraploid, *A. onopteris* diploid). It seems particularly significant that the areas from which such extreme-fronded forms have been recorded are generally in western sites of milder climate and at low altitude often near to the sea. Perhaps, as with the edaphic selection of the blunt frond forms of *A. adiantum-nigrum* to resemble the *A. cuneifolium* parent on serpentine sites (including on cooler northern ones), there is also a parallel selection pressure taking place in milder, western sites, in the absence of known true *A. onopteris* in Britain, for unusual frond forms of *A. adiantum-nigrum* most resembling those of this parent.

Asplenium adiantum-nigrum, Scottish specimens with a morphology closely approaching that of Macaronesian *A. onopteris*: *a*, Midlothian; *b*, Perthshire; *c*, Isle of Arran.

Note added in proof (*Asplenium adiantum-nigrum*).
Exciting new developments in our understanding of the phylogenetic affinities of European *Asplenium* as a whole are indicated through recent molecular studies (e.g. Pajaron *et al.*, 1997; Vogel, 1997; Vogel *et al.*, 1997), from which there is increasing evidence that some of the serpentine and non-serpentine forms of *A. adiantum-nigrum* illustrated here, may, indeed, prove to have distinct and independent evolutionary origins.

Asplenium billotii F. W. Schultz

Lanceolate Spleenwort

Preliminary recognition A small evergreen fern with rather stiff and spreading, untidy-looking spearhead-shaped, bipinnate fronds, with obtusely rounded pinnules.

Occurrence Local and seldom in great abundance, growing in rocky fissures in coastal areas of south-west England, the Scilly Islands, the Channel Islands and south-west and north-west Wales; but also present in a few localities in southern Ireland, the extreme west coasts of northern England and Scotland, and formerly in the Weald of Kent.

Identification Plants have distinctively *ovate-lanceolate outlined fronds*, on which the pinnae are set almost opposite to one another. The lowermost pinna pair is usually markedly shorter than those above it, or, more rarely, of similar length, and *frequently deflexed downwards* (away from the frond apex). The stipes are usually 1/3 *or less* of the overall length of the frond (longer only in plants growing in very deep recesses). The fronds are bipinnate, with the *ultimate oval to orbicular segments distinctly stalked* and minutely serrated on their outer margins into *broad obtuse teeth*, each terminating in a sharply pointed (mucronate) colourless tip. The pinnule midribs are characteristically sinuous, and the whole frond may also be sinuous and asymmetric.

The fronds arise from short, stocky creeping or ascending rhizomes, densely clothed with deep purple-brown shining narrow scales and the persistent brown bases of previous years' fronds. The fronds usually spread closely against adjacent rock surfaces, forming a rather flat rosette, and are of smooth, slightly shining, leathery but not thick texture, with rather stiff stipes. The upper frond surface is of a dull to bright mid-green colour, with a paler green undersurface. The stipe and rachis are green on the upper surface except towards the stipe base, where they become a dark shining chestnut brown. On the undersurface the brown colour extends to about halfway up the rachis, sometimes further, giving it a polished appearance.

The sori are shorter than those of *A. adiantum-nigrum*, and are *set nearer to the frond margins*. Because of this shortness, when they bulge with discharging dark sporangia from late summer to early spring, the soral heaps appear more rounded than linear.

The fronds remain green throughout the winter and usually well into the following spring (later in very sheltered niches). New fronds are flushed very early in the season, usually beginning about late January, and often all are completely expanded by mid-April.

Jan.	Feb.	Mar.	Apr.	May	Jun.	Jly	Aug.	Sep.	Oct.	Nov.	Dec.

Variation This is not a very variable species, except in size.

Possible confusion Most likely to be confused with smaller specimens of *A. adiantum-nigrum*, but the narrower ovate-lanceolate outline with usually very short basal pinnae, shorter stipe, scales to upper frond surface and sori placed nearer the frond margins, separates *A. billotii*. The presence of the good spores distinguishes *A. billotii* from *A.* ×*sarniense* (q.v.).

Technical confirmation Plants are tetraploids, with $n = 72$, $2n = 144$ chromosomes. They are believed to be autotetraploids, as they form bivalents at meiosis in crosses with unrelated diploids, e.g. in ×*Asplenophyllitis microdon* (q.v.). The spores have a wide perispore wing, are very spiny and are slightly larger (43–58 μm) than those of *A. adiantum-nigrum*, and of much paler brown colour. Both frond surfaces as well as stipe and rachis have numerous scattered minute deep-brown narrow pointed scales, which are particularly frequent on the lamina over the veins. Occasional small vegetative plants are known to occur in parts of Guernsey and Cornwall, resembling the diploid *Asplenium obovatum* Viv., but have not yet been found in a fertile condition. This species is present in Brittany, and could yet occur within these islands. Spores of the diploid, in continental material, differ in being about 75% of the diameter of those of *A. billotii*.

Field notes *Asplenium billotii* is a west European–Mediterranean species centred mainly in France, the Iberian Peninsula and the west Mediterranean basin, with stations in the Azores, Madeira and the Canary Islands.

In Britain and Ireland, Lanceolate Spleenwort is a notably south-western, largely maritime species. It is a slow-growing evergreen fern which occurs almost exclusively in fissures and crevices of sheltered rocky sites, often on steep faces such as the upper parts of sea-cliffs, rocky outcrops on headlands, and sometimes in hedgebanks or on inland rocks (e.g. some Dartmoor 'Tors'), although seldom far from the sea. In south-west England and west Wales, it grows mainly on sea cliffs, descending alongside *A. marinum* in cool shady crevices and around cave mouths to within about 5 m of high tide. In such habitats, it is rather a solitary species, seldom occurring in large numbers, on a wide range of mainly non-calcareous rock types, including weathered granites, and fairly hard metamorphic or sedimentary rocks such as slaty shales, hornblende schists, conglomerates, coarse grits and thinly bedded sandstones, which provide sufficient firm crevices for anchorage. It thrives mainly where there is shelter from sun, severe winds and frosts, and where slight seepage of permanent moisture occurs with very free drainage and good air movement.

In sea-cliff habitats it commonly grows with abundant yellow crustose and grey foliose lichens and characteristic cliff-face flowering plants such as Sea Thrift (*Armeria maritima*), Sea Scurvy-grass (*Cochlearia officinalis*), Rock Samphire (*Crithmum maritimum*), Sea Plantain (*Plantago maritima*), Sea Campion (*Silene maritima*), and Wall Pennywort (*Umbilicus rupestris*). It seems intolerant of dense flowering plant competition, which confines it usually to the smallest fissures.

Lanceolate Spleenwort also occurs in lanebank habitats, sometimes in the south-west of England, but particularly in the Channel Islands. Here it grows in steep, often sunny, slopes in banks constructed of granitic rocks set in old clay or turf matrices, where there is some slight surface erosion, and hence continual slight openness of the habitat. The shallow rocky substrates ensure that flowering plant competition is usually not rank, and here *A. billotii* grows with other common lanebank species, including Ivy (*Hedera helix*), Wall Pennywort (*Umbilicus rupestris*), Red Campion (*Silene dioica*), Lesser Celandine (*Ranunculus ficaria*), Herb Robert (*Geranium robertianum*), Wood Sage (*Teucrium scorodonia*), Primrose (*Primula vulgaris*), Violets (*Viola* spp.), Cuckoo-pint (*Arum maculatum*), Butcher's Broom (*Ruscus aculeatus*), Wood False-brome (*Brachypodium sylvaticum*), Cock's-foot (*Dactylis glomerata*), Black Spleenwort (*Asplenium adiantum-nigrum*). Hart's-tongue (*Phyllitis scolopendrium*) and Western Polypody (*Polypodium interjectum*). In a number of Guernsey stations, it hybridises in these habitats occasionally with *Asplenium adiantum-nigrum* and *Phyllitis scolopendrium* to produce the two rare hybrids *Asplenium*

Asplenium billotii: a, Merionethshire; *b–c*, West Cornwall; *d–h*, South Devon.

Atlantic range of *Asplenium billotii* in Europe.

sarniense (q.v.) and *Asplenophyllitis microdon* (q.v.) respectively.

In its south-west Ireland and south-west England stations, *A. billotii* seems to have diminished very considerably in numbers in the last hundred years.

Little is known of the reasons for this, although it seems likely that general air pollution, as well as pollution of seeping groundwater, may very probably have had a particularly significant effect upon such a highly maritime plant.

Asplenium marinum L.

Sea Spleenwort

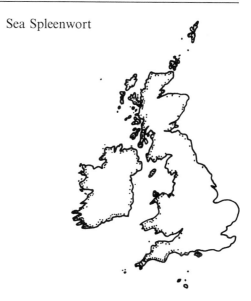

Preliminary recognition A small to medium-sized, robust winter-green fern with distinctive, coarsely divided thick, fleshy shining fronds, growing very near to the sea.

Occurrence Of markedly Atlantic distribution: locally frequent on rocky western coasts of Britain and Ireland, from the Channel Islands to the Shetland Islands; more infrequent and scattered on eastern coasts of Ireland, Scotland and northern England (south to Yorkshire) and absent from the colder coasts of eastern England. It seldom occurs beyond the immediate spray of the sea, and usually grows in rocky fissures and on old harbour walls.

Identification The fronds are mostly 15–20 cm in length (but as much as twice this in very luxuriant plants), and their *lanceolate outlines, simply pinnate or pinnately lobed structure and fleshy rigid texture* are their most distinctive features. The pinnae are ovate-oblong, with (usually) an obtusely rounded apex, and are shortly stalked in the lower part of the frond, with bluntly toothed almost crenate margins. The pinna bases are sometimes developed into a lobe on the upper side, making them markedly asymmetric. The upper portion of the frond is usually pinnatifid with *narrow green wings to rachis connecting between the pinnae*. The *stout, stiff stipes, rarely more than 1/3 of the length of the frond, are of polished dark purple-brown colour, and usually have narrow wings*. The fronds arise in a tuft from short, ascending or decumbent rhizomes, which are densely clothed with narrow, deep purple-brown scales plus the persistent bases of stipes of previous years. The lamina of the frond is a *shining apple-green colour* when young, maturing to a deep glossy green colour above, paler below, and without scales. Abundant sori usually occur throughout the length of the frond.

Variation Sea Spleenwort shows considerable natural variation in size, in degree of cutting and lobing of the pinna margins, and in distance of pinna separation (the pinnae may be overlapping or widely spaced). Forms with trapeziform, highly lobed, deeply incised and

Jan. Feb. Mar.	Apr.	May	Jun.	Jly	Aug.	Sep.	Oct. Nov. Dec.

Asplenium marinum, fronds from juvenile and smaller cliff-crevice plants: *all* West Cornwall.

cuneate-based, wedge-shaped and tapering pinnae can occur. The plant size is greatly influenced by the environment, but its form seems mainly genetically determined.

Possible confusion A distinctive fern, only likely to be confused with the very rare [×]*Asplenophyllitis microdon* (q.v.), but the thick frond texture as well as the plant's abundance and maritime habitat easily distinguish *A. marinum*.

Technical confirmation Plants are diploids, with $n = 36$, $2n = 72$ chromosomes.

Field notes In most of its stations *A. marinum* occurs in a narrow zone seldom higher than 20–30 m above the sea, descending to within 1–2 m of high-tide level. It occurs in crevices and fissures in steep rocky cliff faces, beneath overhangs and around the mouths or hanging from the roofs of cool, moist, sea-scoured caves, where it frequently reaches its most luxuriant development. Small specimens are occasionally abundant in mortar of old, shaded harbour walls.

Sea Spleenwort requires a habitat which is cool, moist, sheltered from full sun in summer and frost-free in winter. It occurs where the sea is warmer than the land in the coldest weather, and where fine spray carries this winter warmth on to suitable rock faces, minimising frost. It is thus

Asplenium marinum, fronds from large sea-cave roof plants: *a–c*, West Cornwall; *d–f*, Bass Rock, Berwickshire.

Atlantic range of *Asplenium marinum* in Europe.

most frequent on cliffs receiving extensive wavebreak from open Atlantic Gulf Stream water, and is infrequent or absent from the coasts most sheltered from the Gulf Stream and from deep inlets on all coasts where there is insufficient wave-break to produce spray. Plants occur higher on cliffs where their slope and exposure help funnel spray-laden winds upwards. They are highly salt-tolerant, but seem only slightly, if at all, salt-requiring.

Asplenium marinum grows on rock types which are sufficiently hard to provide steep rock faces, but which erode to produce the necessary fissures and caves. These include a wide range of volcanic, metamorphic and harder sedimentary rocks, such as basalts, mica schists, hornblende schists, slaty shales, schistose grits, conglomerates, well-bedded red and yellow sandstones, calciferous sandstones and harder limestones.

In its rocky maritime habitats, associated other plants are usually rather few, and include Rock Samphire (*Crithmum maritimum*), Cliff Sand-spurrey (*Spergularia rupicola*), Golden Samphire (*Inula crithmoides*), Sea Scurvy-grass (*Cochlearia officinalis*), and Buck's-horn Plantain (*Plantago coronopus*). In deep crevices and around shaded aspects of sea-cave mouths, closely associated with individual clumps of well-established Sea Spleenwort plants, are often thriving populations of fast-moving, sleek, grey Bristle Tails (*Petrobius maritimus*), which appear to shelter during dried spells, especially within the root-masses and old frond bases of the fern.

Plants are very slow growing, but prothalli and mature plants succeed in conditions of light too low for dense-flowering plant competition. Plants grow mostly in steeply inclined or inverted positions and, in the spray of the sea, water droplets can sometimes be seen to be shed readily along the stipes to drip from the pinna tips. It seems possible that the wings of the stipe and rachis may play a significant role in such drawing of water droplets away from the sensitive crown.

The frequent restriction of *A. marinum* to pockets on headlands probably causes considerable genetic isolation resulting in local inbred populations which can be morphologically distinct.

Fronds of most plants are usually abundantly fertile, and, even on small specimens, from August onwards, the linear sori bulge at maturity with an abundance of brown sporangia. Such highly fertile fronds persist from summer through the following winter, until the early summer of the following year, not finally withering to creamish-white skeletons usually until the succeeding year's flush of new green fronds has fully expanded. The plant thus appears to have an unusually long spore liberation season, and although almost nothing is known about the ecology of its gametophytes, this long availability of spores would seem to coincide with a habitat in which thermal aspects for prothallial growth are correspondingly long, thus maximising the plant's opportunity for colonisation of any newly available habitats which might become exposed through the processes of winter storm erosion.

Outside Britain and Ireland, Sea Spleenwort is an essentially western North Atlantic fern in distribution, occurring elsewhere chiefly along the west and south-west coasts of Europe and in the islands of Madeira, the Canaries and the Azores. Around the coasts of Britain and Ireland, it becomes increasingly infrequent where the minimum sea-water surface temperature is coolest, disappearing altogether from those areas (chiefly eastern English coasts) where this temperature drops below an absolute average minimum of 5.5°C. In a few of its native localities (notably in Ireland) the plant has been recorded also at appreciable distances from the sea. It seems likely to succeed in penetrating to such relatively inland sites mainly where general mildness of winter climate helps to mimic that of its more usual maritime ones.

Sea Spleenwort is of unusual morphological appearance among British and Irish spleenworts but, on a world scale, is most closely approached in overall morphology mainly by an extensive group of species which are characteristic epiphytes and lithophytes of wet forests of the tropics and subtropics, and especially those of the southern hemisphere. It is probably this group with which *A. marinum* has its closest phylogenetic affinities. Against this background, its Macaronesian–Atlantic range, its essentially frost-free ecological requirements within these islands, and its failure to form any known hybrids with other native northern species of *Asplenium*, seem fully in accord.

Asplenium marinum is much less frequent and less luxuriant in many of its habitats today than in the nineteenth century, and this may be general. Although collection of specimens has doubtless had a significant effect, steady pollution of sites through contamination of seepage water, land drainage, and natural changes in climate (against which this species is in an extreme delicate balance) may also be contributing. Plants should on no account be collected.

Asplenium onopteris L.

Acute-leaved
Spleenwort

Preliminary recognition A medium-sized fern with slender triangular fronds like *A. adiantum-nigrum* but with a more finely dissected blade in which the apex, pinnae, pinnules and segments are as if drawn-out into much finer, more slender, more gradually tapering points.

Occurrence A scarce fern of which the few confirmed stations are scattered in the extreme south and south-west of Ireland. It could, however, yet be found elsewhere, especially in south-west England.

Identification Plants are of a size similar to that of *A. adiantum-nigrum*, but differ chiefly from it in having a more narrowly triangular frond, *with all segments more slender, more finely cut and more linear and gradually tapering* (acuminate) in form, the pinnae usually curving quite distinctly forwards towards the tip of the frond, and a *much longer*, thicker, strongly ascending stipe (usually much more than half the length of the frond). They also differ in having fronds which are slightly thicker, more leathery in texture and of *more shining yellow-green* colour. On the stipe the *deep red-brown colouration ascends much higher* – usually to above or well above the lower pair of pinnae on the upper side of the frond and nearly to the top of the frond (i.e. almost throughout the rachis) on the underside. The frond surface and midribs *lack the scattered narrow dark scales* of *A. adiantum-nigrum*. Most of the blade is held in almost the same plane as the stipe, with lower parts of pinnae often bent forwards out of the plane of the frond.

Variation Moderately variable in the degree of all these characters, though perhaps less so in Ireland than elsewhere in Europe.

Possible confusion Most forms of *A. adiantum-nigrum* are easily distinguished, but a few very attenuate forms approach this species closely, and technical confirmation is needed to separate them. *Asplenium onopteris* can be distinguished from *A.* ˣ*ticinense* (q.v.), its hybrid with *A. adiantum-nigrum*, by the high percentage of abortive spores in the cross.

Technical confirmation A diploid species with $n = 36$, $2n = 72$ chromosomes. The spores are good, appear a paler, clearer more straw-brown colour than the mid- to sooty-brown spores of *A. adiantum-nigrum* and are distinctly smaller (30–40 μm mean varies from 31 to 33 μm, i.e. about 75% of the length of those of *A. adiantum-nigrum*). The spores have a wing which is not so wide and is less interrupted than that of *A. adiantum-nigrum* and the spore is minutely spiny, with some spines projecting from the wing (visible under a light microscope with × 400 magnification).

Asplenium onopteris, typical Irish forms: *a*, Co. Kerry; *b*, Co. Down; *c*, *d*, Co. Cork.

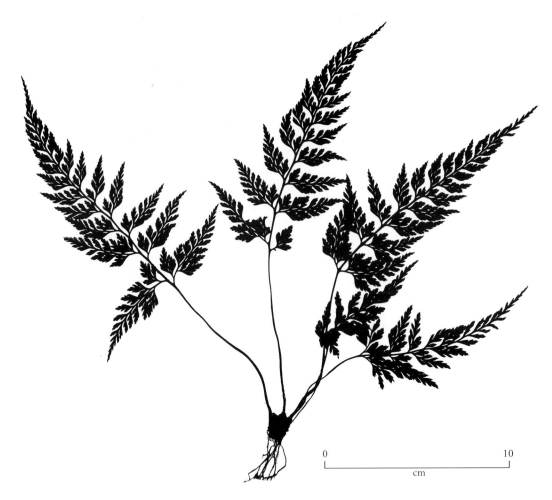

0   10
cm

Asplenium onopteris, whole plant (from preserved eighteenth century herbarium material: Co. Cork).

Field notes *Asplenium onopteris* grows on rather dry, lightly shaded, sunny earth banks and rock faces, especially of limestone. Its sites are all at low altitude, and are mostly within a few kilometres of the sea.

Elsewhere, *A. onopteris* is a common fern all around the Mediterranean, but is most frequent in the west, and is abundant too in the North Atlantic Islands (the Canary Islands, Madeira and the Azores). From the north of the Iberian Peninsula, its range jumps to the south-west of Ireland – a discontinuity shared by a number of other Irish plants of restricted range. The populations in the Mediterranean and the North Atlantic islands are extremely variable in form. Those in Ireland appear, however, to be much more constant and of very much more finely divided frond dissection. The considerable remoteness of the Irish stations suggests that their rather extreme but fairly constant appearance probably results from a considerable period of genetic isolation.

Asplenium ruta-muraria L.

Wall Rue

Preliminary recognition A small tufted fern with finely divided traiangular fronds composed of a small number of stalked, fan-like or spoon-shaped segments.

Occurrence Widely spread through almost all of Britain and Ireland, rather local on lime-rich rocks (especially carboniferous limestones) in natural situations, but frequent and sometimes abundant in old mortar of walls of all types.

Identification Plants have small, wintergreen, ascending fronds, mostly 2–8 cm in length, with *dull dark-green* segments of slightly thick leathery texture. Smaller fronds have trifoliate or simply reniform blades with highly variable stipe length. Larger fronds are more divided, with a small number of ascending alternate pinnae bearing a small number (usually 3–5) of *small, stalked, fan-shaped, wedge-shaped or rhomboidal segments*. The outer margins of all segments are *usually finely serrate*, with a narrow, colourless margin.

The fronds arise in irregular tufts from very short creeping blackish rhizomes, which usually branch to give a small number of adjacent crowns, but never develop to massive clumps. The stipes are slightly wiry in texture, *deep purple-brown at the extreme base, becoming green in colour above*. They are *covered with very numerous small globose glands, readily visible by eye on freshly emerged fronds as a fine, dewy pubescence*, which becomes lost in old specimens. Sori are present on most fronds and their indusia have markedly *ragged-looking, irregularly fimbriate margins* – a character also inherited in hybrids with it.

Variations Extremely variable in frond form. On a single plant, successively older fronds are increasingly divided, whilst there is considerable environmental modification. Plants from moist, shaded situations usually have large fronds with broad segments. Those in dry, exposed situations may be extremely dwarfed.

Possible confusion Despite its variation, most plants remain of distinctive appearance, and are unlikely to be confused with any other common fern. The broader more-divided fronds distinguish it from the vary rare *A.* *×murbeckii*, its hybrid with *A septentrionale*, whilst the glandular stipes distinguish it from all the other *A. septentrionale* hybrids, with which it has some superficial similarity, and all of which are extremely rare.

Jan. Feb. Mar.	Apr.	May	Jun.	Jly	Aug.	Sep.	Oct. Nov. Dec.

Asplenium ruta-muraria: a–c, North Yorkshire; *d*, Midlothian; *e*, Mid Perthshire; *f*, Isle of Arran.

Technical confirmation A tetraploid (autotetraploid) species with $n = 72$, $2n = 144$ chromosomes.

Field notes *Asplenium ruta-muraria* is a principally European species in its centre of distribution, with a probably more fragmented distribution spreading widely into central and northern Asia. In Europe it ranges from northern Fennoscandia to the Mediterranean, with most of its stations concentrated at middle west European latitudes.

In Britain and Ireland, Wall Rue is a strongly calcicolous fern which ranges widely over virtually the whole of these islands. In natural situations it is confined mainly to steep faces of lime-rich rocks, especially limestone, and can be frequent in limestone pavement grykes. It is very widespread and locally abundant in lime-rich mortar of old walls, especially those also constructed of limestone. It often dominates lightly shaded aspects of old country churchyard walls and rural railway bridges and viaducts, and the walls of old walled gardens.

It seems a fast-growing and perhaps rather short-lived spleenwort. In most situations, frequent re-establishment occurs, and its precocious maturity and abundant spore production are presumably considerably advantageous in maintaining a footing on softer mortar and limestone surfaces in the face of steady erosion.

The balance of edaphic and microclimatic factors which determine the success of this species in competition with other mural ferns such as *Asplenium trichomanes* subsp. *quadrivalens* and *Ceterach officinarum* in drier, lighter situations, and *Cystopteris fragilis* and *Adiantum capillus-veneris* in moister, more shaded ones, seems complex and in need of study, as is its occasional occurrence on less lime-yielding rocks in the company of *Asplenium septentrionale* and *A. trichomanes* subsp. *trichomanes*.

Accumulating evidence on this fascinating subject (see e.g. Young, 1985; A. F. Dyer, personal communication) suggests that its gametophyte stages require (or are more tolerant of ?) higher air and substrate temperatures than are those of potentially competing *Asplenium trichomanes* subsp. *quadrivalens* (a mean temperatures difference in the field has been suggested to be as much as 6 deg.). This would also accord with a possibly slightly greater drought-tolerance of its sporophyte (as

Asplenium ruta-muraria, young frond stages (after Orth, 1938).

judged by the occurrence of this species often on higher and lighter aspects of walls compared with *Asplenium trichomanes* subsp. *quadrivalens* (q.v.), which is typically more successful on their lower and more shaded aspects). It also accords with the suggestion (Young, 1985) that temperature cycling is especially effective in the germination of spores of *A. ruta-muraria*. Such spore germination conditions might well occur normally with early-season diurnal temperature fluctuations on more sunny aspects of walls and more natural rock faces, while a related crevice depth correlation (with *A. trichomanes* subsp. *quadrivalens* germinating more successfully in deeper crevices and *A. ruta-muraria*

in more exposed shallow ones) probably also operates.

Today, wall ferns of all types, including both *A. ruta-muraria* and *A. trichomanes* subsp. *quadrivalens*, as well as *Ceterach officinarum* and epiphytes such as *Polypodium* spp., are, through their dependence for water on the immediately incoming precipitation of the district, particularly exposed to the potential effects of sulphurous and nitric oxide pollutants deposited in rainfall, mist and fog (Page, 1988*b*). It is probably as a result of such general acidification of its habitats that *A. ruta-muraria* seems to have declined markedly in the last 50 years in its mainly rupestral habitats in industrial areas.

Asplenium septentrionale (L.) Hoffm.

Forked Spleenwort

Preliminary recognition A small fern with distinctive narrowly wedge-shaped forked fronds, forming compact clumps on steep rocky faces.

Occurrence Rare in a few widely scattered localities, mainly in mountain areas of the British Isles, although not necessarily at high altitude.

Identification Fronds occasionally reach 10 cm or more in length, though are mostly 4–8 cm, their *narrow, lance-shaped segments seldom more than 3 mm in width*. The fronds are simple when young, but when mature, are *irregularly forked a few times, into several narrowly linear segments with acutely tapering tips.* Each division usually has *further small pointed spurs along its outer margin.* At their bases, the blades taper gradually downwards into slender, wiry stipes, as long or longer than the leafy portion, green in colour becoming dark purple-brown near the base. Most fronds bear 3–4 long, linear lengthwise sori, sufficiently crowded on to the backs of the narrow blades to appear confluent at maturity.

The fronds are of a rather stiff, leathery texture, and a deep sage-green colour, which is slightly shining when fresh. They arise from small, short, creeping rhizomes which branch to

Jan. Feb. Mar.	Apr.	May	Jun.	Jly	Aug.	Sep.	Oct. Nov. Dec.

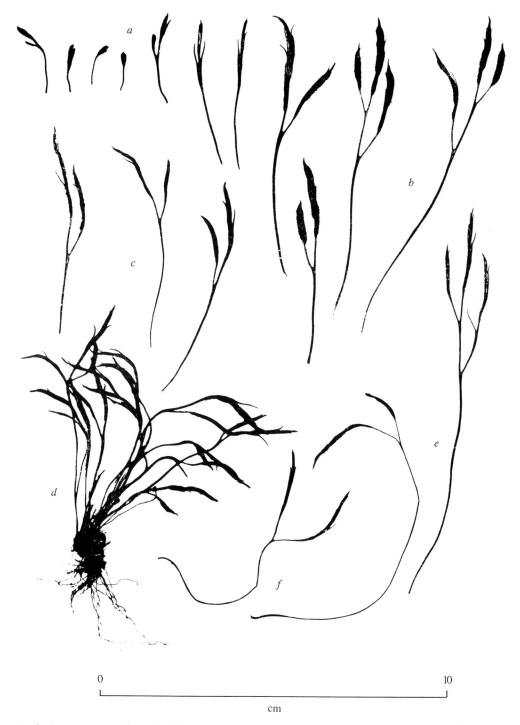

0   10

cm

Asplenium septentrionale: a–b, Mid Perthshire; *c*, Cumberland; *d*, Midlothian; *e–f*, Merionethshire.

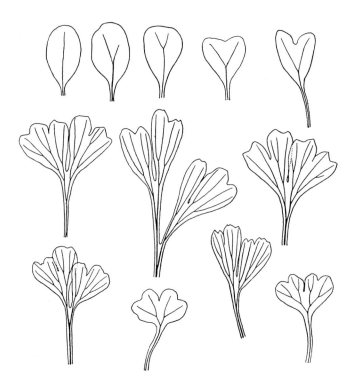

Asplenium septentrionale, young frond stages (after Orth, 1938).

give multiple, tightly packed groups of crowns. On small specimens, the fronds occur in spreading, irregular groups, but as plants mature, the fronds become held more perpendicularly to the rock surface, eventually with older stocks giving rise to densely tufted masses of several hundred fronds per plant, all with their numerous slender tips pointing downwards.

Variation *Asplenium septentrionale* is not a very variable species, except in size of clumps with age.

Possible confusion Unlikely to be confused with any other native fern. The narrowness of the fronds and the compact habit of the plant give a superficial resemblance to a mass of dense, short, curved grass blades, and the plant has also been likened to the leaves of Buck's-horn Plantain (*Plantago coronopus*).

Technical confirmation This is a tetraploid (autotetraploid) species, with $n = 72$, $2n = 144$ chromosomes.

Field notes *Asplenium septentrionale* is a species which is much more abundant on the continent than it is in Britain or Ireland, with a range spreading extensively through the mountainous regions of Scandinavia and central and southern Europe. It is a species of continental climates, tolerant of cold winter conditions and usually demanding relatively hot summers, and habitats in Britain and Ireland, as well as its sole Icelandic station, clearly represent the extreme limit of its oceanic tolerance.

Consequently, in Britain and Ireland, Forked Spleenwort is a very local species, with wide distances between its few localities. Its plants grow gregariously, mainly on steep (often vertical) faces of poor base-yielding, old, hard, dark-coloured volcanic and metamorphic rocks, especially on slates, but occasionally also on granites and basalts.

It is consequently often associated with mountainous areas, but is usually confined to valley sides at only low to moderate heights. In such places, its habitats are almost invariably steep ones, with a generally southerly aspect, and in situations of full exposure, not overgrown by trees. In at least one site in North Wales, it has also successfully spread locally from adjacent cliffs on to low roadside walls of the same slaty rock type.

In cliffside habitats, with much bare rock, other associated plants are often few, but particularly frequently include Gorse (*Ulex europaeus*), which can eliminate Forked Spleenwort by shading if the Gorse bushes become too dense. Many small mosses often occur with the fern, as well as scattered plants of other small crevice-colonising species such as Wood Sage (*Teucrium scorodonia*), Thyme (*Thymus drucei*), English Stonecrop (*Sedum anglicum*), Wall Rue (*Asplenium ruta-muraria*) and Delicate Maidenhair Spleenwort (*A. trichomanes* subsp. *trichomanes*). Very rarely, occasional plants of its hybrids with the latter two species have also been known – *A.* ×*murbeckii* (q.v.) and *A.* ×*alternifolium* (q.v.) respectively. It has been known once to hybridise also with *A. adiantum-nigrum*, to form the hybrid *A.* ×*contrei* (q.v.).

In its sole Irish (Co. Galway) locality, the plant grows on gabbro cliffs, with an associated stunted heath vegetation of lusitanean species of *Erica*. In its most northerly British (west Sutherland) station, it inhabits relatively dry south-west facing exposed rock crevices in epidiorite crags, associated with occasional *Asplenium adiantum-nigrum* (R. E. C. Ferreira, personal communication).

In its exposed, rocky sites, this evergreen species is clearly tolerant of considerable winter cold about its fronds. However, the almost invariably steep and usually south-facing, unshaded situations in which it grows, are also ones which receive considerable direct sunshine. Even in winter, following nights of severe frost, the temperature of the rock in these sites can rise to hand-warmth by mid-morning on a clear sunny day. Such warm situations presumably enable the fern to achieve adequate temperatures for brief periods of active growth, even during severe winter weather. The thermal capacity of its preferred, dark-coloured, hard, siliceous rock types, may well be of ecological relevance in this species' success in oceanic climates on the margin of its geographic range. Further,

although a small-fronded plant, it eventually forms more massive clumps than do most other species of native *Asplenium*. Whilst these undoubtedly serve to absorb and hold a certain amount of moisture, enabling the plant to make a remarkably quick recovery following long dry spells, the sheer sites and the curious structure of the fronds seem to ensure that at the same time, they do not become waterlogged. The dense masses of spreading-descending fronds with winged stipes seem to draw excess moisture away rapidly from the numerous crowns in dense clumps, whilst the curious forking structure of the blades with many acute points directed downwards forms drip tips from which water droplets shed rapidly.

Much play has been made of its frequent association with areas of metalliferous ore mining (especially lead, copper, tin, zinc, manganese) especially in England and Wales, but whether this reflects any essential mineral requirements or merely an ability to tolerate such conditions has often been debated. In such habitats, if plants become shaded by dense establishment of other vegetation such as Gorse, or by upgrowth of trees, then *A. septentrionale* appears to become at a severe competitive disadvantage in the moister and less-airy conditions resulting. In this connection, the nineteenth century observation that plants in cultivation in moist conditions are particularly susceptible to grazing by 'slugs, snails and woodlice' (e.g. Hibberd, 1875: 76) could be significant in the plant's absence from moister situations in the field. Susceptibility to slug depredations might also be related to the plant's apparent success in the naturally moister climates of western parts of the British Isles, in habitats in which highly toxic metalliferous elements surround it.

Clearly, the ecology of this species, rare and on the oceanic edge of the European part of its range in Britain and Ireland, may well prove a complex, although fascinating, one. *Asplenium septentrionale* has become greatly reduced in Britain and Ireland in the last hundred years or more, mainly through collection of specimens involving their total uprooting, but also through upgrowth of shading vegetation in some stations. This is a very slow growing and threatened species in Britain. Adequate conservation measures, founded on a sound understanding of its curious ecology, through detailed research, seem much needed.

Asplenium trichomanes L.

Maidenhair
Spleenwort

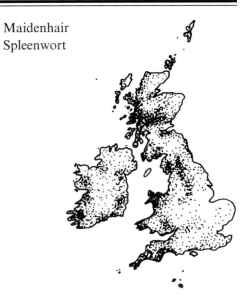

Preliminary recognition A small evergreen fern, with generally long, narrow, tapering fronds either arching from or winding sinuously over rock surfaces and bearing many-repeated similar, mostly rounded and largely uncut pinnae from thin, glossy black midribs.

Occurrence The species is present almost throughout Britain and Ireland in natural rock habitats and often in abundance on walls of all types. Three subspecies are currently recognised in these islands, two of which have much more restricted ranges (see details below).

Identification *Asplenium trichomanes* is a very distinctive and easily recognised species. The fronds are of *slenderly lanceolate outline*, tapering gradually towards the apex, usually 8–20 cm long, occasionally more, seldom over 2 cm wide, and are *divided into a row of many-repeated simple small usually rounded pinnae* on either side, gradually reducing in size towards the frond apex. The pinnae are typically of bright yellow-green colour, small, scarcely stalked, generally rounded or slightly squared and largely uncut. The stipes and rachides are dark throughout their entire length, and two very narrow wings (visible through a × 10 hand lens) run along the full length of the stipe and rachis beneath the pinnae. The fronds arise in tufts, sometimes abundantly, from very short, stocky, decumbent rhizomes.

Variation A quite variable species under differing environmental conditions in addition to the taxonomic variation of the three subspecies. In general, under differing environmental conditions (including luxuriance of growth), all of the taxa may show variation in the degree of development of many of their features as well as overall plant size and numerousness and luxuriance of the fronds. In deep shade, for example, induced differences include thinner lamina texture, flatter pinnae and often wider pinna spacing.

Jan. Feb. Mar.	Apr.	May	Jun.	Jly	Aug.	Sep.	Oct. Nov. Dec.

Asplenium trichomanes subsp. *quadrivalens*

Jan. Feb. Mar.	Apr.	May	Jun.	Jly	Aug.	Sep.	Oct. Nov. Dec.

Asplenium trichomanes subsp. *trichomanes*

Possible confusion The above features should separate the species from other ferns, and those below should separate the subspecies from one another, but technical confirmation may be needed for more doubtful material. The good spores distinguish the species and each of its subspecies from *A.* [×]*lusaticum* (q.v.), the hybrid between the two more widespread subspecies, and from potential *A.* [×]*adulterinum*, and the stipes which are dark throughout from *A. viride* (q.v. for the latter two taxa).

Technical confirmation Plants of subsp. *quadrivalens* are tetraploids, with $n = 72$, $2n = 144$ chromosomes. Their spores are darker and distinctly larger (mostly 30–50 μm) than those of subspecies *trichomanes*. Plants of subsp. *trichomanes* are diploids, with $n = 36$, $2n = 72$ chromosomes. Their spores are paler and distinctly smaller (mostly 25–38 μm) than those of subsp. *quadrivalens*. Plants of subsp. *pachyrachis* and its varieties are presumed to be tetraploids with spores 32–38 μm, but native material is the subject of current research.

Asplenium trichomanes L. subsp. *pachyrachis* (Christ) Lovis & Reichstein (Lobed Maidenhair Spleenwort)

Plants are similar in size to Delicate Maidenhair Spleenwort (q.v.), but with paler coloured, flatter, less arching fronds and longer, more lobed pinnae.

Subsp. *pachyrachis* can be separated from the other two native subspecies of *A. trichomanes* mainly by the following differences. Plants are usually of a size similar to or somewhat smaller than those of subsp. *trichomanes*, measuring about 5–9 cm in length, with a lamina texture which appears as thin and translucent as is that of subsp. *trichomanes*, but with a colouration which is notably paler in the field. Additionally, plants nearly always form *a much flatter rosette of fronds* which *spread very closely over the rock face* on which they are growing, each often adopting a slightly laterally sinuous habit, and upon which rock they tend to *spread more or less equally in all directions*, even on the usual sites of vertical rock faces. Throughout the upper 2/3 or more of each frond, the pinnae are *generally much longer and more distinctly lobed* than is usual in either of the other two subspecies, giving *all specimens of subsp.* pachyrachis *a much greater breadth to the frond* (varying in degree with the variety – see below). The pinnae are borne in a very flat manner in the plane of the rachis, while in the basal quarter or less of the frond, where the rachis arches down into the rhizome crown, the pinnae become rapidly *shorter and more triangular*, with auricled bases, and more distally inclined. In the upper 1/3 of the frond, and especially towards the tip, the pinnae tend to be angled forwards, more

Asplenium trichomanes subsp. *pachyrachis*: *a–c*, West Gloucestershire; *d*, Glamorgan. Note characteristic frequent loss of pinnae from rachides in older specimens.

resembling subsp. *trichomanes* in this respect, but throughout the blade are set more closely together than they are in the other two subspecies, with the distal margin of each pinna tucked below the proximal margin of the pinna above, so that the whole of each frond has a much more congested overall appearance than do those of either of the other two subspecies.

Var. *subequale* Moore has the shorter and less lobed pinnae of the two varieties, usually 4–6 mm in length, and its blade is the least congested.

Var. *trogyense* Lowe has the longer and much more regularly lobed pinnae of the two varieties. Its pinnae are up to 9 mm or more in length, and, especially towards the base of each blade, and often along much of it, typically bear two large, basally pointing thumb-like lobes. These lobes usually also stand markedly above the otherwise flat plane of the rest of the blade, so when the frond is viewed along its length from the base, these lobes stand above it like two regimented rows of leaning fence posts. The structural distinctiveness of this variety from all other native *A. trichomanes*, even in the field, is thus spectacular and impressive.

Subsp. *pachyrachis* was originally known from a very limited number of stations along the southern end of the English/Welsh borderland of Gloucestershire, Herefordshire and Monmouthshire, in steep, shaded aspects of highly lime-yielding substrates. This subspecies could, however, be much more widespread, and its presence in other localities has been suggested. In its originally known localities, plants occur in these regions as at least two structurally distinctive varieties, var. *subequale* and var. *trogyense* (see above).

Asplenium trichomanes L. subspecies *quadrivalens* D. E. Meyer emend. Lovis (Common Maidenhair Spleenwort)

This is by far the commonest of the native subspecies of *A. trichomanes*, present almost throughout the range of the species in Britain and Ireland (see map above), and especially characteristic of both natural lime-rich rock outcrops and often reaching its greatest frequency in mortar of old walls.

Plants of subsp. *quadrivalens* can usually be distinguished by a combination of most of the following features, in which they normally contrast with subsp. *trichomanes*, q.v. Plants typically have a *fairly stout rachis and stipe*, with *rather squared, often asymmetric, closely spaced, thick-textured, scarcely stalked, pinnae* spreading *nearly perpendicularly* (i.e. 'squarely') to the rachis. The pinnae are arranged in nearly opposite pairs, and each pinna varies from flat to concave *downwards* in shape (becoming especially concave in exposed situations). The pinnae are almost without stalks, and in most large specimens, many are fertile on each frond, each fertile pinna bearing numerous (4–9, sometimes 12) quite large sori which are arranged regularly along either side of the pinna midribs. Each also has a conspicuous indusium. On all fronds, the stipes are typically of a polished deep blackish-brown colour, *and remain so with age*. Most pinnae *persist on old fronds until the whole frond is shed*. Plants growing in crevices of vertical rock typically have fronds which *spread rather closely against the rock surface*, many in an ascending direction, and almost always curving as they do so, in a *sinuous 'serpent-like'* manner.

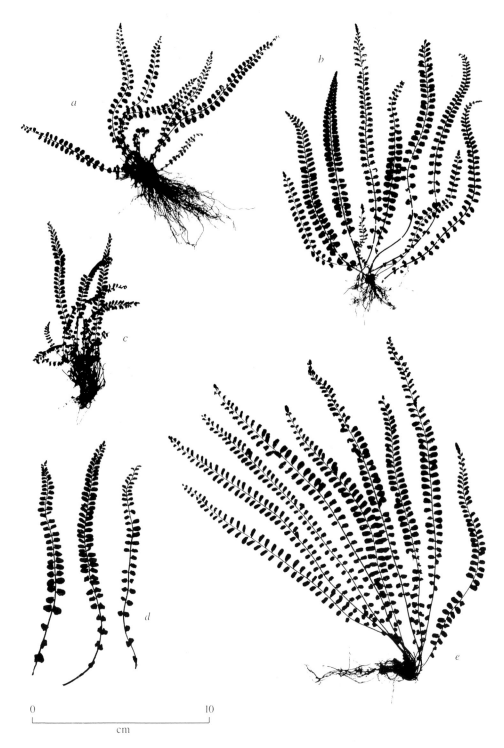

Asplenium trichomanes subsp. *quadrivalens*: *a*, West Yorkshire; *b*, North Northumberland; *c*, Isle of Arran; *d*, Surrey; *e*, Kintyre.

Field notes Subsp. *quadrivalens* is by far the more common and widespread of the three subspecies of *Asplenium trichomanes*. It occurs naturally in fissures of lime-rich rocks, including limestones, at low to moderate elevations, but its frequency in the mortar of old walls (especially in wetter western climates) accounts for much of its general abundance.

It particularly favours walls of old churchyards, interiors of old wells, old walled gardens, disused country railway stations, bridges, viaducts, old castle and abbey ruins, tunnel mouths and walls of disused rural industrial workings. Where these are exposed, plants are often rather stunted, but where these freely-draining lime-rich situations combine with local shelter and high humidity, extensive growths of luxuriant plants may establish.

On walls, it can grow as almost pure stands, or may occur with other wall plants such as Herb Robert (*Geranium robertianum*), Purple Toadflax (*Linaria purpurea*), Wall-rue (*Asplenium ruta-muraria*), Black Spleenwort (*A. adiantum-nigrum*), Western Polypody (*Polypodium interjectum*) and Rusty-back Fern (*Ceterach officinarum*). At low altitude it may occur on natural lime-rich rock in the shelter and shade of ravines along with moisture-demanding plants such as Hard Shield-fern (*Polystichum aculeatum*), Brittle Bladder-fern (*Cystopteris fragilis*), and Golden Saxifrage (*Chrysospenium oppositifolium*), becoming replaced by these species with increasing shade, and in drier habitats it occurs frequently around the tops of fissures of limestone pavements. At higher altitude (above about 900 m, 3000 ft), it becomes mainly confined to crevices in cliff-rocks of mica-schists, where it is hardy alongside Green Spleenwort (*Asplenium viride*), Brittle Bladder-fern, Holly Fern (*Polystichum lonchitis*), and many small alpine flowering plants.

Asplenium trichomanes L. subsp. *trichomanes* (Delicate Maidenhair Spleenwort)

This is a relatively infrequent though fairly widespread subspecies of this fern, present locally mainly in Wales, the Lake District and Scotland, and could yet be found elsewhere. It is present chiefly on natural, usually poorly lime-yielding rocks, extending into wall mortar habitats only more rarely in areas of highest rainfall.

Identification Subsp. *trichomanes* can be separated from subsp. *quadrivalens* mainly by the following differences (although technical confirmation may be needed to separate doubtful material). Plants usually have a *much more slender rachis and stipe*, and have *fewer, much more thin-textured, more rounded, more symmetric, more widely spaced, more distinctly stalked pinnae*, frequently with ornate margins, giving the fronds an overall *much more delicate* appearance in the field. Furthermore, the pinnae are flat or slightly concave *upwards, often have rather scalloped margins*, are usually set more alternately on the rachis, and are *held obliquely forwards* (i.e. angled towards the tip of the frond), increasingly so in the frond's upper half. There are typically fewer sori per pinna (4–6, sometimes 9), which are also smaller and occur mainly near to the outer end of each pinna. They also have a narrower, more delicate indusium. On their upper surfaces, the pinnae usually show fine, radiating striations.

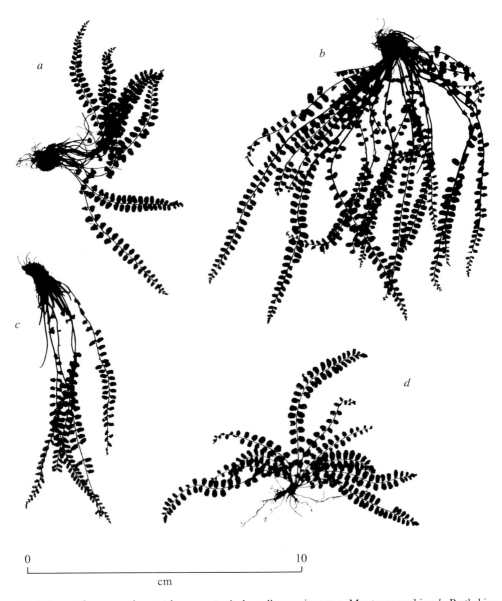

0   10
                    cm

Asplenium trichomanes subsp. *trichomanes*, typical smaller specimens: *a*, Montgomeryshire; *b*, Perthshire; *c*, *d*, Isle of Arran. Note relative ease of loss of pinnae from rachides, both in life and, as here, in herbarium material.

In the field, distinct differences in the habit of the plant from subsp. *quadrivalens* can also be seen: when growing from steep or vertical surfaces, fronds of subsp. *trichomanes* typically *arch away from the rock face and cascade downwards*, rather than adopting the sinuous spreading, 'pressed-to-the-rock' habit typical of fronds of subsp. *quadrivalens*. This difference in growth habit can be distinctive, even from a distance (the habit of subsp. *trichomanes* thus *more closely resembling that of A. viride* than its other subspecies). In addition, marked differences in stipe colour and pinna retention can be seen in the field. In both subspecies, the stipes and rachides on emerging fronds are blackish brown, but they remain this colour only in subsp.

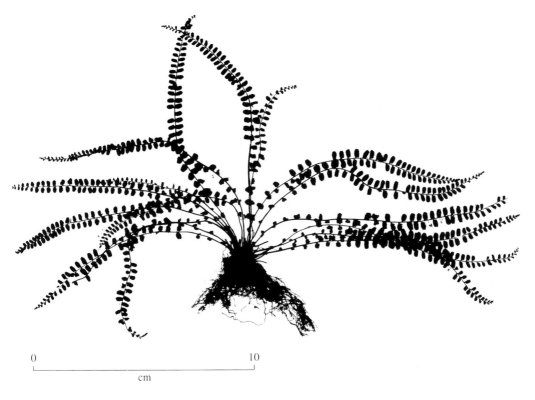

Asplenium trichomanes subsp. *trichomanes*, unusually larger sized specimen: Argyll.

Asplenium trichomanes subsp. *trichomanes*, young frond stages (after Orth, 1938).

quadrivalens: in subsp. *trichomanes* they soon *mature instead to a deep coppery or bronze-red colour, which persists* even into old specimens. Further, a notable feature of subsp. *trichomanes* is that the *pinnae are readily shed from the rachides of ageing fronds* and the resulting *leafless rachides then persist for long periods* (several years in established plants) often to form *dense, leafless tufts below the current year's fronds.* Old frond rachides are seldom retained in this way in subsp. *quadrivalens*. The fronds also unfurl more slowly during early summer than do those of subsp. *quadrivalens*, whilst many of the pinnae become shed during the course of their first winter. The persistent rachides, more delicate appeareance and more cascading frond habit make subsp. *trichomanes* quite distinctive in the field.

Field notes Subsp. *trichomanes* is much less abundant and widespread in the British Isles than is subsp. *quadrivalens*, largely because of the abundance of the latter in mortared-wall situations.

The main habitats of subsp. *trichomanes* are on hard, volcanic or metamorphic rocks, which are not normally highly lime-yielding. Hence plants are mainly confined to mountain areas of the British Isles, but, like *Asplenium septentrionale*, do not necessarily occur at high altitude (the two ferns often occurring at the same sites). *Asplenium trichomanes* subsp. *trichomanes* appears however, more shade tolerant than either subsp. *quadrivalens* or *A. septentrionale*, as well as more susceptible than either to desiccation.

Plants usually grow from fissures where there are small moss and humus accumulations, and although often on southerly exposures, occur especially where these are also lightly shaded from direct sun. In its rocky habitats, this fern seldom meets severe competition from other dense plant growth, but associated species, often in small numbers, commonly include Wood Sage (*Teucrium scorodonia*). Purging Flax (*Linum catharticum*), Mountain Thyme (*Thymus drucei*), Thyme-leaved Sandwort (*Arenaria serpyllifolia*). Herb Robert (*Geranium robertianum*), Wood False-brome Grass (*Brachypodium sylvaticum*), Wood Melick (*Melica uniflora*), Early Hairgrass (*Aira praecox*), Sweet Vernal-grass (*Anthoxanthemum odoratum*), scattered shrubs of Gorse (*Ulex europaeus*), and occasionally Forked Spleenwort (*Asplenium septentrionale*) and Wall Rue (*Asplenium ruta-muraria*). Only rarely does it occur with the other subspecies (such as at some north Wales sites on dolerite rock), creating increased opportunity for its hybrid with subsp. *quadrivalens* (*A. ×lusaticum* – see figures on p. 104) to form.

Asplenium trichomanes subsp. *trichomanes* spreads on to wall mortar only very locally where the walls are sheltered and rainfall is frequent, particularly in western Scotland. Elsewhere, its more delicate, desiccation-sensitive structure as well as different base tolerance compared with subsp. *quadrivalens*, is probably one of the main factors limiting its success on walls.

Asplenium viride Huds.

Green Spleenwort

(*Asplenium trichomanes-ramosum* L.)

Preliminary recognition A small wintergreen fern of similar general appearance to *A. trichomanes*, but with pinnae of softer texture and borne from midribs which are green, not brown.

Occurrence Widely scattered, but often local, in upland regions of Wales, northern England and Scotland (mainly Highland), exclusively on limestone and other base-rich rock. Very local in Ireland.

Identification The lanceolate-outlined fronds, 5–15 cm in length (sometimes more) and under 1.4 cm wide, bearing *numerous entire, rounded to ovate pinnae, attached by short stalks to the green rachis*, are distinctive. The pinnae are well-spaced and tilted forwards in the lower part of the frond, becoming gradually smaller, more crowded and flatter towards its apex. The pinnae have finely but *distinctly toothed margins, which in some specimens become deeply scalloped*. The stipe is normally at least 1/8 to 1/4 the length of the frond (much longer than specimens growing in deep fissures), green in its upper part becoming deep *chestnut-brown only at its extreme base*, and unwinged (cf. *A. trichomanes*).

The fronds are a bright pale-green colour and of delicate, non-glossy texture, and arise in tufts, *sometimes in large numbers*, from short, slender branched rhizomes. Unlike *A. trichomanes* (and especially subsp. *trichomanes*), pinnae are not shed from the rachis of *A. viride*, but old fronds die down as a whole to leave only persistent short frond bases. Sori occur mainly on the upper part of each frond, with 4–8 sori sufficiently clustered near the middle of each pinna to appear confluent at maturity.

Variation Not a particularly variable species, except in size and degree of scalloping to the pinna margins. Forking of fronds often occurs.

Possible confusion Only likely with *A. trichomanes*, but the bright pale-green colour of the rachides as well as the longer stipes without wings easily separate *A. viride*. *Asplenium adulterinum* Milde, a European fertile hybrid resulting from the cross *A. viride* × *A. trichomanes* subsp. *trichomanes*, is known from Scandinavia and central Europe, and could be found in these islands, especially in northern Britain (see silhouette and comment on p. 95). The upper

Jan.	Feb.	Mar.	Apr.	May	Jun.	Jly	Aug.	Sep.	Oct.	Nov.	Dec.

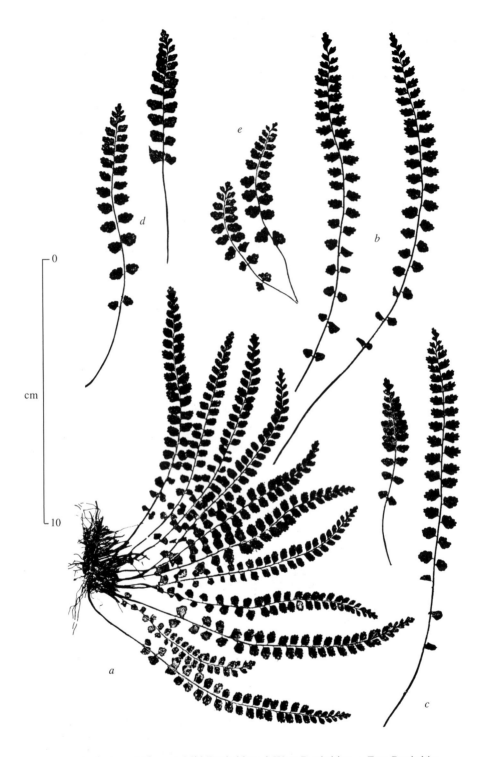

Asplenium viride: a–c, Mid Perthshire: *d*, West Perthshire; *e*, East Perthshire.

part of its rachis has the green colour of that of this species, grading into the purple-brown colour of *A. trichomanes* throughout the lower part.

Recently, the name '*A. trichomanes-ramosum* L.' has been unnecessarily resurrected and proposed to be adopted for the name of this species (e.g. Jermy & Camus, 1991). This trend is *not followed here because of the potential confusion of this new name with the unrelated A. trichomanes* and its several existing subspecies, and the availability and wide usage of the already established, widely known, distinctive and totally unambiguous name *A. viride* for this taxon.

Technical confirmation A diploid species with $n = 36$, $2n = 72$ chromosomes.

Field notes *Asplenium viride* is a widespread species with a geographic range from southern Greenland and Iceland, Britain and Ireland through Europe south to the Iberian Peninsula, the Alps, and the mountains of Greece and Turkey, thence discontinuously across northern Asia to Japan, with outlying stations in mountainous northern regions of the North American continent. In Europe, it occupies the range of a typical arctic-alpine.

In Britain and Ireland, Green Spleenwort occurs in all our more mountainous regions, including south-west and north-west Ireland, Wales and much of northern Britain, and is most frequent in the north of this range.

Plants occur mainly in hilly upland districts and on mountains (to *c.* 1000 m), in habitats with low mean summer temperatures. They grow in moist shaded sheltered crevices of basic rock, especially limestones, base-rich shales and mica schists, but also on rhyolitic lavas, andesites and dolerites, and especially in far northern Scotland, also occur on ultrabasic rock types. In mountains, it is mainly a plant of damp, shaded cliff-faces, screes and rocky sides of ravines, always in intimate direct contact with rock. At lower altitudes it can be frequent in sheltered parts of the grykes of limestone pavements. Its habitats are cool in summer, permanently moist, well drained and sheltered from direct sun and rainsplash. It ascends to higher altitudes than does *A. trichomanes*, where it is clearly tolerant of considerable winter cold. It can succeed in much lower light than *A. trichomanes*, and is much less drought-resistant.

Plants seem long-lived, and old specimens with multiple close crowns can give rise to large masses of arching, descending fronds, forming conspicuous bright green clumps. It seems intolerant of dense competition, and in mountain habitats its main associates include other ferns: Common Maidenhair Spleenwort (*Asplenium trichomanes* subsp. *quadrivalens*), Holly Fern (*Polystichum lonchitis*) and Brittle Bladder-fern (*Cystopteris fragilis*), as well as many small alpine flowering plants, such as Alpine Lady's Mantle (*Alchemilla alpina*), Alpine Penny-

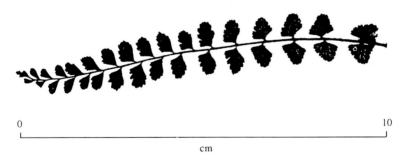

0   10

cm

Silhouette of an undoubted frond of *Asplenium* ˣ*adulterinum* (*A. viride* ˣ *A. trichomanes* subsp. *trichomanes*) in the British Herbarium of the Royal Botanic Garden, Edinburgh. Collected in 1891, the specimen is unlocalised, and could have been from a cultivated source, but bears the inscription 'v.-c. 79'. This is Selkirkshire, and certainly within the general sympatric range of both parents. This hybrid of *A. viride*, known on the continent, could thus occur, and may have formerly been present, in these islands.

cress (*Thalaspi alpestre*), Mountain Sorrel (*Oxyria digyna*), Moss Campion (*Silene acaulis*), Purple Saxifrage (*Saxifraga oppositifolia*), Starry Saxifrage (*S. stellaris*), Yellow Mountain Saxifrage (*S. aizoides*) and Alpine Scurvy-grass (*Cochlearia alpina*). In limestone pavements, *Asplenium viride* is often associated with other ferns, including Wall Rue (*Asplenium ruta-muraria*), Brittle Bladder-fern (*Cystopteris fragilis*), Hart's Tongue Fern (*Phyllitis scolopendrium*) and Limestone Oak-fern (*Gymnocarpium robertianum*), as well as Herb Robert (*Geranium robertianum*), Wood-sorrel (*Oxalis acetosella*), Common Dog-violet (*Viola riviniana*) and Lily-of-the-Valley (*Convallaria majalis*). In lighter, drier situations on limestone cliffs and around the tops of grykes, with increasing light and drier conditions, *A. viride* becomes replaced by *A. trichomanes* subsp. *quadrivalens* and *A. ruta-muraria*, and it is probably its intolerance of drought and of high summer temperatures which prevents it from succeeding, unlike the latter ferns, in the mortar of walls.

Asplenium ×alternifolium Wulf.

Alternate-leaved
Spleenwort

(*A. septentrionale* × *A. trichomanes* subsp. *trichomanes*)

Preliminary recognition A small, tufted fern, with narrowly triangular fronds and slender, forward-pointing, alternate pinnae, growing in steep rocky places in the company of *A. septentrionale*.

Occurrence A very rare plant, entirely restricted to a few very widely scattered stations in North Wales and the Lake District, and formerly also in Northumberland and Scotland.

Identification Plants inherit the narrowness and some of the acuteness of segments of *A. septentrionale*, but have the pinnate habit, more ascending growth and persistence of old leafless stipes of *A. trichomanes* subsp. *trichomanes*. The fronds, mostly 5–8 cm long, occasionally more, have a narrowly triangular-linear or lanceolate outline, with 2–5 narrow, *wedge-shaped, widely spaced, forward-pointing, stalked pinnae arranged in a strongly alternate fashion along either side.* Usually only the lowermost pinnae are extensively cut, sometimes into three portions, whilst most of the upper pinnae are merely *acutely notched, their outer margins trimmed with rather obtuse teeth.*

The short creeping or ascending rhizome branches to give several nearby crowns, but plants form much less-extensive clumps than do those of *A. septentrionale*. Fronds of the hybrid are also of more ascending habit and of paler-green colour. The stipes are wiry, of variable length but usually shorter than those of *A. septentrionale*, and are dark in colour through at least the lower half of their length but with the dark colour occasionally ascending to, or beyond, the lower pinnae. As with both parents, *the stipes have no glands* (cf. *A. ruta-muraria*). The spores are abortive.

Asplenium ×*alternifolium: a–c,* ex-horticulture; *d,* Merionethshire; *e–h,* Cumbria; *i, j,* Merionethshire (*all* from Nineteenth century herbarium material).

Variation There is considerable minor variation in frond shape, degree of cutting, frond size and plant vigour.

Possible confusion The alternate pinna-form, triangular-outlined fronds, shorter, darker stipes, paler-green frond colour and abortive spores distinguish *A.* [×]*alternifolium* from *A. septentrionale*. These characters, plus the narrow pinnules, an entire indusium and absence of glands from the stipe distinguish the hybrid from all forms of *A. ruta-muraria*, to which, although superficially similar, it is unrelated. Two other extremely rare-hybrids involving *A. septentrionale* in their parentage in Britain could also cause confusion: *A.* [×]*murbeckii* (with *A. ruta-muraria*) and *A.* [×]*contrei* (with *A. adiantum-nigrum*). The lack of stipe glands, the more rounded, more obtuse, marginal teeth and the entire indusium separate *A.* [×]*alternifolium* from *A.* [×]*murbeckii* (which is a tetraploid hybrid, thus also differing from *A.* [×]*alternifolium* in chromosome number, q.v.), whilst the much more narrowly triangular frond of *A.* [×]*alternifolium*, smaller size, smaller pinnae and simpler degree of frond (and especially lower pinna) dissection separate it from *A.* [×]*contrei*. Two other continental European hybrids with *A. septentrionale* in their parentage could also cause confusion: *A.* [×]*souchei* (with *A. billotii*) and *A.* [×]*heufleri* (with *A. trichomanes* subsp. *quadrivalens*). The former is very unlikely to be found in Britain through lack of proximity of parents; however, the latter could occur, and would be extremely difficult to separate from *A.* [×]*alternifolium* without the benefit of a chromosome count, as *A.* [×]*heufleri*, unlike *A.* [×]*alternifolium*, is tetraploid.

Technical confirmation Plants are sterile triploid hybrids, with $2n = 108$ chromosomes.

Field notes Plants occur on similar hard volcanic and metamorphic poorly lime-yielding rocks as *A. septentrionale*, and occur only as solitary individuals in the company of this species (those from the Lake District originally grew on Skiddaw slates). Many of the ecological points raised under *A. septentrionale* may also apply to this hybrid.

Although all the hybrids with *A. septentrionale* are rare, *A.* [×]*alternifolium* was clearly formerly less rare than any other. Like *A. septentrionale*, this plant suffered extensively from collection by total uprooting in the nineteenth century, and is now very rare in Britain. On no account should specimens be collected.

Asplenium ×clermontiae Syme

Lady Clermont's
Spleenwort

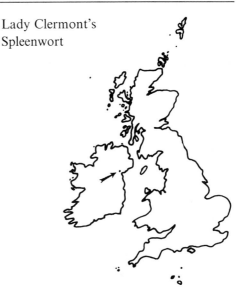

(*A. ruta-muraria* × *A. trichomanes* subsp. *quadrivalens*)

Preliminary recognition Plants have similar overall appearance to *A. trichomanes* subsp. *quadrivalens*, but have a distinctly broader frond outline and at least the upper part of the rachis is green in colour.

Occurrence An extremely rare hybrid, known with certainty only once (mid-nineteenth century in a mortared wall), in Co. Down, Ireland, although a plant later recorded from a similar habitat in Westmorland was perhaps also this hybrid.

Identification Plants have narrowly lanceolate or narrowly triangular fronds, mostly 4–6 cm in length, which are *simply pinnate in structure* in a similar manner to *A. trichomanes*, but are slightly broader, with fewer, more distinctly stalked pinnae. The lowermost 2–4 pairs of pinnae are characteristically trilobed, and all have minutely denticulate outer margins. The stipes occupy up to about 1/4 the length of the frond. At least the upper half of the rachis is green in colour, or the green colour may descend through the upper half of the stipe, which then grades into brown in only its lower half. Only one or a few sori are present on each pinna throughout the length of the frond, and these *inherit the irregular-edged indusium* of the *A. ruta-muraria* parent. Their spores are abortive.

Variation There seems to be considerable variation in the length of rachis and stipe, which is green. The number of pinnae which are trilobed increases with size and luxuriance of the frond.

Possible confusion *A.* ×*clermontiae* is readily distinguished from its parents: from *A. ruta-muraria* by its narrower pinnate frond, and from *A. trichomanes* by the stalked pinnae with the lower ones larger and lobed, and the partially or wholly green rachis. The long, narrow, simply pinnate frond readily distinguishes *A.* ×*clermontiae* from *A.* ×*murbeckii*, which shares the common *A. ruta-muraria* parent. The plant also has a superficial resemblance to a *Woodsia*, but the paucity of hairs and scales as well as the linear sori along the veins of the *Asplenium* readily distinguish it.

Technical confirmation Plants are tetraploid hybrids, with $2n = 144$ chromosomes.

Field notes The plant was originally found growing with both parents in the mortar of a garden wall, and should be searched for again in such situations. The hybrid is nevertheless remarkably rare considering the abundance of both parents and the frequency with which they occur together in quantity in old mortared walls throughout many parts of the British Isles, although amongst a dense growth of quantities of both parents, the hybrid could easily be overlooked.

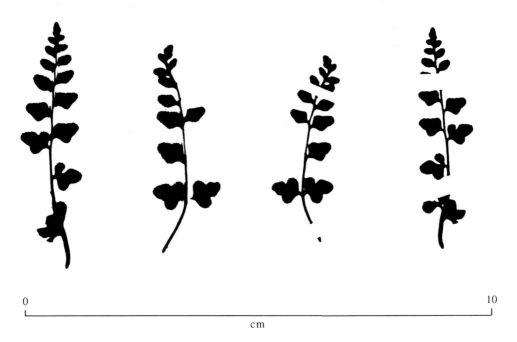

0   10
                                          cm

Asplenium ˣ*clermontiae: all* Co. Down (from nineteenth century herbarium material).

Asplenium ×contrei
Calle, Lovis & Reichstein

Caernarvonshire
Spleenwort

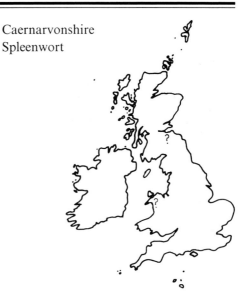

(*A. adiantum-nigrum* × *A. septentrionale*)
(*A.* [×]*souchei* auct., non Litarde)

Preliminary recognition A small fern closely similar to *A.* [×]*alternifolium* in general appearance, but larger in size, with more broadly triangular fronds.

Occurrence An extremely rare hybrid, known only from a single Caernarvonshire locality a century ago. Possibly extinct, but it could recur where its parents grow together.

Identification Plants have a strong intermediacy of appearance between *A. adiantum-nigrum* and *A. septentrionale*. Their fronds can reach about 12 cm long, but are mostly 7–10 cm. The lowermost pinnae are clearly longer than those above, giving the frond a distinctly triangular outline, and at least the lowermost pinnae are themselves divided into three or more narrow main pinnules. All pinnae and pinnules end in acutely pointed segments. The stipes and blades are of nearly equal length, and the stipes green in the upper half and dark brown below. The spores are abortive.

Variation Unknown, but different clones would presumably reflect some of the variability of the *A. adiantum-nigrum* parent (q.v.).

Possible confusion Its strong intermediacy of appearance and abortive spores readily distinguish *A.* [×]*contrei* from both parents. It is most likely to be confused with *A.* [×]*alternifolium* (q.v.), which shares the common *A. septentrionale* parent, but *A.* [×]*contrei* differs in its proportionately longer stipes, broader, more-divided frond with more acute segments, and generally larger size. It can be separated from large specimens of *A.* [×]*murbeckii* by the above characters and absence of the irregular edge to the indusium of the latter hybrid.

Technical confirmation A tetraploid hybrid with $n = 144$ chromosomes.

Field notes In its original locality, the plant probably occurred on slaty rock faces. *A.* [×]*contrei* appears to be a large and vigorous hybrid, and one reason for its rarity is probably the absence of *A. adiantum-nigrum* from most British sites of *A. septentrionale*.

A more recent find of a hybrid which is very probably of this parentage has been from Arthur's Seat, Edinburgh. Here the plant grows on south-facing, steep, basalt crags, associated with locally frequent but scattered plants of *A. septentrionale*, with numerous scattered *A. ruta-muraria* and occasional *A. adiantum-nigrum*. Clearly the plant could be found again in the very few sites in which its parents continue to grow together.

Asplenium ×*contrei: a–e.* Caernarvonshire (1870); *f–i,* horticulture, ex Caernarvonshire (*all* from nineteenth century herbarium material).

Asplenium ×lusaticum
D. E. Meyer

(*A. trichomanes* subsp. *trichomanes* × *A. trichomanes* subsp. *quadrivalens*)

Preliminary recognition A relatively large and vigorous Maidenhair Spleenwort, of frond form intermediate between that of the two parents, and tending to have arching and cascading fronds which are long and borne in considerable abundance on individual plants.

Occurrence A surprisingly rare hybrid, which may be generally overlooked. It has been recorded with certainty only in a few localities in North Wales (Merionethshire) and western Scotland (Arran), and could well yet be found in other districts where both parents grow near together.

Identification Fronds of this relatively vigorous hybrid are typically large, often 10–20 cm or more in length in well-established clumps, and usually arch away from the rock surface and cascade downwards in the manner of those of subsp. *trichomanes* (q.v.), but are usually longer and more robust. The overall blade outline is narrow, more resembling that of subsp. *quadrivalens*, as are the rather squared pinnae which are fairly perpendicularly inserted throughout the greater part of the length of the frond and have the relatively deep green blade colour of this subspecies. The pinnae are, however, relatively conspicuously stalked and have margins which are usually somewhat scalloped, as in subsp. *trichomanes*. In the upper quarter of the blade the pinnae become more obliquely inserted along the rachis, as in this parent, while the stipe also usually possesses a more copper-tinged colouration.

Variation Apart from considerable variation in size with age and presumably habitat, so far as is known, plants from the two known distant sites appear to each show the relatively constant blend of parental characters noted above.

Possible confusion Likely with each of its parents from which the usually greater vigour of *A.* [×]*lusaticum* as well as intermediate morphology serve to distinguish it initially. The abortive spores (see below) confirm its hybrid status. This, plus the lack of lobing of the pinnae, distinguishes this hybrid from the very local *A. trichomanes* subsp. *pachyrachis* (q.v.).

Technical confirmation *A.* [×]*lusaticum* is a triploid hybrid with $2n = c.$ 108 chromosomes (cf. $2n = 72$ in subsp. *trichomanes*, $2n = 144$ in subsp. *quadrivalens*). Its spores are abortive.

Strictly speaking, this plant is an intersubspecific hybrid (since both parents are regarded as subspecies of *A. trichomanes*). Although correct, this is, however, a somewhat technical concept of its status, and since it differs in chromosome number from both of its parents, from which it is hence genetically isolated, I have thus treated it here, for consistency, in the same way as other interspecific hybrids.

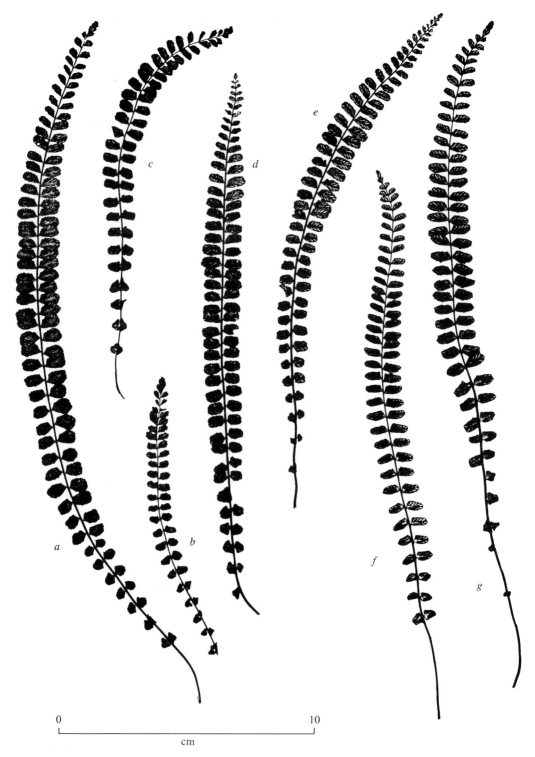

Asplenium [×]*lusaticum: a–e*, Arran; *f–g*, Merionethshire.

Field notes Plants have been found in generally high-rainfall areas of the western British Isles, in north Wales (Merionethshire) by P. Benoit, and in western Scotland (Isle of Arran) by the author. This hybrid may well be more widespread in each of these areas than is currently known, and could be sought in others, such as notably, perhaps, in southern parts of the Lake District. In both areas plants are known from a small number of moderately shady, sheltered and rocky lanebank and roadside sites where a mosaic of rock types assures a close admixture of patches of locally varying base status, adequate for both parents to grow nearby with some frequency. Plants of the hybrid typically ultimately form larger clumps showing greater vigour than either parent locally. Although it seems likely that this hybrid may be largely underrecorded, nevertheless the infrequency with which it is known with certainty, and its occurrence in its known sites in predominantly artificial habitats may indicate a genuine rarity in nature.

Asplenium ˣmurbeckii
Dörfl.

Murbeck's
Spleenwort

(*A. ruta-muraria* × *A. septentrionale*)

Preliminary recognition A very small fern, sometimes 3 cm or less in height, with triangular-outlined fronds composed of a small number (often only 3) of narrowly wedge-shaped, stalked lobes.

Occurrence A very rare hybrid, known formerly in East Perthshire and Midlothian (mid-nineteenth century) and more recently in Cumberland, on hard volcanic (basaltic) and metamorphic rock ledges, with southerly exposure.

Identification Fronds are mostly 3–6 cm in length, occasionally to 10 cm, with usually only 1–2 pinnae plus a single terminal pinna per frond. The pinnae are themselves subdivided only in occasional large fronds. The frond outline is approximately triangular, and the pinnae are merely narrowly wedge-shaped segments with an irregularly rounded outer margin trimmed with numerous fine, acute teeth. The stipes are long, occupying 1/2 to 3/4 the length of the frond, green throughout except at their extreme bases, which are deep brown, and similar glands to those of *A. ruta-muraria* are present (q.v.). The leafy portion of the fronds is of slightly thick, leathery texture, and of dull deep green colour. Sori are few but large, only a single sorus usually occurring on each pinna undersurface, and its indusium has a distinctly irregular edge.

Variation Mainly in frond size.

Possible confusion Smaller fronds approach *A.* ˣ*alternifolium* in superficial appearance, larger ones *A.* ˣ*contrei*, both of which share the common *A. septentrionale* parent. But *A.* ˣ*murbeckii* differs from both in inheriting the irregular-edged indusium of *A. ruta-muraria* and the longer, more extensively green stipe. It also differs from *A.* [×]*alternifolium* in having more acute pinna tips with minutely serrate pinna margins composed of fine acute teeth (cf. more rounded pinna tips with coarser, more obtuse teeth in *A.* [×]*alternifolium*).

Technical confirmation *A.* [×]*murbeckii* is a tetraploid hybrid, with $2n = 144$ chromosomes (cf. *A.* [×]*alternifolium*, which is triploid).

Field notes *A.* [×]*murbeckii* is an extremely rare hybrid, likely to occur as an occasional plant in the presence of both parents, in the small number of sites in Britain where both grow together. Its rock ledge situations are on hard volcanic and metamorphic rock faces and, like *A. septentrionale*, seem to be in exposed patches with southerly aspect. Although recent studies of the only British material have shown it to be highly sterile with a small number of meiotic bivalents, theoretically, as both parents are autopolyploid, a high number of chromosome pairs giving partial fertility could occur; such fertility is known amongst some wild continental examples which have given progeny when sown in cultivation. From these, backcross hybrids, which are always highly sterile, have been produced in cultivation. Such phenomena could occur in the wild.

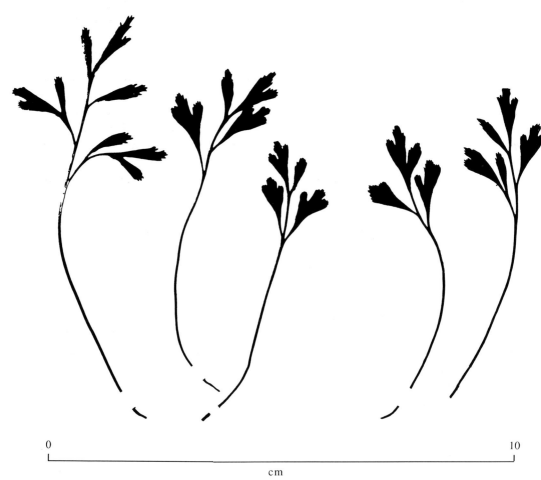

0   10

cm

Asplenium [×]*murbeckii, all* Mid Perthshire.

Asplenium ˣsarniense Sleep

Guernsey Spleenwort

(*A. adiantum-nigrum* × *A. billotii*)

Preliminary recognition A small to medium-sized fern, much like a large and vigorous form of *A. adiantum-nigrum*, but with a narrower frond and blunter segments. The pinnae and pinnules are well spread, giving the fronds an overall rather lace-like appearance.

Occurrence A rare fern known only from the Channel Islands (Guernsey) where it occurs in several localities on sheltered roadside hedgebanks. It was discovered only recently, and could yet be found elsewhere within the overlapping ranges of its parents (particularly south-west England or possibly the Welsh coast).

Identification Plants have fronds mostly reaching about 30 cm in length (but may be as large as 40 cm) and a narrowly triangular frond outline, with the lowest pinna pair (or rarely the next two above it) the longest. The cutting of the frond is distinctly intermediate between that of the parents, and the fronds usually present an *overall more finely cut, blunter-segmented appearance* than is typical for *A. adiantum-nigrum* (the parent for which they are most likely to be taken). The pinnules (especially those in the middle of the frond) are oval to orbicular in shape, rather crowded and sometimes overlapping, and they usually have distinct short stalks. All *pinnules are much more oval in shape than those of A. adiantum-nigrum*, and they have the *obtusely rounded tips* trimmed with the obtuse mucronate-tipped teeth of *A. billotii*. As in *A. billotii*, some of the pinnae (especially in the lower half of the frond) may curve slightly back towards the frond base. The stipe is usually shorter than that of *A. adiantum-nigrum* (mostly much less than half the length of the frond) and slightly stouter, dark brown on the underside and either brown or green on the upper side, with the upper side of the rachis frequently green throughout.

The leafy part of the frond is of a dark green colour and much like *A. billotii* in texture. Sori are produced in abundance almost throughout the frond, and although the spores are mostly abortive, occasional better-looking spores occur which are dark brown in colour and approach those of *A. adiantum-nigrum* in form.

Plants with young fronds can be readily distinguished from *A. adiantum-nigrum* by the immature indusia having distinctive orange edges along one side, whilst the spores also mostly have a markedly orange-brown tinge (A. Sleep, personal communication).

Variation There seems to be considerable size-variation, as well as some variation in details of frond outline and degree of cutting. Occasional individuals with fronds 30–40 cm long forming clumps about 50 cm across are known.

Possible confusion Most likely to be confused with its parents, but the large lower pinnae forming a more triangular blade, and the longer stipe, should separate it from *A. billotii*. The

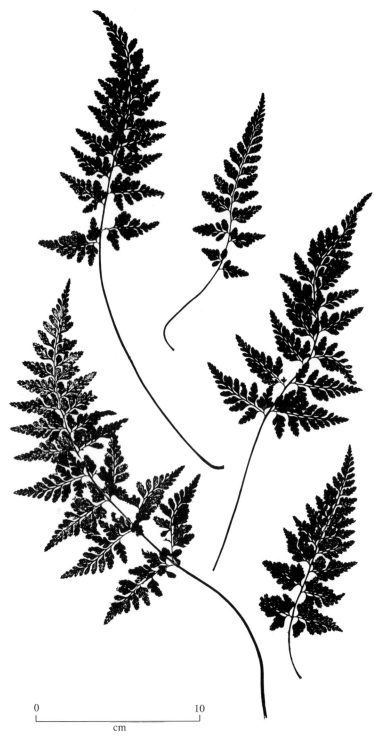

0
10
cm

Asplenium ×*sarniense: all* Guernsey, Channel Islands (courtesy Dr Anne Sleep).

narrowness of the frond and pinnule shape with an overall blunter-segmented appearance and more oval pinnules from *A. adiantum-nigrum*, and the overall more finely cut, rather lace-like appearance as well as colour and time of flushing (see 'Field notes'). should separate it from both.

Technical confirmation Plants are tetraploid hybrids, with $2n = 144$ chromosomes.

Field notes *Asplenium* ˣ *sarniense* was first recognised in the early 1970s (Sleep, 1971*b*), and is now known in a number of separate stations in the western and southern areas of the island of Guernsey, where it grows on steep, earthen hedgebanks, mostly at very low altitude and near to the sea. Plants occur mostly as widely scattered individuals, but in one area, over 20 plants have now been recorded scattered along the banks for about 400 km of a single, sheltered bridle-track (A. Sleep, personal communication).

In Guernsey, its fronds begin to flush as early as late March, usually slightly later than those of neighbouring plants of *A. billotii*, but much earlier than local plants of *A. adiantum-nigrum*. The young fronds long retain a rather pale pea-green colour, more closely resembling that of *A. billotii* in the field than the ultimately rather darker blue-green of the *A. adiantum-nigrum* parent. They do, however, typically have the more erect habit of the latter parent, and it is this combination of features which makes them most readily initially identifiable in the field.

Its habitats are mostly on the south- or west-facing sides of sunny or lightly-shaded, sheltered slopes or sandy-loam or clay soil amongst granite boulders. They are usually slightly more exposed than those of ˣ*Asplenophyllitis microdon* (q.v.) in similar nearby habitats and, although very freely-drained, probably remain just damp throughout much of the year, through slight ground moisture seepage. Their steep position mostly ensures there is also some surface erosion in these habitats, maintaining a mosaic of slightly open patches and denser vegetation. Both parents are usually present nearby, often in some quantity, but the shallow, rocky, steep sites usually prevent the establishment of too

tall and rank surrounding herbaceous vegetation, and associated species are rather few. Besides the parents and Hart's-tongue Fern (*Phyllitis scolopendrium*), these are limited mainly to Ivy (*Hedera helix*), Wall Pennywort (*Umbilicus rupestris*), Celandine (*Ranunculus ficaria*), Red Campion (*Mellandrium rubrum*), Bluebell (*Hyascinthoides nonscriptus*), Violet (*Viola riviniana*), and Wood Falsebrome (*Brachypodium sylvaticum*).

Although all confirmed records in these islands are limited, *Asplenium* ˣ *sarniense* could yet be found in other areas of Britain and Ireland, where the two parents grow together, including especially perhaps in the other Channel Islands, in the Scilly Isles, in coastal Devon and Cornwall and in west Wales, in such habitats as artificial steep earth-and-rock banks.

Jermy *et al.* (1978: 55) record that sterile specimens of similar morphology are also known from the adjacent Brittany coast. In Guernsey, the frequency of this hybrid is probably related to the large number of suitable lanebank sites, and the frequency of both parents over the essentially granitic substrates, and suggests that most plants have probably originated as independent crossings. The wide range of sizes of hybrid plants probably results from variation in individual vigour, as well as the likelihood of their having originated at many different times in the historical past. The habitats in which they occur are virtually all ones along the banked-up sides of footpaths, bridle-paths and lanes leading between fields or along the coast. The origins of many of these tracks and fields must trace back through considerable periods, and it seems likely that these hybrids from time to time form anew, and have perhaps done so in these habitats, through a long period of Channel Islands history.

Asplenium ˣticinense
D. E. Meyer

Hybrid Black
Spleenwort

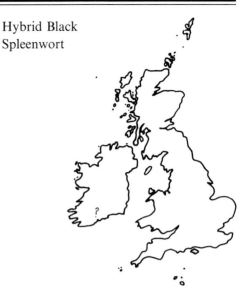

(*A. adiantum-nigrum* × *A. onopteris*)

Preliminary recognition A rare hybrid of intermediate appearance between *A. adiantum-nigrum* and *A. onopteris.*

Occurrence Extremely local, and likely to be confined to localities where both parents grow close together. Known only from southern Ireland.

Identification Specimens which have been recognised in the wild have large fronds (over 30 cm in length) and are of distinctly intermediate appearance in degree of frond dissection between the more coarsely cut, blunter frond of *A. adiantum-nigrum* and the more finely cut and attenuate frond of *A. onopteris.* They inherit the long, acuminate frond and pinna tips of the *A. onopteris* parent, with segments which are narrowly lanceolate, but less so than in *A. onopteris.* The stipe is nearly half the length of the frond, thicker than in *A. adiantum-nigrum*, and the dark colouration ascends higher (to above the lowermost pinna pair).

Variation Crosses made synthetically are known to show wide variation in morphology between that of the parents, in which characters of either parent may be dominant. They generally approach the *A. adiantum-nigrum* parent more closely, but have narrower segments.

Possible confusion The combination of the above characters and especially a high percentage of totally abortive, mis-shaped spores separates *A.* ˣ*ticinense* from *A. adiantum-nigrum* and *A. onopteris.*

Technical confirmation Plants are triploid hybrids, with $2n = 108$ chromosomes, forming 36 pairs and 36 singles at meiosis in synthesised plants. (Wild plants have not been cytologically studied.)

Field notes *A.* ˣ*ticinense* could arise anywhere where the two parents, which are known to be capable of crossing in either direction, grow together. Its Irish station is on a partially shaded grassy and stony bank.

Synthesis of artificial hybrids shows that some plants show hybrid vigour and others do not.

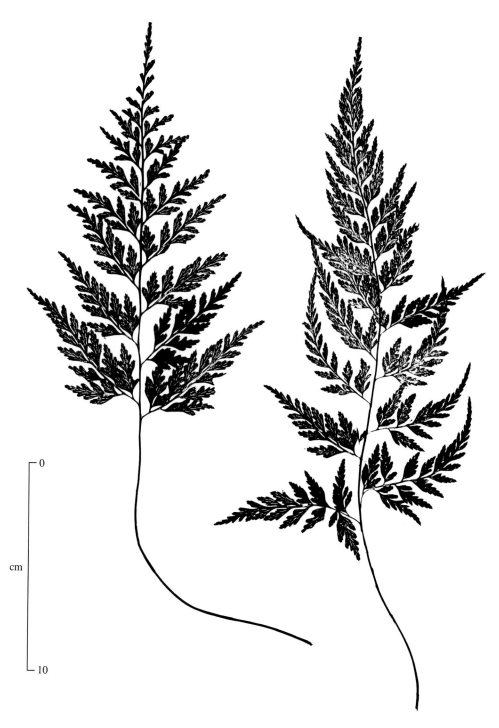

Asplenium ˣ*ticinense: all* Co. Cork.

*Asplenophyllitis confluens
(T. Moore ex. Lowe) Alston

Confluent Maidenhair
Spleenwort

(*Asplenium trichomanes* subsp.
  quadrivalens × *Phyllitis scolopendrium*)

Preliminary recognition A small fern, looking like a rather more robust, broader-outlined form of *Asplenium trichomanes*, with fewer broader and larger pinnae notably overlapping and becoming fused in the upper part of the frond.

Occurrence A very rare hybrid, known wild once in each of three widely separated localities: north-east Yorkshire, Westmorland and Co. Kerry. It could, however, occur within the overlapping ranges of its parents almost anywhere in Britain or Ireland, and has recently been so in Kerry.

Identification Plants have fronds of similar size (10 cm or more) to those of *Asplenium trichomanes*, with a slightly broader lanceolate outline, but with a much thicker stipe and midrib to each frond, and broader, more overlapping pinnae. The pinnae are mostly of somewhat irregular length, the upper ones overlapping and becoming confluent in the upper 1/4 or less of the lamina. Some of the lowermost pinnae are distinctly fan shaped. The leafy portion is green and of thin texture, with the stipe a dark, shining black throughout most of the length of the frond, except in the confluent apex where it becomes green. The spores are abortive. The habit of the fronds is presumably more ascending than that of *A. trichomanes* subsp. *quadrivalens*.

Variation Little known. Considerable variation in the length of the confluent portion might well be expected.

Possible confusion The thicker stipe and rachis, confluent upper pinnae and abortive spores, distinguish this hybrid from *Asplenium trichomanes*, the only parent with which it is likely to be confused. The much narrower frond outline distinguishes it readily from other known *Asplenophyllitis* hybrids (except perhaps from *Asplenophyllitis claphamii* (T. Moore) Alston, of nineteenth century horticultural origin, and unknown parentage).

Technical confirmation Presumed to be a triploid hybrid with $2n = 108$ chromosomes.

Field notes The occurrence of this rare hybrid, of distinctive form, has been both scattered and sporadic. Known records include those of Cumbria (Levens Park, v.c. 69, by Stabler in 1865), North Yorkshire (Whitby, v.c. 62 by W. Wilson *c.* 1870) and Co. Kerry (Killarney, v.c. H2 by F. N. Fraser in *c.* 1875), with other possible, though unauthenticated and unlocalised, Irish records.

Since the nineteenth century, *Asplenophyllitis confluens* had been unknown in any of these former or other localities, until a single plant of it was recently found in Co. Kerry (Rush, 1983). The form of the frond of this plant (see figure on p. 114) matched closely those of the nineteenth century finds, and like the former Kerry specimens (originally found some 24 km distant from the new locality), the habitat of this new plant was that of wall mortar.

The small size of this plant and its low frequency in the wild considering the abundance of its parents and suitable habitats, suggests that this is not a

×*Asplenophyllitis confluens: a–d*, Kerry; *e*, Cumberland.

particularly vigorous hybrid. Nevertheless, it has been successfully synthesised in cultivation (Lovis & Vida, 1969), and it could be found again in the wild in areas where its parents occur together, particularly, perhaps, in limestone districts.

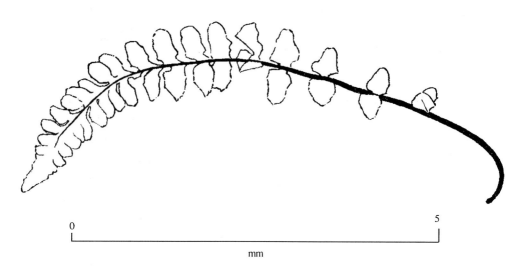

×*Asplenophyllitis confluens*: Co. Kerry (after Rush, 1983).

ˣ**Asplenophyllitis jacksonii** Alston

Jackson's Fern

(*Asplenium adiantum-nigrum* × *Phyllitis scolopendrium*)

Preliminary recognition A small fern resembling a fleshy, less finely cut, simply pinnate form of *A. adiantum-nigrum*.

Occurrence A very rare hybrid recorded in the mid-nineteenth century from widely scattered stations in the Channel Islands, Devon and Cornwall, growing in similar habitats to *A. adiantum-nigrum*. Although it has not been refound since, it could possibly occur anywhere within the overlapping ranges of the parents (virtually the whole of Britain and Ireland), but is perhaps most likely in south-western areas, where both parents reach their greatest abundance.

Identification ˣ*A. jacksonii* has a narrowly to broadly triangular, regular to slightly irregular frond outline, usually 10–16 but up to 20 cm long, somewhat variable in the completeness of division, but usually divided into more or less entire, partly overlapping pinnae, themselves slightly lobed.

The fronds are ascending, mid- to yellowish-green, of fleshier texture than most pure *Asplenium* plants, with a rather undulating frond surface often curving slightly downwards towards the outer edges of the frond (as in *P. scolopendrium*). Their stipes are noticeably shorter than those of *A. adiantum-nigrum* and less black.

The sori are long and frequently twinned, as in *Phyllitis* (q.v.), and occur near to the midrib of each pinna, whilst the spores are abortive.

Variation Insufficient material is known to assess this.

Possible confusion The distinctly triangular frond outlines and the pinnately-cut blades are very different from those of *P. scolopendrium*, but much more closely resemble those of the *A. adiantum-nigrum* parent. They can best be distinguished from the latter by their coarser degree of cutting, thinner frond texture, shorter stipes and abortive spores. ˣ*A. jacksonii* can be separated from ˣ*A. microdon*, which shares the common *P. scolopendrium* parent, by its more ascending broad triangular fronds (cf. more spreading ovate-lanceolate fronds of ˣ*A. microdon*) and normally longer stipes (more than 1/4 the frond length or more, cf. less than 1/4 in ˣ*A. microdon*, q.v.). The ascending triangular fronds of fleshier texture and abortive spores also distinguish ˣ*A. jacksonii* from *Asplenium marinum*, some forms of which it superficially resembles.

Technical confirmation Plants are triploid hybrids with $2n = 108$ chromosomes, showing only single (unpaired) chromosomes at meiosis.

[×]*Asplenophyllitis jacksonii: all* Guernsey, Channel Islands (silhouettes prepared from nineteenth century herbarium specimens).

Field notes The scattered nineteenth century records of ×*A. jacksonii* in the Channel Islands and south-west England suggest that this is a hybrid which can form from time to time in regions where the parents are abundant and can grow closely together. Its habitats seem likely to have been mainly lanebank ones, where there is abundant *A. adiantum-nigrum*, and probably not differing greatly from the habitats occupied by this parent, nor from those occupied by ×*A. microdon* (q.v.) in the Channel Islands today. The frequency of this type of lanebank in all of these areas suggests that this has probably been an important factor in the occurrence of this fern.

The reasons for its presence in the mid-nineteenth century and apparent absence since in these areas is less easily understood, and contrasts with the upsurge in recent records for ×*A. microdon*, at least in Guernsey. The parents of ×*A. jacksonii* would appear to occur as frequently together, if not more so (especially in Devon and Cornwall), as do those of ×*A. microdon*, and it seems no less likely to be overlooked. The original plants appear to have been fairly vigorous. The plant from Guernsey was probably dug up, but thereafter seems to have been successfully maintained in cultivation for over half a century, becoming propagated and listed by at least four fern nurseries between 1865 and 1928 (McClintock, 1975). Artificial synthesis of this hybrid in cultivation (Lovis & Vida, 1969) suggests that it may be more difficult to produce than ×*A. microdon*, hence indicating one likely reason for its rarity in the field, despite the abundance of its parents, compared with the latter.

A further factor of significance in reducing the frequency of ×*A. jacksonii* hybrids compared with ×*A. microdon* may well perhaps be the rather later-season maturation of the spores of both *A. adiantum-nigrum* and *P. scolopendrium* compared with the very early season flushing and maturation of *A. billotii*. It seems possible that prothalli of *A. billotii* may well be more frequently in a condition to act as female parents to either of the others, and it may be significant in this connection, that under conditions of artificial synthesis, Lovis & Vida successfully synthesised hybrid ×*A. microdon* and ×*A. jacksonii* only with the respective *A. billotii* and *A. adiantum-nigrum* parents as the female partners.

×*Asplenophyllitis microdon* (T. Moore) Alston

Guernsey Fern

(*Asplenium billotii* × *Phyllitis scolopendrium*)

Preliminary recognition A plant with a strong superficial resemblance to *Asplenium marinum* (although not related to it), but with larger, thinner-textured, more-ascending fronds, of crinkled, somewhat untidy appearance.

Occurrence A rare hybrid, known wild in the Channel Islands (Guernsey) and as unconfirmed nineteenth century records also from Cornwall, Devon and Wales. It could be refound anywhere within the range of overlap of the parents (mainly the Channel Islands, south-west England, west Wales and southern Ireland).

Identification Plants may have robust fronds, up to 30 cm long, which have a distinctly ovate-lanceolate, somewhat irregular, outline. Each is only 1 × pinnately cut into entire, somewhat lobed, distinctly overlapping, crowded, tapering blunt-tipped, broad pinnae, often of unequal length, *which usually become more or less confluent in the upper part of the frond*. Each pinna usually has a broad base, and its margins are irregularly waved and finely toothed.

0   10

cm

Asplenophyllitis microdon, from specimens cultivated in Victorian times, annotated 'Guernsey and Cornwall'.

The fronds are of bright mid-green colour, glossy above, and have a *soft, fairly thin texture.* They are dished upwards near the centre line with downsweeping outer edges, giving the frond a *rather undulating overall form with distinctly crinkled pinnae.* The fronds are more ascending, especially in their lower halves, than are those of *A. billotii,* and the pinnae are often tilted towards the frond apex, especially on the more ascending fronds. The stipes are fairly short, about 1/4 or less of the length of the frond, but thick and rigid. They are dark on the underside and usually green on the upperside.

The fronds may either be only sparsely fertile, with a few linear sori, or may bear sori throughout much of their length. In the latter case, the sori are numerous, long, twinned, and together form outward-curving sprays along each pinna. They tend to occur on the outer halves of the pinna-segments, as in *A. billotii,* and thus differ from the condition in ×*A. jacksonii,* where they are near the centre of each segment. Their spores are abortive.

The fronds inherit the early-season flushing habit of *A. billotii,* with many of the new

Asplenophyllitis microdon, from specimens cultivated in Victorian times: *all* probably ex Guernsey origin.

season's fronds already fully expanded by late April, whilst several of those of the previous year may be still present. When starting to flush, the young croziers are clothed with whitish scales with buff tips – a feature they inherit from the *P. scolopendrium* parent.

Variation As might be expected of a bigeneric hybrid between two such morphologically differing parents, the frond form is somewhat variable, though nevertheless distinctly recognisable, between specimens. Fronds vary, especially in overall frond size (mostly 14.5–30.0+ cm in length by 4.8–7.8+ cm in breadth) and details of frond cutting, which usually involves the subentire pinnae being varyingly asymmetric, varyingly wide, varyingly elongate and varyingly lobed, and merging upwards into a confluent portion throughout a varying portion of usually approximately the upper 10–25% of the blade (see figures on pp. 118–19).

Possible confusion Plants differ distinctly in appearance from both parents whilst the abortive spores, greener stipes, and less fleshy fronds distinguish them from *Asplenium marinum*, which they superficially resemble. They are distinguished from ˣ*Asplenophyllitis jacksonii* (q.v.), which shares the common *Phyllitis scolopendrium* parent, by the more spreading, distinctly ovate-lanceolate, comparatively narrow frond outline (cf. more ascending, broad and triangular in ˣ*A. jacksonii*) and much shorter stipes (often 1/3 to 1/2 the length of the frond in ˣ*A. jacksonii*, q.v.).

Technical confirmation Plants are triploid hybrids, with $2n = 108$ chromosomes, showing approximately 36 pairs of small chromosomes and 36 single large chromosomes at meiosis (the pairs all deriving from the autotetraploid *A. billotii* parent).

Field notes This somewhat unlikely, but spectacular, cross is thought to have been the origin of a number of sporadically occurring nineteenth century plants recorded in Cornwall, Devon and possibly in Wales (Alston, 1940). More positive records of it (McClintock, 1968) were also made during the latter half of the nineteenth century in Guernsey. The plant thereafter remained unknown, until it was rediscovered in 1966 in Guernsey (Girard, 1967; Girard & Lovis, 1968), where it has since been recorded in more than a dozen separate localities (P. J. Girard, personal communication).

Recently, two further, previously unknown, Victorian herbarium sheets came to light in the herbarium of the Royal Botanic Garden, Edinburgh (Page, 1990a), further corroborating the nineteenth century records of this plant. These sheets include at least five separate gatherings made between 1860 and 1865, as 10 beautifully preserved fronds, with annotations indicating that they are not only from Guernsey, but may also include the previously unsupported finds of nineteenth century Cornish specimens. So far as is known, this hybrid is endemic to the British Isles.

In its Guernsey habitats, where the plant is still known, it grows on the faces of steep, laneside and roadside hedgebanks, and in one place, amongst steep, mortarless roadside stonework. Its sites are all at low altitude, and are well drained, sheltered, moist, humid and usually heavily shaded, mostly with a northerly, north-easterly or north-westerly aspect. It usually grows near the bottoms of tall banks, where they are dampest, steepest, and most sheltered. In most localities, such banks are constructed out of closely spaced, granitic boulders, set in a matrix of clay soil, and *Asplenium billotii* and *Phyllitis scolopendrium* occur on them, intimately mixed together and often in some numbers. As with *Asplenium* ˣ*sarniense* (q.v.), which in some sites also occurs nearby, the habitats of ˣ*A. microdon* are mainly those where their steep position helps ensure that there is some surface erosion, maintaining a mosaic of variously open patches amongst slightly denser vegetation. The shallow, rocky substrate, and the shaded position, usually help to reduce the competition of surrounding herbaceous vegetation, and the main associated species in addition to the two parents are usually rather few, and chiefly Ivy (*Hedera helix*), Wall Pennywort (*Umbilicus rupestris*), Red Campion (*Silene dioica*), Herb Robert (*Geranium robertianum*), Lesser Celandine (*Ranunculus ficaria*), with scattered Primrose (*Primula vulgaris*), Creeping Buttercup (*Ranunculus repens*), Sorrel (*Rumex acetosa*), Bramble (*Rubus* *fruticosus*), Black Spleenwort (*Asplenium adiantum-nigrum*) and Western Polypody (*Polypodium interjectum*) as well as small clumps of various grasses – mainly Red Fescue (*Festuca rubra*), Cock's-foot (*Dactylis glomerata*), Wood False-brome (*Brachypodium sylvaticum*) and Oat-grass (*Arrhenatherum elatius*).

In cultivation, ×*A. microdon* is a fairly vigorous hybrid, and plants of this cross can be made artificially from the parents with greater ease than can those of ×*A. jacksonii* (q.v.) (Lovis & Vida, 1969). This greater facility for crossing must, however, in the wild, be to some extent offset by reduced opportunity resulting from the somewhat different edaphic preferences of the two parents, over much of their range, for *Phyllitis scolopendrium* is generally more calcicolous than *Asplenium billotii*. The frequent occurrence of the parents together along Guernsey lanebanks is probably largely the result of the close mosaic formed by the granitic boulder and clay background. Most such habitats are probably also lightly irrigated by moisture seepage, and, to some extent, influenced by the warm but windy and often foggy maritime climate. The frequency of the hybrid doubtless results from the widespread nature of this lanebank type of habitat, in which the steady effect of light surface erosion continuously opens opportunities for re-establishment of both parents, and thus for hybrids to arise between neighbouring prothalli growing in crevices around the boulders.

This hybrid could well occur again elsewhere in the British or Irish areas of overlap of the parental ranges, especially perhaps in Devon or Cornwall or elsewhere in the Channel Islands, where similar lanebank habitats very frequently occur. On Guernsey, laneside banks in much their present form probably date back through considerable historic periods. For how long this hybrid has been present on them is unknown before its single nineteenth century record, although there is some evidence that its numbers may have increased over a fifteen-year period since the first one was refound this century (P. J. Girard & A. Prain, personal communications).

Athyrium distentifolium
Tausch ex Opiz

Alpine Lady-fern

(*A. alpestre* (Hoppe) Rylands)

Preliminary recognition Usually a medium-sized fern, of graceful appearance, with upright-growing, lance-shaped fronds which are finely divided. They look very similar to those of *A. filix-femina*, but have pinnae extending almost to the stipe base, and have round, not J-shaped, sori, without obvious indusia.

Occurrence An alpine species, of local and scattered range in the Scottish Highlands, in high mountain sites, in moist gullies and hollows where there is appreciable winter snow-lie.

Identification Plants are very similar in general appearance to *A. filix-femina*, but usually of smaller size, with suberect, elliptical or lanceolate outlined fronds, mostly 20–30 cm high but

Jan. Feb. Mar.	Apr.	May	Jun.	Jly	Aug.	Sep.	Oct. Nov. Dec.

occasionally as much as 70 cm, arising in shuttlecock-like groups from crowns of ascending or erect rhizomes. The fronds are as finely divided as those of *A. filix-femina*, but the *pinnae are usually more ascending*, more crowded along the rachis (especially in its upper part), are less acutely pointed and of less variable, usually dull pale yellow-green colour, translucent with a pink or greenish-brown stipe and rachis. The stipe is shorter and thicker than that of *A. filix-femina* (usually about 1/4 or less the length of the frond), and is sparsely clothed with rather limp, pale buff-brown scales. In smaller specimens, the pinnae are often more rounded than those of *A. filix-femina*, with fronds adopting a flatter-looking appearance.

Fertile fronds are similar to the vegetative ones, with sori either distributed over the whole of the lower surface, or occurring mainly in its upper half. The sori differ from those of *A. filix-femina* in being *much smaller and of circular (not J-shaped) plan, with either an extremely rudimentary indusium or no indusium at all.* Fronds less than about 15–25 cm in length are seldom fertile (cf. *A. flexile* q.v.).

Fronds arise in spring following snow-melt, and die and turn brown in autumn, although the old red-brown fronds may remain largely undecayed beneath the snow until exposed again the following season. Beneath them, the rhizomes spread and branch decumbently, gradually building up large clumps, each containing many nearby crowns.

Variation Mainly in size. In form and colour, it seems a less variable fern than *A. filix-femina*. However, many dwarf forms occur throughout much of the Scottish range of this species, and appear to be distinct ecotypes.

Possible confusion Likely to be confused only with plants of *A. filix-femina*, which can grow as high as the lower part of the altitudinal range of this species (see 'Field notes'). The combination of the above features, and especially the round, naked sori, should separate *A. distentifolium* from *A. filix-femina*. Smaller specimens might be confused with *A. flexile*, with which there is also an altitudinal overlap, but the more erect frond habit, different frond outline, and presence of sori on fertile fronds from the tip downwards (from the base upwards in *A. flexile*, q.v.) distinguishes *A. distentifolium*. The lanceolate frond outline (never triangular) distinguishes *Athyrium* from finely divided infertile fronds of *Dryopteris expansa*, at similar altitudes. The more finely divided fronds distinguish them from *Oreopteris limbosperma*.

Technical confirmation Plants are diploids, with $n = 40$, $2n = 80$ chromosomes. The spores have an irregularly folded and ridged perispore, giving the spore surface a reticulate appearance under a light microscope (hence differing from the unridged, granular perispore of *A. filix-femina*). The distribution of sori, their shape and virtual lack of an indusium, confirm *A. distentifolium*.

Field notes *Athyrium distentifolium* is present in southern Greenland, Iceland and Europe, with scattered stations in northern Asia as far as Japan, Sakhalin and Kamchatka. What is probably a further related form occurs in mountainous regions of North America. This overall range shows close similarities to that of *Cryptogramma crispa* (q.v.). In Europe, *A. distentifolium* shows a generally arctic-alpine distribution, with scattered station stretching south to the Pyrenees, and Macedonian and Anatolian mountains to the Caucasus. Within Britain *A. distentifolium* is always a high-mountain species of the Scottish Highlands seldom occurring at all below about *c.* 600 m (2000 ft), and usually seen between about 550 and 1100 m (1800 and 3600 ft). Although the typical form of the plant has tall, erect fronds, many apparently genetically determined more dwarf forms exist which can produce spores from a small frond size, especially in the Scottish mountains. These dwarf forms of

Athyrium distentifolium, typical populations: *a*, *b*, Perthshire (Bredalbane); *c*, Perthshire (Meall Ghaordie); *d*, Aberdeenshire (Cairntoul); *e*, Forfarshire (Glen Clova); *f*, Perthshire (Lawers Range).

A. distentifolium, which differ distinctly from *A. flexile* (q.v.), occur notably both in the central and north-west Highlands. All forms of Alpine Lady-fern are generally calcifuge plants, growing mostly on acidic rock types, but extending on to more basic rocks (such as mica schists) where they are locally overlain by acidic surface layers. It is characteristic of rocky mountain gullies and damp pockets in boulder slopes, and where ravines or local topographic detail provide some protection, large, almost pure local stands may occur. Its sites are characteristically ones where there is regular accumulation of appreciable winter snow, which every year covers the plant for long periods lasting well into spring. Such snow-cover not only protects them from damage by severe cold, but presumably also minimises winter desiccation.

Ratcliffe (1977) records that following snow-melt, the dead overwintering fronds give a characteristic red-brown colour to patches of this fern on the hillsides and that its habitats are ones which are irrigated by spring meltwater, presumably bringing a supply of mineral downwash. The heavy frond-fall at the end of the season, and the fronds' slow rate of decay, result in the gradual accumulation of a dense, acidic, raw humus around the plant clumps. It seems possible that these dark-coloured fallen fronds also form a heat-absorbing blanket, assisting with the normal thermal balance of the plant, especially in early spring sunshine. Associated species in these clumps are often few, but may include Bilberry (*Vaccinium myrtillus*), Alpine Lady's-mantle (*Alchemilla alpina*), Heath Bedstraw (*Galium saxatile*), and Stiff Sedge (*Carex bigelowii*).

Alpine Lady-fern generally replaces Parsley Fern (*Cryptogramma crispa*) on similar rock types at higher altitudes and in edaphically moister situations, away from the most oceanic west. It also replaces Alpine Buckler-fern (*Dryopteris expansa*), Holly Fern (*Polystichum lonchitis*), and Alpine Bladderfern (*Cystopteris montana*) on more acidic rocks.

A. distentifolium is limited to areas of alpine topography, to which it is confined probably by a requirement for a low maximum summer temperature of around 27°C, coupled with adequate year-round precipitation. It is presumably the increasing dryness of climate which limits it from more continental eastern areas. In the west, it is probably the less persistent snow-lie, removed readily by early warm rain, that limits it from descending to low altitudes where its temperature maxima might otherwise be met. In the lower part of its altitudinal range, at about 600 m (2000 ft) there is usually a small overlap with the upper part of the altitudinal range of *Athyrium filix-femina*, whilst in the upper part of its range, there is a larger overlap – especially above about 900 m (3000 ft) – with *Athyrium flexile*.

As with *A. flexile*, *A. distentifolium* seems confined to altitudes above the tree limit of the old Caledonian forests of Scotland, and may well be a fairly old, early post-glacial member of our flora. Like *A. flexile*, the trigger for its spring flush of fresh frond-growth is probably the sudden rise in temperature following spring snow-melt, rather than daylength. At the other end of the season, rhizomes appear to become completely quiescent, until they have received the appropriate period of winter chilling after which they again become temperature sensitive. Such a temperature-induced response is probably of considerable advantage to an alpine species, allowing maximum benefit to be taken of the short growing season, once that season has firmly arrived. It is perhaps this physiology which confines it, like *A. flexile*, to alpine snow-beds, where a sudden single temperature change marks the beginning of the season. At lower elevations, in this oceanic climate, such a temperature-triggered species would probably suffer badly at a vulnerable early-flushing stage, were it to be stimulated into spring growth by an early mild spell, only to be followed later by exposure and severe frost.

Note added in proof (*Athyrium filix-femina* – see entry opposite).
Athyrium filix-femina is now known to be one of our native ferns capable of forming natural soil spore banks in the wild (Lindsay & Dyer 1997). These, when re-exhumed, are capable of germination into new gametophyte and, potentially, new sporophyte plants. New developments in our understanding of the genetic variation of soil-banked spores of European *Athyrium filix-femina* (Schneller & Holderegger 1997a) has shown increasing evidence that these have the potential to contain considerable amounts of genetic variability, even within small soil samples, and that thus amounts of genetic variation may become potentially 'stored' within (and perhaps beyond?) existing populations.

Athyrium filix-femina (L.) Roth   Woodland Lady-fern

Preliminary recognition A moderate to large sized fern, with lance-shaped, finely divided, rather delicate-looking, ascending fronds, arising in shuttle-cock-like clusters. In woodland, they often form luxuriant plume-like masses of bright green colour.

Occurrence Widespread and common throughout most of Britain and Ireland, especially in moist woodlands, along streambanks and in ravines, and to a considerable altitude on mountains.

Identification Plants have fronds up to 120 cm or more in height, which are of elliptical or lanceolate outline, broadest about the middle of the frond, and finely bipinnately or tripinnately divided throughout. The fronds are usually steeply ascending, and arise in a shuttlecock-like manner from the crowns of stout heavy, woody rhizomes, which are ascending or erect and covered with the persistent bases of old stipes. The rhizomes occasionally branch, and in old plants, often develop into large clumps with numerous closely packed crowns.

The fronds have stipes which are usually at least 1/4 of the length of the blade, and are fairly thickly clothed with soft, rather limp, pale-brown scales. The pinnae taper gradually to an acute point, and the pinnules have finely crenate or serrulate margins. In some specimens, the pinnae are held more or less perpendicularly to the rachis, and the pinnules perpendicularly to the pinnule midribs, giving the fronds an elegant, geometrically constructed appearance.

The lamina is usually of soft texture, and varies from a bright yellowish-green to a rather blue-green colour. In shaded conditions, the lamina is fairly flat, but with increasing light, the pinnule edges bend downwards along their midribs and, in full light, all the pinnae themselves may be held at a steep downward angle. Between different plants, in almost all populations, the rachis and stipe vary much in colour, from a pale green to dark green, plum-red or purple, and stipes of more than one of these colours may occur.

On most plants, fertile fronds occur abundantly, with sori distributed over much of their lower surfaces. The curved sori can be very variable in shape even on a single frond, but are mostly J-shaped, occasionally more linear, or sometimes C-shaped or horseshoe shaped. They are, however, always covered by a distinct indusium, the shape of the sorus, and hinged along its inner side. The indusia persist until maturity.

Jan. Feb. Mar.	Apr.	May	Jun.	Jly	Aug.	Sep.	Oct. Nov. Dec.

Athryium filix-femina, habit of plant in life.

Most fronds emerge in a single spring flush. They reach maturity by about late July, and die down in autumn, usually in conjunction with the first frosts.

Variation Beside the variation in lamina, rachis and stipe colour, fronds are also variable in both pinna-spacing and cutting of pinnules. Much of this variation is maintained in cultivation, and may be the result of segregation in most populations of a small number of highly distinctive genes.

Possible confusion The elegant appearance and large size of most specimens in lowland forest makes them unlikely to be confused with any other fern. On mountains, *A. distentifolium* is of very similar appearance, but plants of *A. filix-femina* always retain the curved sori and persistent indusia. On wet rocks, smaller specimens of *A. filix-femina* can be confused with *Cystopteris fragilis* (q.v.), but fronds of *Athyrium* differ in their cutting (see illustrations) as well as in their more tapering pinnae and more perpendicularly-held pinnae and pinnules. *Athyrium* differs from *Oreopteris limbosperma* (q.v.), which has a similar outline, in its much finer frond dissection as well as distinct stipe. Its frequent association, in woodland, with often similar numbers of individuals of the much more coarsely divided *Dryopteris filix-mas*, is

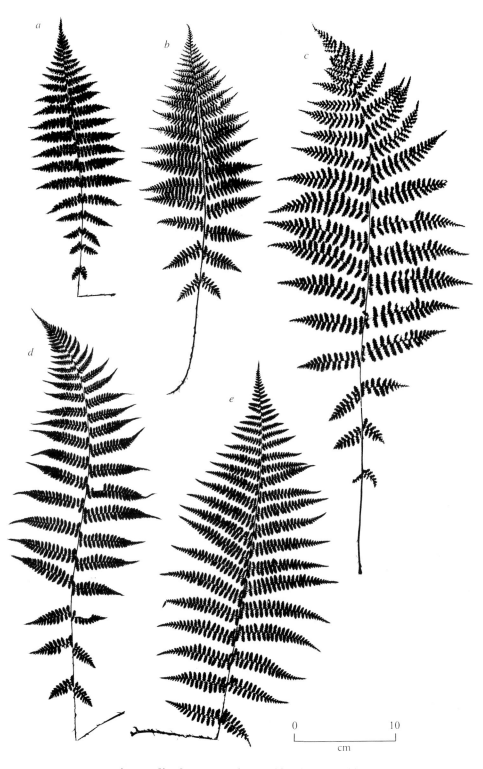

Athyrium filix-femina: *a*, Wigtownshire; *b–e* Perthshire.

Athyrium filix-femina, young frond stages (after Orth, 1938).

presumably the reason for the *Athyrium* having been formerly regarded as the female of the pair, although both are good species in their own right.

Technical confirmation Plants are diploids, with $n = 40$, $2n = 80$ chromosomes. The spores have a non-ridged perispore, with a granular appearance under a light microscope (cf. an irregularly folded and ridged perispore of reticulate appearance in *A. distentifolium*). The shape of the sori and the persistence of the indusium confirm *A. filix-femina*.

Field notes *Athyrium filix-femina* is a very widespread species of middle latitudes of the northern hemisphere, with a natural range from Europe across northern Asia to Japan and across North America. Within this overall range, a large number of segregates are generally recognised. Its overall range compares closely with that of *Cystopteris fragilis* (q.v.). Within Europe, it is widespread almost throughout moister northern and middle latitudes of the continent, thinning more in Mediterranean latitudes, but spreading to the Azores, Madeira and the Canary Islands.

Within Britain and Ireland, Woodland Lady-fern is very widespread, especially in western counties and in most wetter and more mountainous districts, becoming more local only in East Anglia and the English and Irish Midlands. Throughout most of its range, it is a frequent member of the ground vegetation of deciduous forests and woodland, especially on damper soils near to streamcourses

and in moist, rocky situations. But it also occurs in a wide range of other habitats, including the edges of drainage ditches in lighter patches of coniferous plantations, along rocky gorges, on moist sides of upland glens, in sheltered hollows on mountain flanks and along the courses of small mountain streams and rills, from sea-level to around 915 m (3000 ft). Throughout its range, it associates with a very wide range of other herbaceous, generally woodland species, but especially with other ferns, such as Hard Fern (*Blechnum spicant*), Broad Buckler-fern (*Dryopteris dilatata*) and Male Fern (*D. filix-mas*).

It is a calcifuge species, avoiding base-rich soils, except where these are overlain by acidic surface layers. It occurs mostly in soils with good drainage, but which are kept permanently moist. Most of its sites are on sloping ground, where there is constant light irrigation through groundwater seepage. Its greater abundance in more western and more mountainous areas seems to correspond to a general increase in rainfall frequency, increase in resulting peaty acidic surface layers and decrease in summer droughts.

In more permanently exposed situations, and with very high rainfall, increasing altitude, and increased leaching and acidity of the ground, *A. filix-femina* becomes largely replaced by Sweet Mountain Fern (*Oreopteris limbosperma*). In woodlands, however, Woodland Lady-fern seems more tolerant of shade than is *O. limbosperma*, and to a large extent, the occurrence of Woodland Lady-fern in other habitats may well reflect distribution of former woodland vegetation.

Athyrium filix-femina is, however, a vigorous species, and is able to also colonise a wide range of open, artificial, semi-natural situations, especially along hedgerows and ditches. Of particular interest too are some local strains which appear tolerant of high concentrations of certain metalliferous soil contaminants, such as lead and tin. In association with various rushes (*Juncus* spp.), plants of *Athyrium* may pioneer edges of pools and rills across old mine tailings and around settling ponds, where few other plants can succeed. In such habitats, its rapidity of growth and density of frond-cover may be of particular ecological importance.

Athyrium flexile (Newm.) Druce   Flexile Lady-fern

Preliminary recognition An important small alpine fern with finely divided, mostly tapering fronds, held in a spreading rather than erect position, and looking, perhaps, superficially more like a spleenwort than a lady-fern.

Occurrence Very local and rare in a few high-mountain screes and alpine corries in the central Scottish Highlands, to which area this fern seems entirely restricted (endemic).

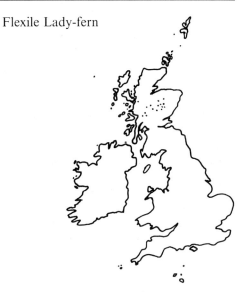

Identification Plants have small, finely divided, rather rigid fronds, mostly 8–20 cm long, occasionally to 30 cm, of *rather irregularly obovate-lanceolate outline.* The fronds arise in a *flattened shuttlecock-like fashion* from small, decumbent rhizomes, which branch occasionally to give multiple closely packed crowns. The blades are borne on *very short stipes*, about 1/8 or less the length of the frond. The *stipes are mostly bent rather sharply backwards at a distinct elbow*, and hold the blade in a spreading, often nearly horizontal, plane. This habit contrasts strongly with the upright habit of *A. distentifolium* (q.v.), the bend in the stipe providing the source of the name *flexile*.

The structure of the blades is also very distinctive. Throughout the frond, the *pinnae are quite short and distantly spaced in most specimens*, and the blades consequently have a

Jan. Feb. Mar.	Apr.	May	Jun.	Jly	Aug.	Sep.	Oct. Nov. Dec.

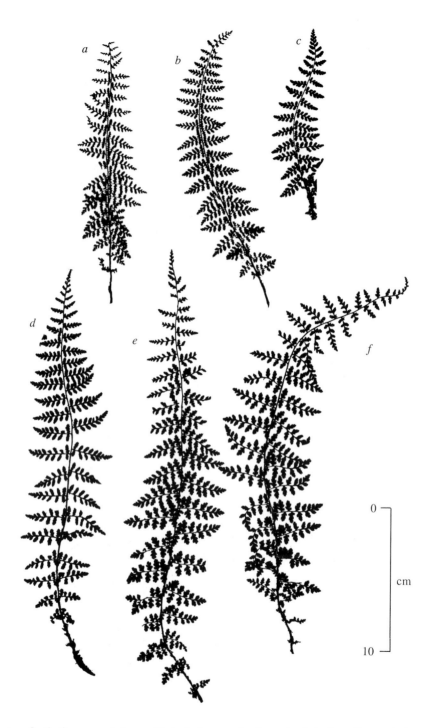

Athyrium flexile, from populations with typical, more slender and relatively well-spaced pinnae and pinnules: *a–d*, Argyll (Ben Alder); *e, f*, Forfarshire (Glen Prosen).

Athyrium flexile, from populations with typical stout-textured fronds: *a–d*, Argyll (Ben Alder); *e, f*,
Forfarshire (Glen Clova and Glen Prosen, respectively); *g*, Inverness-shire (Beinn Eibhinn).

Athyrium flexile, from populations with fronds possibly intermediate between this species and *A. distentifolium* (see the text): *a–d*, Perthshire (Breadalbane); *e–g*, Forfarshire (Glen Clova).

markedly narrow outline. The widest point is about 1/3 of the way from the base of the blade, below which the pinnae shorten abruptly to become very short basal pinnae, but above which the pinnae shorten only gradually to give the frond a *long, attenuately tapering apex.* In the upper part of the blade, the pinnae are quite spreading, but in the lower half they become gradually more deflexed, so that the *lowermost ones are angled variously backwards towards the frond base.* Along the pinnae, the pinnules are usually each narrowed at their base, resulting in wide gaps between bases, giving the fronds a characteristic, somewhat lace-like cutting. The frond texture is thin but slightly rigid, whilst the lamina, stipe and rachis are of a uniform yellow-green colour throughout.

Fertile fronds occur sparsely and are similar to the vegetative ones but have sori which, very unusually amongst ferns, are normally present *only over the central or basal part of the underside of the frond, not near its tip.* Each sorus is of circular shape, like those of *A. distentifolium,* but smaller and more distant, and contains fewer sporangia. Each has only a minute and inconspicuous indusium. Fronds as small as 7–10 cm long can be fully fertile, in contrast to those of *A. distentifolium* and *A. filix-femina,* which are seldom fertile at this size.

Variation The characteristic features of this fern, especially its small size and spreading frond habit, as well as somewhat congested frond form and soral distribution over the lower parts of fertile fronds, are totally retained under conditions of uniform experimental cultivation, and thus genetically determined. Recent research (H. McHaffie, personal communication, and see also 'field notes') has shown that populations in its few known localities vary somewhat in features of overall vegetative structure mainly between sites: some populations (e.g. especially Glen Prosen) have typically spreading and generally the most elongated frond form and relatively well-spaced pinnae and pinnules; others (e.g. Ben Alder) have typically spreading and much more congested pinnae and pinnules; and ones from some other localities (e.g. Bridge of Orchy, Breadalbane and Glen Clova) have vigorous individuals with more ovate and more ascending fronds, which are least congested and may be hybrids with or backcrosses to *A. distentifolium* (see 'Field notes').

Possible confusion The small, spreading fronds, with tapering tips, deflexed lower pinnae, pinnules with narrowed bases and stipes which are elbowed, give plants a distinctive appearance, unlikely to be confused with any other fern.

Technical confirmation Plants are diploids, with $n = 40$ chromosomes. The circular sori with rudimentary indusium confined only to the lower half of the frond, confirm *A. flexile.*

Field notes This very local species, endemic to Scotland, is confined to a few widely scattered high alpine corries in the central and western Scottish Highlands, where it occurs either as scattered individuals or occasionally in some numbers. It was first described by Newman in 1853 from Glen Prosen, and is now known to have a number of scattered stations mostly in the central and eastern Scottish Highlands, nearly all between 1040 and 1140 m (3400 and 3750 ft), but descending occasionally as low as 750 m in Glen Prosen. In total, plants have now been recorded from up to 20 separate Scottish sites, with more outlying stations in Glen Doll and Lochnagar, and it has recently been recorded as far west as Knoydart. Its classic site at Ben Alder, where it has been known since 1867, has long been recognised to have the largest population of several hundred plants (H. McHaffie, personal communication). In addition to these populations of the typical form of the plant, a very vigorous and varied form which may involve some hybridisation with *A. distentifolium* occurs in some localities, including especially in three corries near Bridge of Orchy. *Athyrium flexile* grows in damp rocky places, on similar, non-calcareous rock types to *A. distentifolium*, especially in moist fissures of acidic rock scree, spreading occasionally on to more base-rich rock types, chiefly where there is surface leaching.

Athyrium flexile is sympatric with *A. distentifolium* across its entire range in the Scottish Highlands (Page, 1988*b*; 345), but, although generally growing with *A. distentifolium*, plants of *A. flexile* are overall much less abundant, occurring as discrete patches or even a few or sometimes single individuals, often in moist pockets amongst boulders. Its habitats are mostly in acidic sheltered corries that benefit from medium to late winter snow-lie, and on aspects that are usually north-east or north-west facing, although the Glen Prosen colony is on a south-facing slope beneath overhanging rocks. Its habitats are, however, ones that are mostly cool, moist and frequently shaded, and plants seem particularly closely attuned to the short growing season and windy exposure of their montane situation. Spring frond-flushing appears to be triggered initially by a rapidly rising temperature following snow-melt. The small fronds then expand quickly, with their small number of sori on the early-expanded basal part of the frond maturing rapidly. The rigid texture of the stipe and rachis holds fronds firmly against battering by winds, whilst their remarkably spreading habit gives even fully flushed fronds a very low wind-profile.

The exact nature of the ecology, life history and phylogenetic affinity of *A. flexile* is the subject of current active research (H. McHaffie, personal communication). Field observations suggest a higher degree of micro-ecological distinction from *Athyrium distentifolium* than had hitherto been appreciated. It is possible that *A. flexile* succeeds only where it is not outcompeted by *A. distentifolium*. Where the two taxa occur together, *A. flexile* can typically dominate the more marginal edaphic niches, though this may vary somewhat between sites, while plants of apparent morphological intermediacy appear to be much more local in certain areas. Further, plants maintain their form in cultivation, breed true when selfed, and have a morphology maintained from its earliest juvenile phases which is distinctive in the genus as a whole. There is clear opportunity for plants of *A. flexile* to establish as true-breeding self-replicating populations where appropriate environmental conditions are met, and this situation appears relatively stable in time, at least since the plant was discovered a century and a half ago.

No such comparable situation has been identified elsewhere within the extensive range of *A. distentifolium* outside these islands, which clearly indicates that, whatever its genetic status, this fern is a Scottish endemic with a distinctive structural, geographical and ecological identity which is not limited (as are some *Asplenium*, q.v.) to the occurrence of specific soil types. *Athyrium flexile* thus appears to behave as an incipient insular species in evolutionary terms, which is how I thus prefer to recognise it. The exciting issue is the existence of a new montane fern taxon at all, apparently in the process of actively evolving, and the opportunity that this gives for furthering our understanding of the mechanisms of this process against an observable ecological background.

H. McHaffie has recently collected much valuable first-hand data on this enigmatic endemic, on which she is to be congratulated. I am grateful for permission to summarise some of her observations quoted above, although the conclusions derived remain entirely my own.

Blechnum spicant (L.) Roth   Hard Fern

Preliminary recognition A small to medium-sized fern, with leathery, dark green, herringbone-like fronds, arising in dense masses or rather flattened rosettes.

Occurrence An abundant species in all western, wetter and upland parts of Britain and Ireland in a wide range of acidic peaty places, from sea-level to near mountain tops, especially in acidic woodlands and in open heathy places.

Identification Plants have quite numerous, hard, *coarse-textured and leathery, glossy, pinnately divided, herringbone-like fronds*, commonly 20–50 cm or more in length (much smaller when in very exposed conditions), and much larger (occasionally to about 1 m) when exceptionally luxuriant. The fronds arise from the crowns of short, stocky, ascending or erect rhizomes, which are dark brown or blackish in colour, and covered in slender, pointed dark purple-brown scales. Similar scales also clothe the bases of the stipes.

The fronds are of two types. The vegetative (sterile) ones usually form a rather flattened basal rosette. They are of long, linear-elliptical outline, and are pinnately divided in their lower parts and pinnatifidly divided in their upper parts *into very numerous, simple, oblong, parallel-sided, bluntly round-tipped pinnae.* Throughout the greater part of the frond, the pinnae are long, closely spaced and usually rather upward-curved towards the tip of the frond. Towards the frond base, they become progressively shorter, straighter and more widely spaced. Most fronds have only a short length of stipe, often 1/4 or less of the length of the blade. The stipe is rigid and wiry, and of a polished, deep purple-brown colour. The dark colour extends some distance up the rachis, which eventually becomes green in the upper part of the frond, whilst the upper side of the rachis is distinctly channelled throughout.

The fertile fronds are usually much larger than the vegetative ones, up to twice as long, and are held in a strongly erect manner, arising from the centre of the rosette of the vegetative ones. They are similar to the vegetative ones in structure, but have numerous, *very slender, long, acute pinnae*, which are widely spaced along the rachis, giving the fertile fronds a particularly gaunt, herringbone-like appearance. They also usually have much longer stipes, sometimes as long as the blade length. The pinnae have slightly dilated bases, attaching them to the tall, wiry

Jan. Feb. Mar.	Apr.	May	Jun.	Jly	Aug.	Sep.	Oct. Nov. Dec.

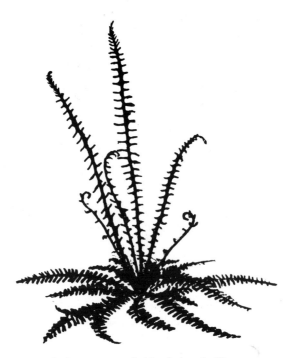

Blechnum spicant, habit of plant in life.

rachis. Each pinna bears a pair of long, linear sori which run as a continuous line along each side of its midrib. The continuous indusia are attached along the outer edge of the pinna, opening inwards, and at maturity the sori bulge with masses of dark-brown dehiscing sporangia.

All the fronds are of pale-green colour when flushing, but rapidly turn to a shining deep green once they have fully expanded. The fertile fronds appear later in the season than the vegetative ones, and die down sooner. The vegetative fronds remain wintergreen, and usually persist in this condition until after the vegetative fronds of the following year have expanded. Plants in very exposed situations often eventually build up extensive vegetative masses by branching of the rhizome to form many interconnecting crowns.

Variation Plants vary widely in size with shelter of the habitat. Most specimens are of fairly constant appearance. Occasional monstrosities and small aberrations of incomplete dimorphism between vegetative and fertile frond states can occur.

Possible confusion The simply pinnate structure of the fronds makes them unlikely to be easily confused with other ferns, except perhaps for having a superficial resemblance to *Polypodium vulgare* and *P. interjectum*. The dark-coloured stipes arising from single crowns, the dimorphic fronds and linear sori readily confirm *Blechnum*.

Technical confirmation Plants are diploids, with $n = 34$, $2n = 68$ chromosomes.

Field notes Blechnum spicant is a predominantly European fern, with highly discontinuous range elsewhere at generally middle latitudes of the northern hemisphere, especially locally in northern Asia, Japan, the Aleutian Islands and the Pacific north-west of western North America. Within this overall range, various segregates are sometimes recognised. Within Europe, it is widespread almost throughout moister northern and middle latitudes of the continent, thinning more in Mediterranean

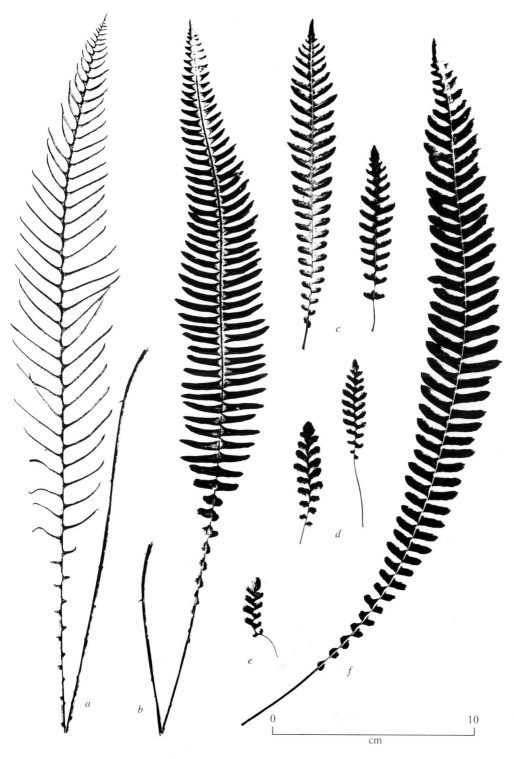

Blechnum spicant: a–b, West Inverness-shire; *c–e*, Argyll; *f*, Mid Perthshire.

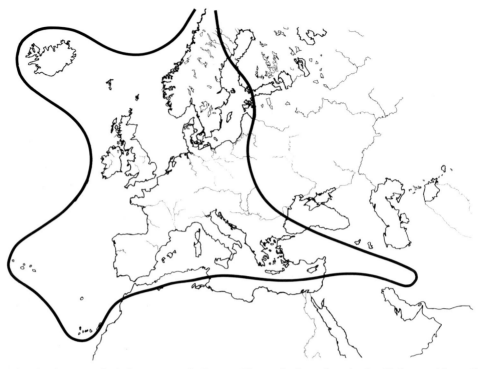

Sub-Atlantic range of *Blechnum spicant* in Europe. The species has other closely allied taxa with smaller ranges in western North America and Japan.

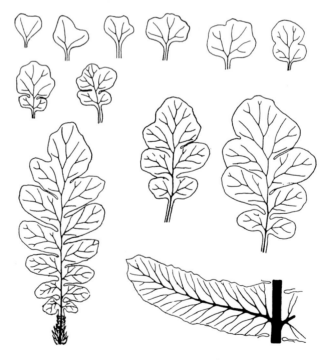

Blechnum spicant, young frond stages (after Orth, 1938).

latitudes, but spreading to the Azores, Madeira and the Canary Islands and eastwards to more scattered localities in the Balkans, Turkey and Asia Minor.

Within Britain and Ireland, Hard Fern is a frequent and generally well-known species in wetter, western climates throughout western Britain and western Ireland. It grows in a very wide range of moist, acidic sites from sea-level to nearly 1220 m (4000 ft). It is particularly characteristic of a wide range of acidic woodlands, such as those of oak and pine developed over poor, sandy soils, of hedgerows in acidic areas, and of moorlands, heaths, acidic river valley sides, steep, damp acidic upland slopes, along peaty overhangs, damp mountain hollows, beside rills and streamlets on mountain sides and along drainage ditches in forest plantations. Most sites occur on peaty soils, but it also colonises damp mineral-rich soil slopes wherever these are lime-free and sufficiently moist.

It is a strongly calcifuge species and thrives in highly acidic ground, with many sites having a pH between about 4.0 and 5.0, and sometimes as low as 3.5. It demands edaphic conditions which are permanently moist but freely drained. Plants are intolerant of drought, and generally seem to avoid topographic situations exposed to desiccating easterly or northerly winter winds. They thrive best where conditions of reasonable local shelter combine with light shade and there is also frequent high humidity.

In woodlands at low to moderate elevations, it associates with a number of common acid-woodland understorey herbs, usually becoming most abundant in locally damper patches, or alongside small streams and drainage ditches. On the woodland floor it often grows with Bilberry (*Vaccinium myrtillus*), Broad Buckler-fern (*Dryopteris dilatata*) and Woodland Lady-fern (*Athyrium filix-femina*). It can dominate locally on damp sheltered slopes of mineral soil, along steep river-banks forming a fringe below festooning masses of Woodrush (*Luzula sylvatica*), where there is continual light irrigation by moving acidic water from above.

In plantation woodland, indigenous plants often survive at least temporarily and re-establish along the ridges and breaks of coniferous forest, where the effects of increased drainage are to a large extent offset by increase in shade and humidity, as well as reduced herbaceous competition.

Blechnum becomes increasingly abundant in western oceanic climates, where there is a higher humidity, more continuous cloud cover and frequent light rain. Under such conditions, it extends into more-exposed situations especially on to slopes of moorland streambanks in ericaceous dwarf-shrub heath communities. Here plants grow alongside Heather (*Calluna vulgaris*), Bell-heather (*Erica cinerea*), Cross-leaved Heath (*E. tetralix*), Bearberry (*Arctostaphylos uva-ursi*), Crowberry (*Empetrum nigrum*), Wavy Hair-grass (*Deschampsia flexuosa*) amd Sheep's Fescue (*Festuca ovina*), with scattered other associates such as Heath Bedstraw (*Galium saxatile*) and Field Woodrush (*Luzula campestris*). It frequently occurs near Sweet Mountain Fern (*Oreopteris limbosperma*) and Woodland Lady-fern (*Athyrium filix-femina*), both of which the Hard Fern tends to replace in edaphically more freely draining situations.

At higher altitude, Hard Fern becomes increasingly restricted to moister soil pockets. It occurs particularly in communities of Alpine Crowberry (*Empetrum hermaphroditum*), Bilberry (*Vaccinium myrtillus*), and Wood-sorrel (*Oxalis acetosella*), with *Hypnum cupressiforme* and *Pleurozium schreberi* mosses and *Cladonia* lichens, where these mark areas of long winter snow-lie. In such habitats, the fern benefits not only from winter frost protection by the covering snow blanket, but is also protected from winter desiccation, and benefits from the increased edaphic moisture once the growing season resumes following spring snow-melt.

Although Hard Fern is still a common species throughout many parts of Britain and Ireland, it has nevertheless probably become considerably reduced in lowland habitats through the removal of natural woodland cover and increase in agriculture. It is also rare in most industrial areas, presumably due mainly to air pollution to which it seems especially sensitive. Like many other British and Irish pteridophytes, it is also particularly sensitive to burning and, on upland moorlands in Scotland, has become greatly reduced where the land is managed by regular heather burning.

Botrychium lunaria (L.) Swartz　　Moonwort

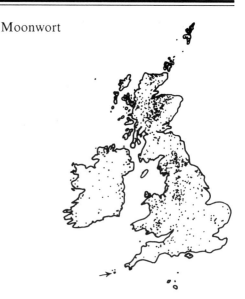

Preliminary recognition A usually small and fleshy plant, with a single leafy lobe of fan-shaped segments borne on one side of a taller, fleshy shaft, ending in a spike of numerous spherical sporangia, like a bunch of miniature grapes.

Occurrence Present throughout Britain and Ireland, more frequent in northern upland areas, and becoming scarcer southwards and in Ireland. It grows mainly in grassy uplands and old meadowland, in open places, and sometimes in old sand-dune slacks.

Identification Plants arise usually as *solitary, scattered, individual shoots.* They are erect, smooth and succulent, about 5–25 cm in height, with each plant merely a single shoot bearing a divided leafy segment and a fertile spike. The shoots arise from a deeply buried, slender, ascending or creeping rootstock, bearing abundant, spreading, long fleshy roots. At the surface of the ground, the base of the shaft arises out of a brownish, membranous portion, formed from the remains of a previous shoot. The shaft bears the sterile, leafy portion, asymmetrically to one side, usually at some distance above the ground.

This sterile, leafy portion is of oblong or slightly tapering outline, about 2–8 cm long and 0.7–3.0 cm wide, and of a bright, rich green colour when young, becoming a duller, yellower green with age. It is subdivided into a small number (usually about 3–8 pairs) of *broadly crescent-shaped segments with fan-shaped or wedge-shaped bases.* These segments arise from a stout, green midrib and connect together at their bases by a narrow leafy wing. Their size decreases upwards, and their edges frequently overlap and they are also often tilted into a rather vertical plane.

The fertile portion is a vertical continuation of the shaft, which ends in a *pinnately divided, slightly loose, broad spike,* about 1–5 cm in length, and of similar width. The shaft is at first short but, in luxuriant plants, eventually extends to become held high above the leafy portion. The spike has a triangular outline, with the lowest divisions the longest and spreading horizontally. All the branches are *covered in globular, leathery, sporangial capsules, each about the size of a poppy seed,* borne along the margins of one side only of each division of the spike. The capsules are green when young and orange-brown when mature. Each eventually opens by a horizontal split, to release masses of bright yellow spores.

Jan. Feb. Mar.	Apr.	May	Jun.	Jly	Aug.	Sep.	Oct. Nov. Dec.

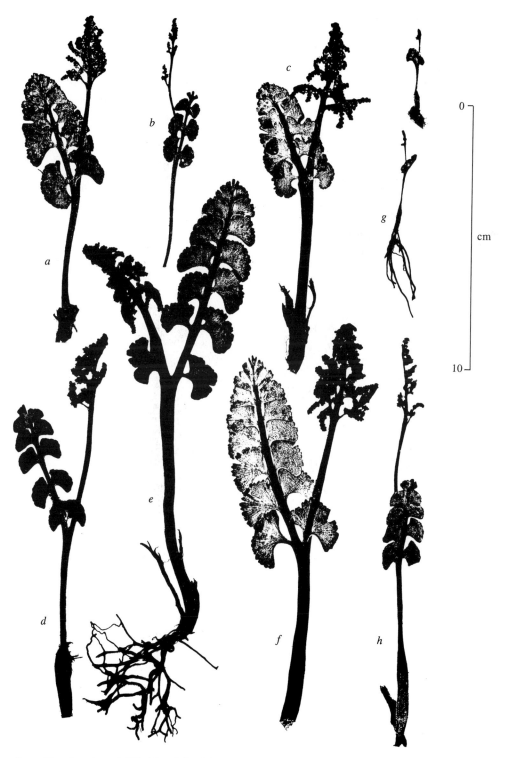

Botrychium lunaria: a, Midlothian; *b*, Mainland, Orkney Islands; *c–f*, East Inverness-shire; *g*, Forfarshire; *h*, Fife.

Botrychium lunaria, typical grassland plants: Perthshire, May. Note hand lens for scale (*right*).

New shoots arise about early May, from within the withered sheath-like base of the previous year's stem. Emerging shoots are *not circinately coiled*, but before fully expanded, the segments of the sterile portion of the frond are usually folded inwards over the fertile portion. The sporangia mature about mid-June, and shoots start to die down from early summer.

Variation Plants vary considerably in size with luxuriance and shelter of the habitat, with sand-dune forms often the smallest. Particularly luxuriant specimens may have more profoundly branched spikes. Additionally there is some variation between populations in shape of sterile segment, the margins of which sometimes can be serrated or quite deeply incised. Occasional monstrous forms also occur.

Possible confusion The peculiar structure of the plant, its fleshy texture and conspicuous crescent-shaped leafy segments, make confusion with any other plant unlikely.

Technical confirmation Plants have $n = 45$, $2n = 90$ chromosomes.

Field notes *Botrychium lunaria* is a widespread species of moonwort, with a geographic range from southern Greenland and Iceland, Britain and Ireland through nearly the whole of the northern and central Europe south to the Iberian Peninsula and the Alps, with more outlying Mediterranean stations across very much of northern Asia to Japan and Kamchatka, and throughout much of the higher latitudes of the North American continent from Alaska to Labrador.

In Britain and Ireland, Moonwort is particularly widely spread. It is a less-gregarious species than Common Adder's-tongue (*Ophioglossum vulgatum*), to which it is related, seldom forming such dense colonies, and occurring more often as scattered individual shoots through turf. Although predominantly a northern and upland species, it occurs in a wide range of open, exposed, short-grassy habitats throughout its British and Irish range. At low altitude, these include old meadow grasslands and pastures, grassy banks, downs, sandy commons, dune pastures, grassy links and old stable sand-dune slacks. At higher altitudes it grows in grassy moors and heaths, hill pastures, gritstone edges, old turfy screes, alpine meadows and cliff ledges. In such habitats, it ascends to about 425 m (1400 ft) in Devon, to 700 m (2300 ft) in the northern Pennines, and to around 1020 m (3350 ft) in Perthshire. Plants can also colonise old industrial spoil heaps and shale bings (tips), especially where natural Moonwort populations are frequent in surrounding hills, such as in the Forth–Clyde valley of central Scotland.

Plants generally grow under less damp and better-drained ground conditions than are typical for Common Adder's-tongue, and on rare occasions when the two grow nearby one another, Moonwort usually occupies the better-drained hummocks and knolls. It grows mainly on soils of circum-neutral to strongly basic character. On heathlands, it is usually limited to areas where there is some lime present in the ground, and where acid heath plants such as *Calluna* and *Erica* are absent. Moonwort is sometimes present in soils directly over limestones, especially in areas of higher rainfall such as in limestone pastures of Galway and Co. Clare, or over chalk in south-east England. But it generally thrives best in deeper soils such as those developed over alluvium, calcareous sands, basic clays, marls and basic shales. In mountain areas, it usually grows where peaty soils on steep slopes are flushed with bases from lime-rich veins. Deep, well-aerated soils probably benefit the plant by giving adequate scope to its deeply running root system, and presumably provide conditions conducive to the success of appropriate mycorrhizal fungi.

The prothalli of *Botrychium* are subterranean brownish-yellow bodies with rhizoids, which lack chlorophyll, and have an essential association with endotrophic (internal) mycorrhizal soil fungi. These provide the prothallus with nutrients saprophytically derived from decaying matter in the soil. It seems likely too, that mycorrhizal fungi are also essential for the long-term well-being of the sporophyte plant, and that the occurrence of other higher plants in the surrounding turf may be essential in providing a steady supply of decaying organic matter for the continued success of the Moonwort.

Notable amongst its associates in most habitats are a wide range of grasses, often including Fescues (*Festuca* spp.) and Bents (*Agrostis* spp.). Other frequently associated plants in lowland sites include common turf species such as Self-heal (*Prunella vulgaris*), Birdsfoot-trefoil (*Lotus corniculatus*), Restharrow (*Ononis repens*), Common Speedwell (*Veronica officinalis*), Eyebrights (*Euphrasia* spp.), Harebell (*Campanula rotundifolia*), Bedstraws (*Galium* spp.), Purging Flax (*Linum catharticum*), Common Dog-violet (*Viola riviniana*) and Field Woodrush (*Luzula campestris*). In old meadow pastures of southern England, the presence of Cowslip (*Primula veris*) is often particularly indicative of sites suitable for Moonwort. In mountain-ledge habitats, various small alpine plants, such as alpine Saxifrages (*Saxifraga* spp.) and Roseroot (*Sedum rosea*) are often present, whilst Lesser Clubmoss (*Selaginella selaginoides*) often grows in moist pockets nearby.

On mountain ledges, Moonwort survives in spots where there is low grazing pressure, and in lowland pasture, it may benefit from a moderate amount of grazing which maintains the overall herbage short. In very short turf its shoots seem particularly readily eaten by rabbits. Plants also have a reputation for disappearing for a number of years from places in which they have been formerly seen, although whether this is due to grazing or more inherent properties of the plant is not known.

Perhaps because of its rather sporadic appearance and disappearance, the plant has long been known in folklore and accredited with all sorts of magical properties, including loosening locks and shackles, and dislodging horseshoes from horses' feet. The former attribute is perhaps an association with the key-like structure of the sterile branch. Clearly plants were very widely known and recognised in former times and presumably frequently encountered. Today, in lowland England and almost throughout Ireland, the plant seems to have declined very considerably in frequency, especially in recent years. Much of this decline seems likely to be associated with ploughing and agricultural improvement of old, formerly long-established pasture land, whilst other factors such as fluctuations in rabbit populations, increased use of herbicides and spread of bracken into many upland habitats may all have had a substantial effect.

Although a certain amount of recolonisation by spores appears to take place, as shown by the appearance of Moonwort in tips of mining debris, much of the plant's spread in established turf is probably by vegetative means. Both re-establishment by spores and vegetative spread in existing turf seem likely to be very slow processes, with existing patches of established plants perhaps often being of considerable age.

Ceterach officinarum D.C.

Rusty-back Fern

Preliminary recognition A small, rather stout plant, with numerous slightly fleshy fronds which are only coarsely divided into a few, bluntly rounded, alternate lobes, giving them a wavy curving outline like a short length of ricrac braid, with the back of the frond densely covered with overlapping scales.

Occurrence Present throughout almost the whole of Britain and Ireland, on natural limestone and mortar of old walls, most abundant in the south and west, becoming much rarer in eastern and northern districts.

Identification Plants have numerous, strongly spreading fronds, about 4–15 cm in length and up to 2 cm broad (but up to 20 cm long and 5 cm wide in some western Irish material – see 'Variation'). The fronds arise in a *densely tufted manner* from the crowns of very short, ascending or erect, usually much-branched rhizomes. They are of overall oblong-lanceolate outline, tapering towards both ends, but are conspicuously divided into *a few, short, broad*, rounded, obtuse, *markedly alternate* lobes. Each lobe is *attached by the whole of its base* to the frond rachis, and their *bases connect between the lobes*, especially towards the apex of the frond. The blades are borne on short stipes, usually about 1/4 of the length of the frond or less. Above, the blade is of *slightly fleshy and somewhat leathery texture*, smooth and a rich deep-green colour, with a slightly glaucous tinge. Beneath, it is *densely clothed over the whole of the undersurface with numerous small scales*, which overlap closely like tiles on a roof, but are pointed. The scales are silvery-white as the frond first expands but later become a pale reddish-brown colour. They project slightly beyond the frond margins, and so can be seen fringing the edges even from above.

Most fronds on mature plants have sori towards their upper ends, which are linear along a vein like an *Asplenium*. They are well hidden beneath the dense scales whilst young, erupting as dark-brown elongate masses near maturity. The fronds are wintergreen, and after emerging in late spring or early summer, last about 18 months or more. On some plants, fronds appear uniquely to produce two generations of mature sporangia in succeeding years (Dyce, 1979). During continued dry weather, fronds tend to inroll, exposing the lower, scaly surfaces, mostly opening again when rain returns.

Variation Plants become somewhat more yellow green in colour when exposed to strong sun, but are relatively constant in appearance within any one population. They do, however, show a strong geographic pattern in variation. Most Scottish, English and Welsh plants are of generally smaller stature and have more or less entire margins to the frond. In some south-west England and west Welsh populations, plants become larger and more crenately lobed along the margins. Such crenate-margined plants become the usual form over most of Ireland. In the extreme west of Ireland, fronds on most plants are regularly extremely crenated, of very large size and often proportionately broader. Preliminary experiments suggest that these characteristics are genetically determined. These Irish forms have been referred to var. *crenatum* Moore.

Possible confusion The very distinctive frond outline and rusty scale-clad frond undersides are unique in the British and Irish flora, and make confusion with any other plant unlikely.

Technical confirmation Plants are tetraploids, with $n = 72$, $2n = 144$ chromosomes.

Field notes *Ceterach officinarum* is a mainly European species in overall range, occurring from Britain and Ireland through south-western Europe and the west Mediterranean basin where it has its principal centre of distribution, and eastwards to the Crimea, Turkey, the Caucasus and Asia Minor. Other species of this small and unusual genus occur in Africa and the Himalayas.

In Britain and Ireland, 'Rusty-back' Fern is widespread and is a strongly calcicolous species, growing in lime-rich crevices in natural rock and also in mortar of well-matured old walls. It occurs mainly at low elevation, from sea-level to about 180 m (600 ft), but scattered plants have been recorded from as high as c. 550 m (1450 ft) in Wales. It is a local fern, but usually highly gregarious, occurring where it does in some quantity. It gains its greatest luxuriance where lime-rich rock occurs at low altitude in oceanic, west-coast climates.

Plants grow mainly directly in fissures of limestone rock, especially Devonian and Carboniferous limestones and Jurassic Oolite, and in walls constructed of these stones. It occurs only occasionally on other, harder, lime-rich rocks, but has been recorded on Ordovician slates in Co. Wexford. It often occurs on village walls in limestone districts, especially in the north Yorkshire and Derbyshire Dales, the Cotswolds, south Devon, and in many parts of western Ireland, especially in Co. Kerry, Co. Limerick and Co. Clare, where Praeger (1934) suggests it to have spread extensively from original restricted areas of limestone. In Ireland, it can become particularly abundant along many low-altitude walls and dyke tops, forming a densely capping mass. The plant is always less frequent in the mortar of walls made of brick.

Plants are tolerant of very strong light, occurring on sunny as well as shaded aspects of both walls and natural cliffs. They seem highly intolerant of tree overgrowth, and usually occur in completely exposed or only extremely lightly shaded situations. In sunny conditions, plants seem tolerant of considerable summer baking, especially on more massive limestone cliffs where adequate reservoirs of stored moisture from within the stonework probably help to maintain a degree of edaphic moisture during spells of dry weather. In such habitats, as well as on walls, plants seem considerably more drought tolerant than either Maidenhair Spleenwort (*Asplenium trichomanes*) or Wall Rue (*A. ruta-muraria*). In milder, moister climates too, the denser growth of *Ceterach*, when present, probably partially excludes these other two species from the same habitats.

Rusty-back Fern demands edaphic conditions which are extremely free-draining. Plants usually establish initially in small, sheltered fissures in limestone rock, and as they slowly build up large rhizome masses, expand over more-exposed faces. As they do so, they accumulate a certain amount of rudimentary soil about their roots, partly through degradation of the rock itself and partly from incoming wind-blown particles. But as the crevices become shaded and humid, they also seem to

Ceterach officinarum; a–d, South Devon; *e*, Shropshire; *f–g*, Co. Clare.

provide a temporary habitation for insects, especially foraging ants. Such insect activity may help to bring in further mineral and humus particles and doubtless also helps to add to the nitrogen available to the plants.

Plants are fairly slow growing, and probably also long lived, and each year their fronds also expand rather slowly. Despite its somewhat xerophytic habit, *Ceterach* undoubtedly thrives best where there is year-round high humidity and a long growing season, in sites where absence of desiccating winter winds and mild winter temperatures allow the plant to benefit most fully from its evergreen condition.

Ceterach officinarum, first fronds, adult venation, young plant and detail of mature scale (after Orth, 1938).

Cryptogramma crispa (L.) Hook.   Parsley Fern

Preliminary recognition A small, bright-green-coloured fern, with finely cut fronds looking very like the leaves of Parsley, and arising in dense clumps.

Occurrence Locally frequent in upland Wales, northern England and throughout upland Scotland, on rocky slopes and screes of acidic rocks on mountains. Absent from lowland districts of Britain and rare throughout Ireland.

Identification Plants have finely dissected vegetative and fertile fronds which are somewhat different from one another, the vegetative ones up to about 15–20 cm tall and of arching habit, *the fertile* up to about 25 cm tall, *standing more erectly from the centres of established clumps*. The fronds all arise from the crowns of short decumbent or ascending scaly rhizomes, which branch freely to eventually build up large masses, with multiple crowns grouped closely together.

On well-established clumps the fronds of both types can be very numerous. All have blades of *oval-triangular or nearly triangular outline, which are finely and somewhat irregularly nearly tripinnately divided*. They are borne on rather slender, brittle stipes, which are mostly green but become brown near the base, where the stipe bears scattered, sparse, brown scales. The vegetative (i.e. sterile) fronds are borne on stipes about as long as the blades or slightly longer, whilst those of the fertile fronds are much longer – usually twice or more the length of the blades.

The vegetative fronds have blades with ultimate segments which are unstalked and look thin textured and flat. Their segments are of broad, cuneate or wedge-shaped outline, with obtuse outer margins which are finely toothed, and the segments often overlap. The fertile fronds, by contrast, have thick, short-stalked segments which are narrowly linear-oblong or almost linear in shape, with their lengthwise margins recurved tightly downwards, forming indusia which almost completely cover the developing sori. The slender-segmented structure gives the fertile frond a finely dissected appearance, with non-overlapping pinnae.

All fronds are of a *vivid, conspicuous, bright light-green colour*, and have a *somewhat leathery, slightly fleshy texture and waxy feel*. They emerge in late spring, and die down rapidly with the first severe frosts in autumn, to dry to a bright rusty-brown colour by early winter. These usually lie incompletely decayed until well into the following season, still attached to the parent clump.

Jan. Feb. Mar.	Apr.	May	Jun.	Jly	Aug.	Sep.	Oct. Nov. Dec.

Variation This is not a very variable species.

Possible confusion The finely dissected, vivid green fronds of remarkably parsley-like appearance, are unlikely to be confused with any other native fern.

Technical confirmation Plants are tetraploids, with $n = 60$, $2n = 120$ chromosomes.

Field notes *Cryptogramma crispa* is present in all the more mountainous districts of western Europe, and the same or other closely related species disjunctly in scattered stations at mainly mid-latitudes in northern and central Asia, Japan, Sakhalin and Kamchatka, and in mountainous regions of North America. This overall range shows close similarities to that of *A. distentifolium* (q.v.). In Europe, *C. crispa* shows a generally arctic-alpine distribution, with its main centres of distribution in western Scandinavia, northern Britain and Ireland and the Alps, and scattered stations stretching south to the Iberian Peninsula and Macedonian mountains.

Within Britain, plants are strongly calcifuge, with a predominantly northern and western range. Parsley Fern forms characteristically bright-green summer patches, up to a metre or more in diameter, distinctive by their colour, when thickly studding steep mountainsides. It is a strongly calcifuge pioneer species, virtually confined to steep scree slopes and tumbled boulder fields of non-calcareous rock on steep mountain slopes, occasionally extending onto cliff ledges around them. It usually avoids the small-fragment-sized central and upper parts of scree slopes, when these are very mobile, but establishes around areas of slightly greater stability, particularly along the flanks of screes and around their lower talus fans. It thrives especially where these reach upland valley bottoms, where larger boulders accumulate and rock movement is slight. *Cryptogramma* occurs in such habitats through a wide range of altitude, descending to about 100 m (300 ft) in a few places and ascending to around 1200 m (4000 ft) in others.

Plants are present on screes on a wide range of acidic, predominantly hard rock types, including ones of sandstones, gritstones, shales, and lime-free

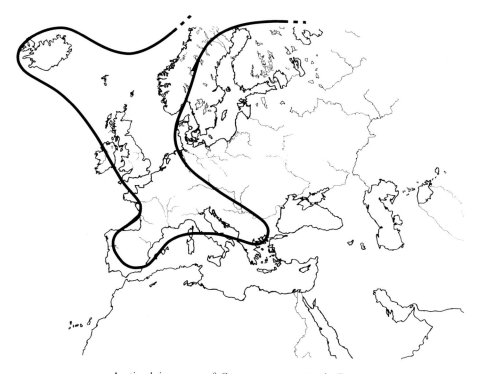

Arctic-alpine range of *Cryptogramma crispa* in Europe.

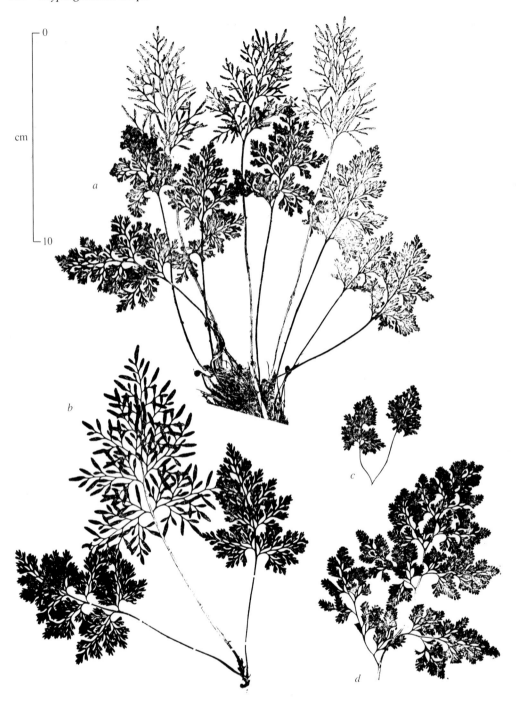

Cryptogramma crispa: a, Argyll; *b*, Cumberland; *c*, Westmorland; *d*, West Inverness-shire.

basalts, but are especially frequent on granitic rock or on slates. In Wales and in the English Lake District, where plants are abundant on nearby hills, individuals frequently appear also in more artificial situations, especially in walls, retaining banks and on bridges, where these are made of local rock. Measurements suggest the acidity of the plant's sites to range usually between about pH 4.0 and 5.0, sometimes as high as 6.1 or as low as about 3.8.

Parsley Fern is a relatively oceanic species, avoiding climates of very severe (and especially dry) winter cold, but also avoiding conditions of high summer maximum temperatures (Connolly & Dahl, 1970). It is, however, tolerant of long snow-lie, and at its higher altitude situations (above about 915 m (3000 ft)) is mostly confined to fern-beds where a thick blanket of regular winter snow adds considerably to the protection of the plants against both cold and winter desiccation.

Plants probably initially colonise screes by establishment of prothalli in their sheltered crevices, where there are also establishing bryophyte cushions. As sporophyte plants develop and build up rhizome masses, so the densely tufted foliage decaying annually but remaining long-attached to the parent clump slowly helps to build up root-bound raw humus masses. Clumps so formed probably often survive to a great age, but eventually they themselves provide habitats in which mosses and flowering plants establish, marking the beginning of the end of the *Cryptogramma* dominance. Such plants often include Heath Bedstraw (*Galium saxatile*), Crowberry (*Empetrum nigrum*), Bilberry (*Vaccinium myrtillus*), Brown Bent-grass (*Agrostis canina*), Wavy Hair-grass (*Deschampsia flexuosa*) and Sheep's Fescue (*Festuca ovina*). With establishment of dense flowering-plant competition, the *Cryptogramma* declines in vigour and usually dies. From such areas, it re-establishes again in more uncolonised areas of scree, thus playing a significant role in eventual vegetation development over such unstable upland habitats.

Cystopteris dickieana Sim

Dickie's Fern

Preliminary recognition A small fern, similar in general appearance to *C. fragilis*, but with smaller, more spreading fronds, with shorter stipes and broader, more crowded, slightly crisped pinnae.

Occurrence Known with certainty only very locally in sea-caves in Kincardineshire, eastern Scotland (its type locality), although there have been sporadic records elsewhere in Scottish mountain glens.

Identification Plants have fronds mostly 8–20 cm long, arising in irregular, tufted groups, from the apices of much-branched, insignificant, decumbent and creeping rhizomes. The fronds mostly adopt an arching habit, giving plants a *lower-growing, more spreading appearance* than

Jan.	Feb.	Mar.	Apr.	May	Jun.	Jly	Aug.	Sep.	Oct.	Nov.	Dec.

0   10
                    cm

Cystopteris dickieana, frond variation in sea-cave habitat of type locality: *all* Kincardineshire.

Cystopteris dickieana, habit of plant in life.

is typical for *C. fragilis*. The short stipes are brittle, are about 1/4 the length of the frond, and are of a greenish-brown or straw colour, becoming darker brown at the base. The blades are of oblong-lanceolate, often rather irregular outline, and are *less finely dissected* than those of *C. fragilis*, with *broad, bluntly rounded, and only slightly notched, ultimate segments*. The base of each pinna is particularly broad, and its tip abrupt and obtusely pointed. The broad pinnae are *crowded along the rachis*, usually becoming confluent towards the frond tip, and are often slightly twisted, giving the whole blade a *congested and slightly crisped appearance*, which is very characteristic of this plant.

The lamina is often a rather deep-green colour with a dull surface, and is of a thin but less delicate texture than *C. fragilis*. Its venation is often conspicuous as darker lines.

Fertile fronds occur abundantly, are similar to the vegetative ones and have sori which are nearly marginal in position. The sori are small and green when young, but shining black when mature, before shedding their sooty black spores. Plants are vegetatively vigorous and produce spores abundantly.

Fronds flush early in spring, and the first fronds mature their spores by early June, much earlier than those of *C. fragilis*, whilst a succession of fronds, with maturing spores arise throughout the summer until autumn. Fronds begin to die down with the first frosts, but appear rather less frost sensitive than those of *C. fragilis*.

Variation The above description refers to plants in the Kincardineshire (type) locality, and to the features maintained in cultivation in material from this locality, originally distributed by Dickie in the mid-nineteenth century, and maintained in cultivation by fern growers since. Plants possessing the *C. dickieana* non-spiny spores (see 'Technical confirmation') but more closely resembling *C. fragilis* in frond form, are known elsewhere in Europe, and possibly also occur in Scotland, as might hybrids between *C. dickieana* and *C. fragilis*.

Possible confusion The much more crowded pinnae as well as non-spiny spore type (see 'Technical confirmation') distinguish *C. dickieana* from *C. fragilis*. Plants of *C. dickieana* also have a superficial similarity to *Woodsia* plants, but lack the hairs and dense scales of this genus.

Cystopteris dickieana, east facing sea-cave type locality: Kincardineshire, June. Note two shadowy figures to the near right for scale, and hanging fronds from the high cave roof.

Technical confirmation Plants are tetraploids, with $n = 84$, $2n = 168$ chromosomes. The spores differ distinctly in surface sculpturing from those of *C. fragilis*. Spores of *C. dickieana* have an outer coat (perispore) which is wrinkled into irregular folds and ridges and low projections (described as rugose-verrucose), in contrast to the very spiny (echinate) spores constantly found in all *C. fragilis*.

Field notes Like the other species of *Cystopteris*, Dickie's Fern is a strongly calcicolous species, in its type locality growing on easily weathered, base-rich andesitic lavas. Its sites have a pH range of about 6.3–7.7, and the fern is associated with a number of other calcicolous plants. It occurs in deep shade in east-facing caves, which are cool in summer and kept permanently moist by trickling seepage of water throughout the year.

Cystopteris dickieana could yet be found in other, particularly more montane, areas of Scotland. Its occurrence in a coastal habitat seems at first unusual, but it parallels the distribution of several other base-loving, essentially montane species which also have coastal stations. Like montane sites, these habitats are cool in summer and often lack excessive competition from rank herbaceous vegetation. The survival of Dickie's Fern in such calcicolous maritime conditions, in contrast with the uncertainty of its montane occurrence, is of particular interest because of the notable absence of *C. fragilis* from such coastal sites. The apparent ability of *C. dickieana* to survive salt-laden air, may well have enabled it to survive locally on coastal rocks in a genetically relative 'pure' form, after possible widespread displacement by (or partial introgression with?) plants of the *C. fragilis* group in other cool, moist sites on Scottish mountains. Wild plants of *Cystopteris dickieana* are specifically protected by law (Wildlife and Countryside Act, 1981).

Cystopteris fragilis (L.) Bernh.

Brittle Bladder-fern

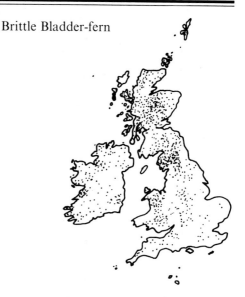

Preliminary recognition A small, delicate fern of variable form, but always with lance-shaped, finely dissected, rather fragile blades borne on slender, brittle stipes.

Occurrence Widespread throughout Britain and Ireland, but very local in the south and east of England and in most of Ireland. It is frequent in upland districts of Wales, northern England and Scotland, especially where there is lime-rich rock, on damp shaded cliff faces and springing from rocky crevices. It also grows in mortar on damp, shaded walls.

Identification Fronds arise in irregular, tufted groups, from the apices of occasionally branched, very short and rather insignificant creeping or decumbent rhizomes. The fronds are commonly 10–20 cm long (occasionally to 30 cm or more in luxuriant specimens), with *slender brittle stipes* forming 1/3 or more of the length. The stipes are a pale, *translucent straw-green colour*, usually brown only near the base (occasionally higher, to the middle of the blade), and bear only a few, sparse, soft pale-brown scales. The blades are of oblong-lanceolate outline, narrowing slightly towards the base and gradually upwards to a sharply tapering apex, and are *finely bipinnately or more divided.* The pinnae are usually non-overlapping and frequently quite widely separated on the frond, and set in nearly opposite pairs. Each is widest at the base, has a tapering tip, and is pinnately divided, but the degree of dissection and shape and spacing of the ultimate segments are very variable (see 'Variation'). The lamina is usually of a bright, pale- to mid-green colour, and of *thin but delicately firm*, non-shining and completely glabrous texture, borne on brittle, translucent green midribs. In strong light, plants become yellow green in colour, in very dull light, a deeper green.

Fertile fronds are produced abundantly and are similar to the vegetative ones, and bear submarginal sori. The sori are green and each covered by a *thin, membranous, translucent indusium which is hood shaped* and slightly inflated when young, with a denticulate margin, but shining black with an inconspicuous indusium when mature, before shedding their sooty black spores. The inflated, bladder-like appearance of the young indusium is the source of the generic name of 'bladder-fern'.

Fronds arise early and rapidly in spring, and mature quickly, with their first spores usually completely mature and shedding by early August. A succession of further fronds usually follows, but all die down early in autumn with the first frosts.

Jan.	Feb.	Mar.	Apr.	May	Jun.	Jly	Aug.	Sep.	Oct.	Nov.	Dec.

Cystopteris fragilis, habit of plant in life.

Variation Plants are highly variable in details of frond form, especially pinna spacing, pinna shape, degree of cutting, and shape of ultimate segments, which are sometimes obtuse and rounded and sometimes acute and attenuate. Much of this variation seems genetically determined.

Plants with smaller fronds with blunter pinnae and more rounded, more obtuse segments which are bluntly toothed have, in older literature, often been referred to as 'var. *dentata* Hook.'. Plants with larger fronds with quite widely spaced, narrow, acutely tapering pinnules which are deeply cut and serrated have been referred to 'var. *angustata* Moore'. Both these types seem fairly widespread and it may eventually prove possible to recognise distinctive traits in correlation with known differences in ploidy level (see 'Technical confirmation'), although the genetics and genomic constitution of these high polyploids is likely to be extremely complex (see, for example, Vida, 1974).

In addition, and of very much more local and more doubtful status, is a form with finely cut fronds with blunt, distantly separated segments, recorded from Upper Teesdale in 1872 as 'var. *alpina* Hook.', and probably now extinct, whilst intermediate types between these rather extreme forms also seem to occur.

Further, in recent years, a number of specimens have been found in Highland Scotland (D. Tennant, personal communication) whose frond type looks like that of *C. fragilis*, but which have the rugose-verrucose spore type normally associated with *Cystopteris dickieana*

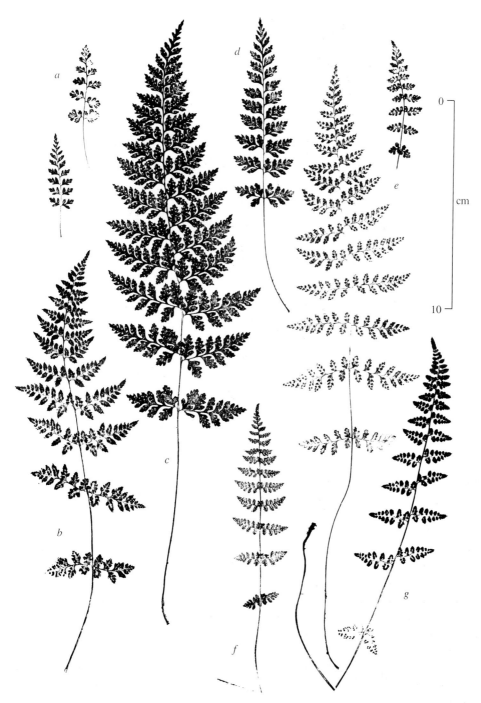

Cystopteris fragilis: a–c, West Argyll; d, Wigtownshire; e–g, Mid Perthshire; silhouettes showing soral distributions.

0

10

cm

Cystopteris fragilis or *dickieana*? Scottish inland populations with a frond morphology of the *C. fragilis* type and a spore morphology of the *C. dickieana* type (specimens courtesy D. J. Tennant); see comments in text.

(q.v.). Much careful work has been put into the examination of these specimens by the finder, who is to be congratulated on this work and their discovery. It is consequently very tempting to interpret these specimens as true *C. dickieana* rather than *C. fragilis*, as a result of this spore type, and this may be the correct interpretation. However, an alternative interpretation could also be that these plants, often occurring as scattered individuals or groups among wider populations of *C. fragilis* with the 'normal' echinate spore type, represent particular genotypes (and presumably cytotypes) of variable *C. fragilis*, in which

Cystopteris fragilis, young frond stages and venation (after Orth, 1938).

such a spore type has perhaps arisen independently. Such individuals might yet rank as a separable and independent taxon.

Possible confusion Mature plants of *C. fragilis* can be very readily confused with young plants of all species of *Athyrium*, from which the best characters for separation are the early fertility of fronds of *C. fragilis* and the generally more widely spaced, more forward-swept pinnae and the usually much more slender and longer stipes.

Technical confirmation Plants are either tetraploids, with $n = 84$, $2n = 168$ chromosomes, hexaploids with $n = 126$, $2n = 252$ chromosomes, or pentaploid hybrids with $2n = 210$ chromosomes. (Octoploids are also known in Europe, and could yet be found in Britain or Ireland, raising also the possibility of hybrids of other ploidy grades.) Spores of *C. fragilis* have an outer coat (perispore) which is very spiny, and gives the spores a variously echinate appearance, in contrast to the rugose-verrucose spores typical of *C. dickieana* (q.v.) (but see note at the end of 'Variation').

Field notes *Cystopteris fragilis*, in the broadest sense, is a very widespread species of middle latitudes of the northern hemisphere, with a natural range from Europe across northern Asia to Japan and across North America. Within this overall range, a number of segregates have been recognised, and the species is also a polyploid complex, whose overall range compares closely with that of *Athyrium filix-femina* (q.v.) Within Europe, *C. fragilis* is widespread almost throughout moister northern and middle latitudes of the continent, thinning more in Mediterranean latitudes, but spreading to the Azores and with related forms in Madeira and the Canary Islands.

Within Britain and Ireland, Brittle Bladder-fern is very widespread, especially in western counties and in most wetter and more mountainous districts. It is a fast-growing and rather gregarious fern, often occurring either in small groups or in quite large numbers over limited areas of moist, rocky habitats. All forms of it are markedly calcicolous, thriving usually where there are exposures of lime-rich rock, and these are moist and shaded. It is consequently most frequent in limestone areas and upland hilly districts, in rocky woodland and shaded, sheltered, wet ravines, but it ascends to well over 915 m (3000 ft) in sheltered, moist spots on mountain cliffs, especially in deep fissures and beneath overhangs, or in crevices of mountain scree.

In limestone districts, it often penetrates more deeply than other ferns into the mouths of caves and sink-holes, and into darker, sheltered, grykes of limestone pavements. Plants also occur, sometimes in considerable numbers, in old mortar on shaded aspects of walls, such as basements of old abbey ruins, old walled-garden brickwork and beneath arches of old stone-built bridges, especially over streams and rivers. In such habitats, its tolerance of low light level enables it to succeed in moist habitats, which, except for mosses, are often nearly competition free.

Although Brittle Bladder-fern often grows abundantly on limestone, it can occur on exposures of a wide range of other rocks containing calcium-rich minerals, such as calciferous sandstones, mica and hornblende schists, rhyolites, andesites, dolerites and calcium-bearing seams in granophyre volcanics. In most of its niches, it occurs in small pockets of rudimentary rocky soils, often with a pH range of about 5.7–6.7. Its small, but very extensive, wiry root system probably often ramifies deeply into small cracks and comes into intimate contact with surfaces of rock.

In better-illuminated situations, associated species often include Herb Robert (*Geranium robertianum*), Harebell (*Campanula rotundifolia*), Wild Strawberry (*Fragaria vesca*), Opposite-leaved Golden Saxifrage (*Chrysosplenium oppositifolium*), Hard Shield-fern (*Polystichum aculeatum*) and Green Spleenwort (*Asplenium viride*).

The scarcity of plants in coastal situations, even where lime-rich rocks outcrop, may indicate poor tolerance by *Cystopteris* of salt spray. In such situations in the south and west, *C. fragilis* seems to be ecologically replaced by Maidenhair Fern (*Adiantum capillus-veneris*). The scarcity of *Cystopteris* in the south and east of both England and Ireland may be associated with its poor tolerance of summer drought.

Cystopteris montana (Lam.) Desv.

Mountain Bladder-fern

Preliminary recognition A small, elegant, delicate-fronded fern, with finely divided, broadly triangular fronds, arising in colonies.

Occurrence Strictly a high-mountain plant, with its main headquarters in the Breadalbane ranges of mid and west Perthshire and north Argyll in the central Scottish Highlands, on cool, moist, north-facing cliffs, where there is lime-rich seepage. Even in these areas it is rare.

Identification Fronds arise individually from short side-branches of extensively creeping, slender, 1–2 mm thick, dark-brown, subterranean rhizomes, bearing fronds at rather distant intervals.

The fronds have tall, erect stipes, and finely cut, spreading, triangular blades, with finely toothed ultimate segments. The stipes are twice or more the length of the blade, usually 10–15 cm long, sometimes 30 cm, are a pale straw-green colour and slightly translucent, becoming dark only towards the base. They are grooved, and carry a few pale, gland-fringed scales. The stipes are *notably thick in relation to the size of the blade* (and comparable proportionately with those of *Dryopteris aemula*).

The blades are usually about 3–12 cm long and of similar width, *triangular-deltoid in outline*, and have green midribs and *finely cut pointed pinnae*. Frequently, the tips of all the pinnae curve towards the apex of the frond in a sickle-like manner, whilst the frond tips all point in a similar direction, which is constantly downslope. The lowermost pair of pinnae, which are set oppositely, are *much more strongly developed and more compound than those above, and asymmetric*, with the more developed and the more divided side towards the base. Just above the point where the lower pinnae are inserted, the rachis curves over to spread the upper part of the blade in a more or less horizontal (or often slightly descending) plane. Arising from before the curve, the lower pair of pinnae are usually inclined upwards compared with the rest of the blade, with only their pinnules held horizontally. The whole arrangement gives the blade a somewhat upward-dished overall shape, similar to that seen in *Dryopteris aemula* (q.v.).

The lamina of the frond is of *firm and crisp, pellucid texture, and of a fresh pale-green colour*

Jan.	Feb.	Mar.	Apr.	May	Jun.	Jly	Aug.	Sep.	Oct.	Nov.	Dec.

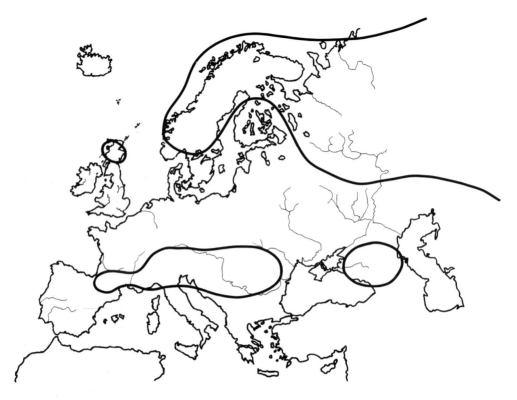

Arctic-alpine range of *Cystopteris montana* in Europe.

when newly emerged, but later in the season turns pinkish or purple from the midribs outwards. The frond surface usually has a few, sparsely scattered, fine-stalked glands.

Fertile fronds are similar to the vegetative ones and carry sparsely scattered small sori approximately centrally on their ultimate segments, each protected when young by a thin, transparent, ovate indusium with an irregularly toothed margin.

Fronds flush rapidly in the increasing warmth in spring, and die down early, along with the first frosts, usually from late August, and disappear by late September.

Possible confusion The triangular-outlined fronds distinguish *C. montana* from all other *Cystopteris* and *Athyrium*. Small alpine forms of *Dryopteris expansa* also have triangular blades and finely-cut fronds, but those of *Cystopteris montana* arise singly, not in tufts.

Technical confirmation Plants are tetraploids, with $n = 84$, $2n = 168$ chromosomes. The much smaller spores (*c.* 29–39 μm), which have only low, rounded surface protuberances, separate it from the large spores (*c.* 38–54 μm) with spinose surface protuberances of *Cystopteris fragilis*.

Field notes *Cystopteris montana* is present in all the more mountainous fairly high-latitude districts of western Europe, and the same or other closely related species disjunctly in scattered stations at mainly mid-latitudes highly discontinuously in northern, central and eastern Asia, and in mountainous regions of North America. In Europe, *C. montana* shows a generally arctic-alpine distribution, with its main centres of concentration in Fenno-scandia, northern Britain and the Alps, with

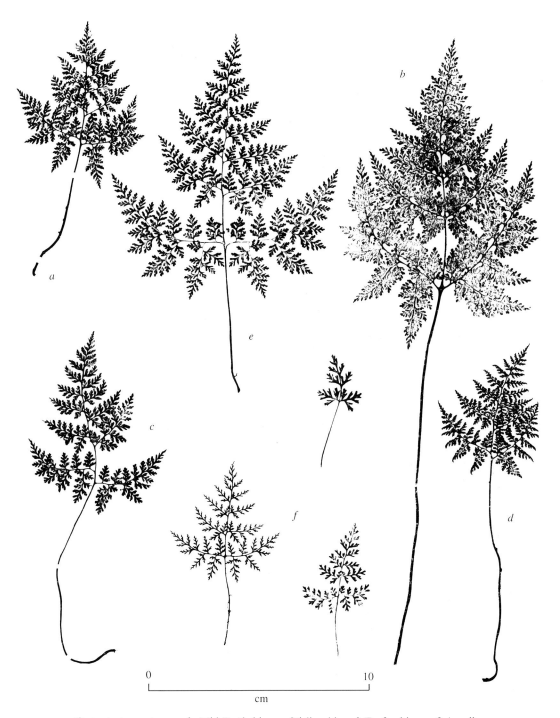

Cystopteris montana: a–b, Mid Perthshire; *c*, Stirlingshire; *d*, Forfarshire; *e–f*, Argyll.

scattered stations stretching south to the Pyrenees and Central European mountains east to the Caucasus.

Within Britain, Mountain Bladder-fern is a distinctly high-altitude species, seldom descending below *c.* 700 m (2300 ft) and mostly growing much higher than this. It occurs very sparingly on cool, shaded, mainly north-facing slopes, in areas where the summer maximum temperature does not usually exceed about 27–28 °C, especially on more western mountains where the climate is oceanic.

Plants occur mainly on dripping rock ledges, on steep sides of gullies, and on steep, loose, unstable, shaly micaceous wet rocky screes irrigated by moving seepage water, at the foot of dripping, shaded crags. They occur in sites where the relatively large fronds are protected from direct sun and strong winds, and where there is constant water percolation and movement of humid air. It is a strongly calcicolous species, restricted to outcrops of calcium-rich veins, almost exclusively on mica-schist rock.

Its moist, basic, alpine habitats are frequently floristically rich ones, and large numbers of other species often occur within and around *C. montana*. These include especially alpines such as Alpine Lady's mantle (*Alchemilla alpina*), Moss Campion (*Silene acaulis*), Mountain Sorrel (*Oxyria digyna*), Alpine Scurvy-grass (*Cochlearia alpina*), Alpine Bistort (*Polygonum viviparum*), Yellow Saxifrage (*Saxifraga aizoides*), Purple Saxifrage (*S. oppositifolia*), Mossy Saxifrage (*S. hypnoides*), Starry Saxifrage (*S. stellaris*), the mosses *Ctenidium molluscum, Bryum alpinum, Rhacomitrium languinosum, Fissidens* spp., and several thallose liverworts. On surrounding rocky fissures there is often also Green Spleenwort (*Asplenium viride*), and sometimes Brittle Bladder-fern (*Cystopteris fragilis*) and Holly Fern (*Polystichum lonchitis*). This is the '*Saxifraga aizoides* association' of McVean (1964), and the number and dominance of mosses and flowering plants of moss-like habit is notable. *Cystopteris montana* often spreads its rhizomes through or beneath the cushions formed by these species, which probably act as additional reservoirs of moisture for it during occasional dry spells. The occurrence of such a proportionately large and delicate-fronded fern contrasts strongly with the low-growing habit of these alpines, and it is perhaps of interest in this connection that around its sites there often also survive scattered individuals of other plants which seem more characteristic elements of river banks and flushed slopes in upland valley woodland, such as Wood Anemone (*Anemone nemorosa*), Common Dog-violet (*Viola riviniana*), Globe Flower (*Trollius europaeus*), Wood-sorrel (*Oxalis acetosella*), Meadowsweet (*Filipendula ulmaria*) and Water Avens (*Geum rivale*).

Suitable conditions for Mountain Bladder-fern seem to occur only over very limited areas. Large stands of it are rare, and it is possible that in most of its sites, each colony represents a single clonal individual, underlining that this species in Britain may already have a very limited genetic base.

Dryopteris aemula (Ait.) Kuntze

Hay-scented
Buckler-fern

Preliminary recognition A graceful, medium-sized, triangular-bladed, winter-green buckler-fern, with vivid green fronds, drooping frond-tips, and distinctive overall crimped or crisped frond texture.

Occurrence A species of markedly Atlantic distribution, occurring sometimes in local abundance near the west coast of Ireland and in western Scotland, present more sparsely in west Wales, south-west England and the Weald of Kent, and elsewhere only in a few widely scattered localities.

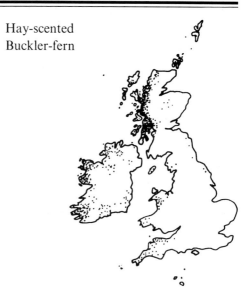

Identification Plants have spreading and drooping-tipped fronds, mostly about 15–60 cm long (occasionally nearly 100 cm), arising in a broad, shuttlecock-like manner from the crowns of short, erect or ascending rhizomes, which eventually branch to build up several closely packed crowns. The blades of the fronds are of distinctly triangular-ovate or triangular outline, borne on *rather thick, slightly succulent stipes*, often as long as the blade and of *dark-brown colour tinged purplish near the base*, and bearing scattered, rather limp, uniformly pale orange-brown scales. The fronds bear about 10–20 pairs of opposite or subopposite pinnae, each of which is of asymmetrically deltoid outline, with the *lowermost side particularly well developed on the lowest and largest pair of pinnae.*

The pinnules on the lowermost pinna are nearly pinnate, whilst those on the rest of the frond are less divided. The edges of all the pinnae are rather coarsely serrated into slightly spinulose teeth, *the tips of which are variously straight, outcurving or incurving.* The edges of all the pinnae *curve markedly upwards, giving each pinna a concave shape and the whole frond a markedly crimped or crisped appearance*, characteristic of this species even from a distance.

Most fronds on mature plants bear sori scattered in rows along each pinnule over the whole of the lower surface, and because of the strongly upturning pinnule margins, occasionally show in oblique view from above.

The fronds are of a very distinctive, *rather vivid, bright mid-green colour*, darkening slightly with age, as they persist through the winter and well into the following season. The blades are of *crisp, turgidly firm, but never leathery, texture.* When lightly bruised, they have a slightly sweet smell in the field, which turns to a stronger smell of new-mown hay (that of the chemical coumarin) when freshly dried.

Jan.	Feb.	Mar.	Apr.	May	Jun.	Jly	Aug.	Sep.	Oct.	Nov.	Dec.

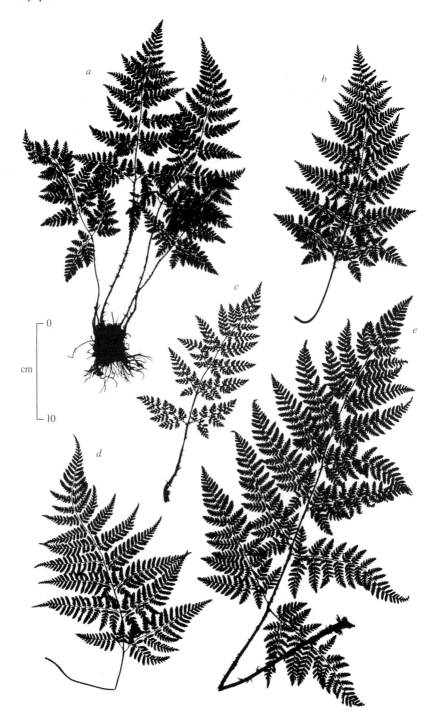

Dryopteris aemula, mature frond stages, Scottish populations: *a*, Isle of Arran; *b*, *c*, Argyll; *d*, Western Ross; *e*, Isle of Bute.

Dryopteris aemula, habit of plant in life.

Variation Mature fronds vary very little between individuals or populations, except in ultimate size, which is probably largely environmentally determined. Fronds of juvenile plants are usually much more narrowly triangular during their first one or two seasons than adult ones, but show, even at this stage, the characteristic upturning margins and crimped appearance.

Possible confusion A very distinctive fern in the field which, once seen, is unlikely to be confused with anything else. Pressed (herbarium) material loses the crimping and is then much less obvious, but the thick stipes and pattern of frond cutting usually remain clear.

Technical confirmation Plants are diploids, with $n = 41$, $2n = 82$ chromosomes. The rachis, stipe and pinnule mid-veins are all quite densely covered in small, sessile glands, which are conspicuous as minute, rounded, colourless, shining spheres under a $\times 10$ hand lens in fresh material. These glands also spread less densely over the undersides and sometimes the upper sides of the lamina, and to the scales and industrial margins, which they fringe. The indusial margin is also distinctly toothed.

Field notes *Dryopteris aemula* is an entirely North Atlantic species in distribution, occurring in Britain and Ireland, the coastal periphery of Brittany and the northern Iberian Peninsula, the Azores, Madeira and the Canary Islands. Further eastwards in the Mediterranean, *Dryopteris liliana* Golcin of Turkey is reputedly related.

Hay-scented Buckler-fern has its main centre of range in Britain and Ireland, where it is a plant of strongly oceanic distribution, becoming most abundant and vegetationally luxuriant only at low altitudes in habitats more or less directly influenced by the winter-ameliorating warmth of the Atlantic Gulf Stream. Most of its habitats occur within about 30 m (100 ft) of sea-level, although it occasionally ascends higher, especially in Ireland, where scattered plants reach exceptional heights of 440 m (1430 ft) in Co. Down and 640 m (1430 ft) in Co.

Dryopteris aemula, young frond stages.

Dryopteris aemula, dorsal view of pinnules in life showing generally upcurved margins (cf. those of *D. dilatata*, q.v.).

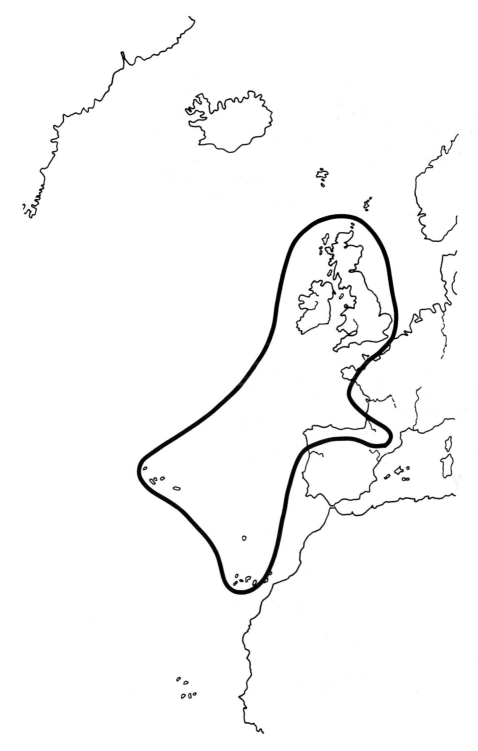

Atlantic range of *Dryopteris aemula* in Europe.

Kerry (More, 1898). It grows mostly in rocky woodland sites, often by streams, and especially where steep, mossy, boulder-strewn slopes are overgrown by a light, low canopy of oak–ash–rowan woodland. Under such conditions, *D. aemula* can form the dominant groundcover vegetation, each plant rooted on the tops and shoulders of mossy boulders, or on mossy faces of rocky slopes, forming curtains of cascading fronds. Although its habitats are ones which are very freely draining, they are mostly also ones which are permanently moist, have year-round high humidity, frequent light precipitation, and are screened from exposure to full sun and desiccating winds. In such habitats, plants can tolerate remarkably low levels of light.

Hay-scented Buckler-fern is a rather slow-growing species. Its fronds also unroll rather slowly each season, with an initial flush followed by further fronds unfurling in a steady succession throughout the summer. Most fronds mature by autumn, and stand overwinter until about May of the following season. One of the main factors confining the plant geographically to highly oceanic areas is probably the long duration of the growing season required for the slow-growing wintergreen fronds to harden fully. Hence, this fern is mostly absent from areas where winter frosts can occur at an early date.

With its main centre of range in Britain and Ireland, *Dryopteris aemula* is a particularly characteristic North Atlantic and especially western British and Irish species of fern, which is distinctly rare elsewhere in Europe. Although not rare locally in these islands, it is a particularly significant native species, with important conservational implications.

Dryopteris affinis (Lowe) Fraser-Jenkins

Golden-scaled Male-fern

(*D. pseudomas* (Wollaston) Holub et Pouzar, *D. borreri* Newm.)

Preliminary recognition A large, handsome fern with robust, tough and rather leathery, lance-shaped, yellow-green fronds, arising in shuttlecock-like clusters, their stout stipes densely clothed with shaggy golden-brown scales. Plants look generally like a more yellow-green, tougher, more scaly version of *D. filix-mas*, sometimes with more square-edged pinnae.

Occurrence Present throughout most of Britain and Ireland. Especially frequent in more oceanic areas of higher rainfall, in woodlands and mountain valleys, from sea-level to moderate altitude, mostly on acidic soils.

Identification Plants have variably winter-deciduous, pinnate-pinnatifid fronds, mostly 50–80 cm in length, but sometimes 150 cm or more, of *ovate-lanceolate outline*, which arise in steeply ascending shuttlecock-like clusters from stout, clump-forming or ascending or erect rhizomes. The fronds are widest about the middle, and taper to the apex. Below, the outline narrows less, and is eventually truncated by the pinnae stopping short of the frond base, with the lowermost pinna pair about half the length of those in the central part of the frond or

Jan.	Feb.	Mar.	Apr.	May	Jun.	Jly	Aug.	Sep.	Oct.	Nov.	Dec.

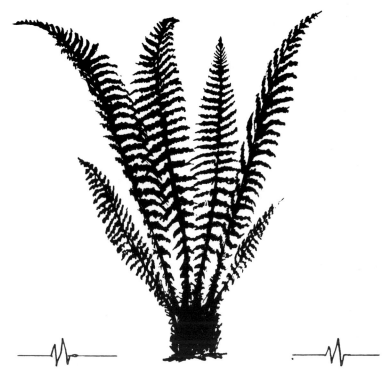

Dryopteris affinis, habit of plant in life.

sometimes less. The stipes are about 1/4 or less of the frond length, *are particularly strong and stout, and are very densely clothed with usually strongly coloured, shining golden-brown or light orange-brown chaffy, shaggy scales*, with bases which usually become darker as the season progresses. These scales remain dense throughout the lower half of the rachis, thinning progressively in the upper half. They are, however, much denser throughout than in *D. filix-mas*.

The individual pinnae are incompletely pinnatifidly divided, cut nearly to or only part-way to the pinna mid-ribs, with nearly entire or somewhat toothed margins. The fronds are of firm, usually tough and leathery texture, and of *much yellower-green colour, with more shining upper surfaces* than those of *D. filix-mas* when newly emerged in spring, but often darker green when fully mature. At the junction of each pinna with the rachis are patches of *dark leaden-grey pigment forming a distinctive black patch or black spot around the pinna-midrib base*, never seen in *D. filix-mas*.

Fertile fronds are similar to the vegetative ones, and bear two lines of *large (1–2 mm in diameter or more), conspicuous kidney-shaped sori along the underside of each segment*, producing a domino-like appearance.

Fronds flush considerably (2–5 weeks) later in spring than those of *D. filix-mas* growing in the same environment, and the unfurling croziers unwind their tips so that they quickly open out into *laxly hanging, open hooks, like a loose shepherd's crook*. This is even more open and lax than that of *D. filix-mas*, and contrasts strongly with the tightly coiled croziers of *D. oreades* when unfurling.

0   10

cm

Dryopteris affinis, medium-sized frond specimens from wild individuals developed under comparable environmental conditions in the same rocky woodland habitat, Scotland, Inverness-shire: *a*, subsp. *cambrensis; b*, subsp. *affinis.*

Variation The satisfactory morphological identification of all of the British and Irish variants of this taxonomically very complex apogamous species usually requires a symphony of characters to be taken into simultaneous account. Its taxonomy and nomenclature in these islands and in Europe as a whole has been, and continues to be, the subject of extensive study (and not inconsiderable debate), with taxonomic concepts and resulting nomenclature intermittently changing. There is probably much further research yet to be undertaken, and some of the problems of nomenclature of the variants undoubtedly result from the inherent difficulties of dealing satisfactorily with an apogamous taxon. Perhaps because of this, some published approaches (e.g. Jermy & Camus, 1991) have adopted a non-specific morphotype concept for the segregates within the *D. affinis* complex, while others (e.g. Fraser-Jenkins 1980, 1982, 1984, 1987*a*,*b*) have developed a more traditional subspecies concept and resulting nomenclature. The latter approach is followed in broad outline here for consistency within this book, and, with the benefit of valuable discussion with C. R. Fraser-Jenkins, A. C. Jermy and A. C. Pigott, the nomenclature and descriptions of the currently most widely recognised subspecies are outlined accordingly.

Subsp. *affinis* has fronds which, when mature, usually form a 'shuttlecock' on an erect rhizome, the fronds *remaining more or less green almost throughout most winters.* Most fronds have an ovate-lanceolate outline, narrowly tapering at the apex and usually more or less tapering at the base, *the stipe thick and stocky*, sometimes short, *with a very dense and conspicuous covering of chaffy deep golden-brown scales.* The pinnae are usually held relatively perpendicularly to the rachis throughout much of the fronds, are relatively long, flat, and held in the plane of the blade, and the whole frond has *a firm and leathery texture and usually a very glossy upper surface* that is rich green above and slightly glaucous below. The pinnules have dorsiventrally nearly flat margins, which are *unlobed to only shallowly lobed along their adjacent edges, and have rounded tips with sparse obtuse teeth.* The lowest basiscopic pinnule of the lowest pinna is partly attached at its base (i.e. not fully stalked and sometimes described as semi-adnate). Technically, the sori each have an indusium that is thick, steel grey in colour when young, browning with age, and *particularly doughnut-shaped. Its margins remain well tucked under but usually readily split radially in more than one place under the pressure of the ripening sporangia, lifting slightly as the sporangia mature: they are long persistent in this condition.* The spores are generally smaller than in the other two subspecies, but more regular in shape and of more similar size range. Var. *palaceo-lobea* Fraser-Jenkins with more lobed and twisted pinnules and more toothed margins is a variant of this subspecies, with the same indusial characters.

Subsp. *borreri* (Newman) Fraser-Jenkins has fronds which, when mature, usually arise irregularly from much-branched rhizome clumps or which, especially in luxuriant specimens in moist and shaded woodland situations, may form a shuttlecock on a more erect rhizome, *the fronds usually dying back early in winter.* Most fronds have a more or less oblong outline, *tapering at the apex but usually less tapering (sometimes quite abruptly truncate) at the base.* The stipe is long and sometimes relatively slender, with a sparse to moderately dense covering of rather pale grey-brown scales. The pinnae are usually held at a perpendicular to somewhat, forward-swept angle along the rachis throughout much of the frond, are relatively *long, flat to somewhat undulated and held in the plane of the blade, and have a less-firm texture and usually dull (non-glossy) upper surface.* The pinnules are usually crowded along the pinnae and have dorsiventrally nearly flat margins which are typically *shallowly notched along their adjacent*

Dryopteris affinis subsp. *affinis*: West Cornwall, July.

edges into low rectangular lobes with acute teeth, and have somewhat squared ('truncated') tips in younger specimens (as though the pinnules have been cut with a pair of scissors), more pointed in more adult ones, with sharp, acute, usually incurved teeth, especially at both corners. The lowest basiscopic pinnule of the lowest pinna is typically fully stalked (i.e. not usually partly attached at its base). Technically, the sori each have an indusium that is thin, of low profile and has margins that are relatively flat. The indusia become rapidly creamish-white as they lift to become funnel-shaped and then shrivel rapidly at maturity of the sporangia beneath, only partly persisting thereafter. The spores are relatively large in size (more so than in subsp. *affinis*), and may contain many misshapen ones. Var. *robusta* (Oberholzer et von Tavel ex Fraser-Jenkins) Fraser-Jenkins & Salvo-Tierra (= subsp. *robusta* Oberholzer et von Tavel ex Fraser-Jenkins) is a variant (and particularly luxuriant growth form) of this subspecies which generally forms large shuttlecocks of large and wide fronds (to 150 cm or more in length) with long, well-lobed pinnules, in moist woodland habitats. It has the same indusial characters and is similar in chromosome number (see under 'Technical confirmation').

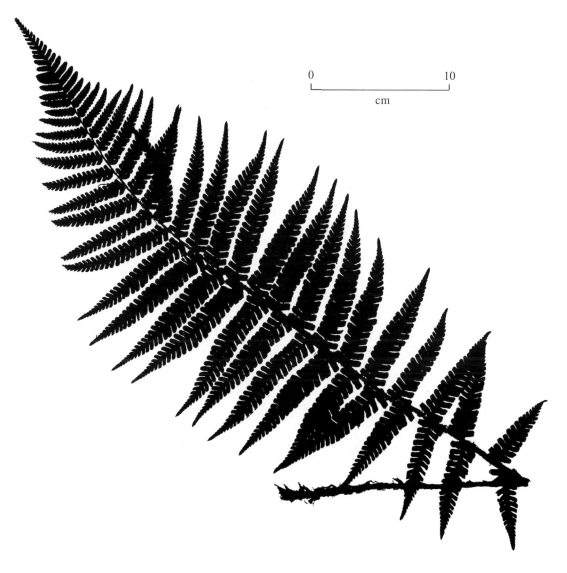

Dryopteris affinis, medium-sized frond of subsp. *borreri:* Mid Perthshire.

Subsp. *cambrensis* Fraser-Jenkins has fronds which, when mature, usually form a dense mass on a much-branched rhizome, the fronds mostly withering rapidly with the first frosts of winter. Most fronds have *a markedly narrow oblanceolate to elliptic outline, tapering at the apex and usually tapering gradually and extensively at the base*, the stipe moderately thick *with a fairly dense covering of chaffy golden-brown to reddish-brown twisted scales.* The pinnae are usually held at a forward-swept angle throughout much of the frond, are relatively short, *and have most basiscopic pinnules usually curved upward to give concave pinnae.* The pinnae are somewhat upwardly rotated with respect to the overall plane of the blade, and have a firm texture and rather glossy surface. Uniquely, the pinnules also have notably downturning adjacent longer margins, with dorsiventrally flatter apices, giving the whole of each pinnule a somewhat slender body (not usually apparent when pressed as herbarium material) and an

Dryopteris affinis subsp. *borreri*: West Cornwall, July.

overall somewhat spoon-like form in life. Most pinnae are unlobed to only shallowly lobed along their adjacent edges, with rounded-truncate tips bearing small, slightly obtuse teeth. The pinnae nearest to the rachis are often the most deeply cut, with the lowest basiscopic pinnule of the lowest pinna typically fully stalked (i.e., not usually partly attached at its base). Technically, the sori each have an indusium which is of medium thicknss with thin margins, which are doughnut-shaped when young, and the margins of which flatten and lift as the sporangia mature, occasionally splitting but more usually shrivelling, thereafter to be shed irregularly with a few relatively long persistent in this condition. The spores are relatively large in size (more so than in subsp. *affinis*), and may contain many misshapen ones. (As the evolutionary origin of this plant is almost certainly from initial hybridisation between *D. affinis* subsp. *affinis* and *D. oreades*, giving rise to many reproducing plants each with an apogamous life-cycle, an alternative view of its taxonomic status suggested by A. C. Jermy (personal communication) would be to give it merely a hybrid rather than subspecific binomial

Dryopteris affinis subsp. *cambriensis*: West Cornwall, July.

name, in which case the combination would be *Dryopteris* ˣ *cambrensis* (Fraser-Jenkins) Beitel & Buck.)

Possible confusion The generally more shining frond which is more yellow green on flushing, the denser, generally more golden-brown scales to the stipe and rachis, the stiffer, more leathery frond texture, the narrow pinnae with usually more truncated segments, the laxer crozier when flushing, and the dark patch to the base of each pinna distinguish most specimens of all subspecies of *D. affinis* from both *D. filix-mas* and *D. oreades*, the only other species with which they are likely to be confused. *D. affinis* subsp. *borreri* var. robusta is clearly likely to be easily confused with *D.* ˣ *complexa*, although *D. affinis* tends to occur as populations rather than as scattered individuals. But technical confirmation may prove essential for firm diagnosis, as the genetic basis of their difference is quite small (see under 'Technical confirmation'). But don't give up yet!

Dryopteris affinis subsp. *borreri*, expanding crozier: May.

Technical confirmation All plants of *D. affinis* have an apogamous life-cycle: prothalli reared from their spores totally lack archegonia, with new sporophytes growing directly from gametophyte tissue.

Plants of subsp. *affinis* are apogamous diploids, with $n = 82$, $2n = 82$ chromosomes. The spores are rather small compared with those of the other subspecies, mostly about 37–51 μm (hence, as a general guide, mostly less than 50 μm in size). Subsp. *affinis* is believed to contain two different genomes, one from *D. oreades* and one from a hypothetical 'pure' *D. affinis* ancestor.

Plants of subsp. *cambrensis* are apogamous triploids, with $n = 123$, $2n = 123$ chromosomes, believed to contain a pair of identical genomes, possibly both from *D. oreades* plus one from a hypothetical 'pure' *D. affinis* ancestor.

Plants of subsp. *borreri* are apogamous triploids, with $n = 123$, $2n = 123$ chromosomes. The spores are larger than those of subsp. *affinis*, mostly about 40–50 μm, with some up to 58 μm (hence, as a general guide, with many spores larger than 50 μm in size). It probably contains

Dryopteris affinis subsp. *affinis*, expanding crozier: May.

three different genomes, which are possibly one each of *D. oreades, D. caucasica* and the hypothetical 'pure' *D. affinis* ancestor (cf. *D. filix-mas* which has those of *D. oreades* and *D. caucasica* only).

Field notes All forms of *D. affinis* are long-lived plants, which become most common in oceanic, high-rainfall areas of western Britain and western Ireland, and become more infrequent and confined mainly to heavier, moister soils in more eastern districts. At low altitude, they occur in a wide range of mainly acidic woodlands, but also extend into more open conditions in the lower parts of mountain valleys. They also ascend in mostly open conditions to a considerable altitude on mountains, where they usually occur in scree, often alongside *D.*

oreades. In woodlands, they usually grow as scattered individuals, with more erect, less branched rhizomes. In open situations, they often occur as much more gregarious populations, with much more freely branched masses of more ascending rhizomes.

The exact distribution and ecology of the different subspecies are very imperfectly known, and must await detailed study. However, it seems likely that all the subspecies are fairly widespread, and follow the general pattern of the species in being most frequent in western, oceanic climates. Manton

(1950) cytologically examined populations of *D. affinis* in Britain, and reported triploids to be generally the commonest of those analysed, with diploids occurring as more local populations. This, indeed, may prove to be the general pattern, as, on morphological grounds, triploid subsp. *borreri* appears to be the most abundant of the subspecies generally, and diploid *D. affinis* subsp. *affinis* the next.

Of the subspecies, plants of subsp. *affinis* occur most often in woodland habitats and on open, rocky, roadside banks, amongst the lower hillsides and valleys of more mountainous parts of Britain, such as the Scottish Highlands, the English Lake District and North Wales. They remain, however, frequent in western districts as far south as Cornwall. In Wales, it has been noted to be absent or scarce in base-rich areas (Hutchinson & Thomas, 1996, p. 128). In many habitats they often intermix with subsp. *borreri*. Their rhizomes branch freely, often eventually building up massive old clumps. Their fronds are typically slightly later flushing than those of subsp. *borreri* by about one week.

Plants of subsp. *cambrensis* share a similar upland but possibly more western distribution than subsp. *affinis*, and are probably most frequent in Scotland. Ironically (in view of its name) it is reported to be the least common of the three subspecies in Wales (Hutchinson & Thomas, 1996, p. 129), and a similar situation appears true in Cornwall (Miss R. J. Murphy, personal communication). In its main areas, plants usually grow over exposed, steep, well-drained hillsides, usually in acidic situations where there is a thin, peaty covering over old, semi-stable scree slopes. In such situations, they can grow in local profusion, sometimes forming large, nearly pure, stands. Their rhizomes branch fairly freely, and eventually build up large, multi-crowned clumps. Their fronds seem later flushing by about 1–2 weeks than those of subsp. *borreri*, seldom beginning to expand much before the beginning of June, when those of subsp. *borreri* are already about half-expanded. Although such plants can ascend to considerable heights on mountains, they seem particularly characteristic of the lower parts of mountain glens in the west Scottish Highlands, where their late-flushing fronds in early summer produce extensive bright golden-green patches along the lower reaches of the valleys and extending up screes.

Plants of subsp. *borreri* occur frequently and usually much less locally in a very wide range of habitats across almost the whole of Britain and perhaps Ireland, and it has been noted to be the most common and widespread of the subspecies in Wales (Hutchinson & Thomas, 1996, p. 129). Its habitats overall most closely parallel those of *D. filix-mas* and include open rocky places, stream valleys, roadside banks, and woodland of a wide range of types. From such habitats, plants readily spread often to establish in old brickwork in cities, railway lines around canals and in disused industrial areas. They also commonly grow in coniferous plantation woodlands along tracks and rides, and along the well-drained sides of gullies and ditches. Although apparently the most abundant subspecies, plants seem much less gregarious than those of subsp. *affinis*, occurring more often as scattered individuals, with rhizomes less branching and seldom building up such massive, or probably such old, clumps. Their fronds flush about two weeks earlier than do those of subsp. *affinis*, and because they are less golden and less gregarious, seldom have such an impressive effect. Their occurrence in such a wide range of habitats from old mortar to fairly acidic forests suggests they tolerate a particularly wide range of pH conditions, and their frequent occurrence with *D. filix-mas* suggests that they may be the most common *D. affinis* parent of the pentaploid hybrid *D. ×complexa* (q.v.).

Plants of probably all of the subspecies can also occur on moister valley-bottom or moist valley-side conditions, especially in valley woodland. In such habitats they usually grow as rather solitary individuals, with rhizomes which are much more massive and stocky than in other sites, and they seem usually more erect and seldom, if at all, branched. The fronds thus tend to arise in regular, massive 'shuttlecocks,' with plants usually much larger than even *D. filix-mas*. This appears to be the habitat in which *D. affinis* subsp. *borreri* var. *robusta* is a particular specialist.

Populations of all subspecies are genetically isolated amongst themselves and from one another, as a result of their non-sexual apogamous life-cycle, which effectively puts all of them into an evolutionary cul-de-sac. Each is true breeding, and probably each can act as a male parent in crosses with *D. filix-mas*, and potentially with other species of *Dryopteris*.

Dryopteris carthusiana (Vill.) H. P. Fuchs   Narrow Buckler-fern

(*D. spinulosa* Watt)

Preliminary recognition A medium-sized fern, with rather delicate bipinnately divided, finely cut, narrow erect fronds, arising usually in sparse, irregular groups, and never in shuttlecock-like crowns.

Occurrence Present throughout Britain and Ireland, but in quantity only locally in low-lying fens and wet woodlands.

Identification Plants have bipinnately divided, *rather stiffly erect*, but gently tip-tilted fronds, mostly 40–80 cm high (sometimes more), which arise *in irregular, rather sparse, groups from buried, pale-brown, often shining, rhizome crowns*. The rhizomes are either short and decumbent or longer and more slenderly prostrate, and branch occasionally to form further nearby crowns. The frond blades are of linear-lanceolate outline, *rather narrower than those of* D. dilatata, *with the lower half approximately parallel-sided*, tapering gradually in the upper half to a pointed apex. The blades are borne on fairly slender, but *strongly erect*, stipes which are usually about as long as the blade, and *green in colour, becoming reddish-brown towards their base. The stipes are sparsely clothed with rather lax, uniformly coloured, ovate, pale-tan or biscuit-coloured scales, which are smaller and paler than those of* D. expansa, *but lack the dark central stripes of those of* D. dilata. The fronds bear about 8–20 pairs of nearly opposite, tapering acutely pointed pinnae, which are usually held at a rather ascending angle throughout the frond, with the lower ones also tilted nearly horizontally. The lower pinnae are rather distantly spaced and unequally triangular, with the basal side of each the largest. The pinnae are pinnately divided and their pinnules pinnatifid, with the margins always *conspicuously serrated into distinct, incurving, sharp, spinulose-tipped teeth*, their saw-like points easily visible, even without a hand lens.

Most of the taller fronds become freely fertile, but in each frond clump there are usually also a few non-fertile fronds, which arise somewhat earlier in the season, are smaller than the fertile ones, and have a more spreading habit. On fertile fronds, the sori are small, with rather thin indusia that shrivel at maturity.

All the fronds are of a firm but delicate texture, of a yellowish mid-green colour, and *usually*

Jan.	Feb.	Mar.	Apr.	May	Jun.	Jly	Aug.	Sep.	Oct.	Nov.	Dec.

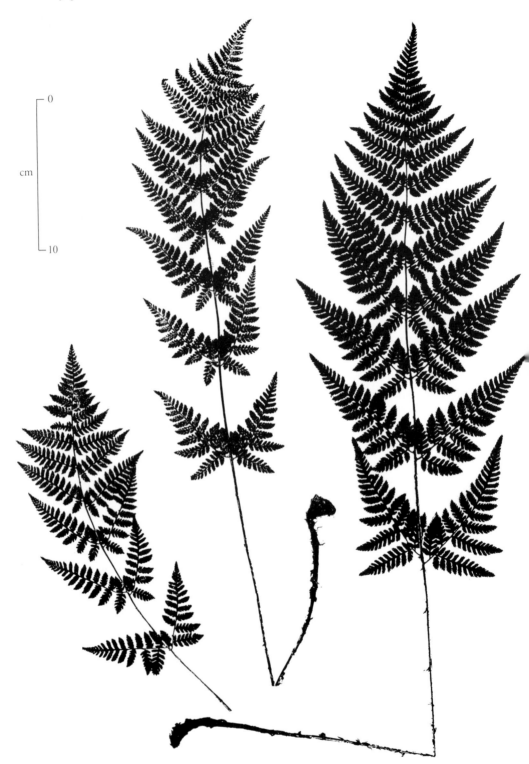

Dryopteris carthusiana: *all* East Cornwall.

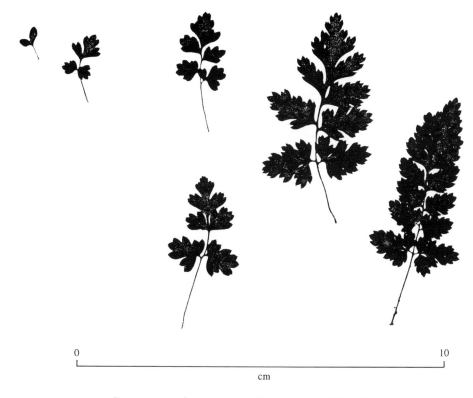

0   10

cm

Dryopteris carthusiana, young frond stages: all Argyll.

appear paler and yellower than those of D. dilatata in the field. The fronds are much more cold sensitive than those of *D. dilatata*, and usually die down rapidly with the first frosts.

Variation Plants vary mainly in size, usually with luxuriance of the habitat and degree of shelter.

Possible confusion Plants differ from *D. dilatata* mainly in their narrower frond outline, more erect frond habit, more prostrate, often more slender, rhizome, paler scales lacking the dark central stripe, and paler frond colouration. They differ from *D. expansa* mainly in having a less triangular and usually narrower frond outline, narower, more widely spaced pinnae, thinner stipe, paler, smaller, fewer, less-gingery stipe scales, and more prostrate rhizome habit. Large *D. carthusiana*-like plants of darker frond colour, showing a median dark stripe to the scales, probably involve hybridity with *D. dilatata* (= *D.* ×*deweveri*, q.v.). Plants differ from *D.* ×*uliginosa* (q.v.), which also shares the *D. carthusiana* parent, in having a broader frond outline and more divided pinnae. Both hybrids also have mainly abortive spores.

Technical confirmation Plants are tetraploids, with $n = 82$, $2n = 164$ chromosomes. The indusium is smooth, and without glands.

   Dryopteris carthusiana is an allotetraploid species, derived from the cross between the *D. intermedia* aggregate (which includes North American *D. intermedia* as well as Madeira and Azores *D. maderensis* and *D. azorica*) and a second diploid species, possibly *D. ludoviciana*, also of North America (Gibby & Walker, 1977).

Dryopteris carthusiana, habit of plant in life.

Field notes *Dryopteris carthusiana* is a European and western Asian species which is widespread at middle latitudes of the continent, and which also occurs widely (but decreasingly) throughout Britain and Ireland. Despite its slightly frail appearance in comparison with the much more robust Broad Buckler-fern, Narrow Buckler-fern is a rapidly growing species, and succeeds under much wetter ground conditions than can be tolerated by *D. dilatata*. It occurs in a wide range of damp, mainly lowland habitats, including wet heaths, fens, mires and wet woodland, especially those developed on flat boggy ground over rich alluvial soils. It seldom ascends far into upland conditions, except into flatter, mountain valley bottoms, and in contrast to *D. dilatata* and *D. expansa*, is usually absent from the most mineral-poor, acidic boggy sites.

In exposed situations in northern England and Scotland, Narrow Buckler-fern often occurs as scattered plants in *Molinia–Myrica* mire vegetation, but it reaches its greatest abundance in low-lying acidic wet woodlands on the edges of lakes, where the groundwater table is kept high by slight seepage of acidic water, with, perhaps, some additional mineral enrichment. Here it occurs usually beneath an alder–willow–birch canopy, amongst a sodden, spongy undergrowth of hummocky *Sphagnum* bog mosses (especially *S. recurvum* and *S. capillaceum*) and rushes (usually *Juncus effusus* and *J. conglomeratus*), sometimes with scattered plants of Heath Bedstraw (*Galium saxatile*), Tormentil (*Potentilla erecta*), Wood-sorrel (*Oxalis acetosella*), and numerous other mosses such as *Thuidium tamariscinum, Dicranum majus, Rhytidiadelphus loreus, Pleurozium schreberi, Plagiothecium undulatum*, and *Polytrichum commune*. In such situations, the thin, creeping rhizome of *Dryopteris carthusiana* usually extends horizontally through the moist, humus-rich ground beneath the mossy carpet. Broad Buckler-fern (*Dryopteris dilatata*) frequently also occurs in the

Dryopteris carthusiana, young frond stages (after Orth, 1938).

same woodlands, but is usually confined to the better-drained tops of hummocks, on raised patches around the tree boles, on old tree stumps and amongst mosses on fallen wood. Where both species are present, their hybrid, *D. ˣdeweveri* (q.v.), sometimes occurs in ecologically intermediate spots. In west coast Scottish sites, Mountain Buckler-fern (*D. expansa*) may also grow with the two other species, and occurs in niches closely similar to those of the *D. carthusiana* plants. In such situations the hybrid between *D. expansa* and *D. dilatata* (= *D. ˣambroseae*, q.v.) sometimes also grows, and the third possible hybrid, that between *D. carthusiana* and *D. expansa* (= *D. ˣsarvelae*) known in continental Europe, has also been found in Scotland (Corley & Gibby, 1981). The occurrence of all three of these variable species together, and the possibility of three combinations of hybrids between them, can provide a daunting task for field separation.

Narrow Buckler-fern also occurs as a locally frequent species in a range of wetland habitats associated with East Anglian fen and carr vegetation, in which Common Reed (*Phragmites communis*) and Saw Sedge (*Cladium mariscus*) play a substantial role. In such communities, it may grow alongside Fen Buckler-fern (*Dryopteris cristata*) in actively growing *Sphagnum* hummocks, and their hybrid, *D. ˣuliginosa*, may form locally.

Despite its widespread occurrence. Narrow Buckler-fern has undoubtedly diminished considerably in abundance in the last 100 years. It is a plant which was formerly quite well known to Victorian naturalists, but today, in many districts where it was formerly frequent, it is now quite local and sometimes surprisingly difficult to find. The plant seems quite sensitive to the right degree of ground moisture and atmospheric humidity to succeed well, and progressive drainage of wetland conditions throughout both Britain and Ireland has undoubtedly contributed much to its rapid decline.

Dryopteris cristata (L.) A. Gray   Fen Buckler-fern

Preliminary recognition A medium- or large-sized fern, with bipinnate, very narrow-outlined fronds and broad, blunt pinnae, standing strongly erect in close clusters.

Occurrence Rare and local in fenland conditions in a very few stations, mostly in East Anglia.

Identification Plants have fronds mostly up to 30–60 cm, but occasionally to about 100 cm, which arise *in close groups in small numbers, mostly standing strongly erect*, from the crowns of fairly stout prostrately creeping, rhizomes, submerged beneath the surface of surrounding vegetation. The rhizomes usually branch frequently to produce several nearby crowns, which are clothed in broad, pale-brown, thin, shining scales. The fronds have a *narrow, markedly parallel-sided outline throughout the greater part of the blade, much narrower than that of* D. carthusiana, *and with much shorter, blunter, less-divided pinnae.* The blade reaches about 8–15 cm in width, and may taper suddenly in its upper portion to a somewhat blunt apex, but narrows only slightly towards its base. The stipe is rather stout, of pale greenish or straw colour, becoming darker chestnut brown towards its base. It is about 1/3 to 1/2 the length of the frond, and is sparsely covered with broad pale-brown scales, which become denser only towards the stipe base. Within each group of fronds, the outermost ones formed earliest in the season, are usually non-fertile and of much lower-growing, more arching, spreading habit, while those in the centre of the group are usually fertile, and of much taller, strongly erect growth.

All fronds are pinnately divided, with deeply pinnatifid pinnae (about 10–20 pairs on either side on the largest fronds). The pinnae occur on alternate or subopposite pairs. They are very short stalked and each broadest at the base, tapering outwards to *a rather blunt tip. In the lower part of the frond, the pinnae are particularly broadly triangular.* They are subdivided into a small number of oblong segments, each attached by the whole width of its base to the pinna midribs except for some of the innermost, which may be partly separated. The segment margins are finely serrulated into sharply pointed teeth, some of which bend inward, but all of which end in only a short spine-like point – not in the long-spined apex of *D. carthusiana* (q.v.). Throughout

Jan.	Feb.	Mar.	Apr.	May	Jun.	Jly	Aug.	Sep.	Oct.	Nov.	Dec.

Dryopteris cristata, young (mostly vegetative) frond phases: *a–d*, Cheshire; *e*, Yorkshire.

Dryopteris cristata, adult fertile frond phases: *a*, Nottingham; *b*, Yorkshire; *c*, Cheshire.

the most erect fronds, most of the pinnae are tilted from their bases, so that they are held in a nearly horizontal plane.

All fronds are of a light-green colour, and the fertile ones of slightly shining and more leathery texture than the lower-growing vegetative ones. On the fertile ones, the sori are quite large and round, covered by a kidney-shaped indusium, produced only in the upper part of the blade, but extending nearly to the apex of the pinnae and pinnules. On uncoiling croziers in spring, the pinnae often partly expand whilst still within the crozier, giving this species a characteristic appearance when flushing. In autumn, fronds die down rapidly with the first frosts.

Variation Not a very variable species.

Possible confusion The much narrower frond outline, with much shorter, broader, blunter pinnae, helps to separate *D. cristata* from *D. carthusiana*. *D. cristata* differs from *D. ×uliginosa* (q.v.), its hybrid with *D. carthusiana*, in having a narrower frond outline, less-divided pinnae, good spores and lack of spine-tipped teeth to the segments.

Technical confirmation Plants are tetraploids, with $n = 82$, $2n = 164$ chromosomes. The stipes, midribs and frond surfaces are completely without glands, and the indusium without a gland-fringed margin.

Dryopteris cristata is an allotetraploid derivative possibly between *D. ludoviciana*, a North American plant and another, currently unknown, diploid species (Gibby & Walker, 1977).

Field notes *Dryopteris cristata* is a European and western Asian species which is a widespread at middle latitudes of the continent, and which has an amphi-Atlantic distribution similarly at middle latitudes of the North American continent, but especially in the east.

In Britain and Ireland, Fen Buckler-fern has a somewhat restricted (and probably decreasing) natural range. It is a lowland species, seldom occurring above about 100 m (300 ft) altitude, in wet heaths, marshes and fens. It thrives in more continental climatic conditions than are usual over much of Britain and Ireland, and has probably always been mainly confined to the eastern half of England and absent from the cloudier, more summer-cool west, including the whole of Wales and Ireland.

It was formerly a locally frequent species in East Anglia, occurring also locally elsewhere, especially in the north English Midlands to Cheshire and Yorkshire, with outlying stations as far north as central Scotland (Renfrewshire). Today it has become mostly confined to the margins of the Broads of east Norfolk, where it grows in damp, acidic, base-poor but perhaps slightly mineral-flushed soils, sometimes in open, exposed conditions, but most luxuriantly in light shade.

Much of its overall reduction in range in Britain has probably been due to pollution and widespread

drainage of lowland wetland habitats during the last century or more, although it is possible that climatic shifts may also have played a significant role. Even in East Anglia, it has disappeared from many former stations. In addition to suffering from drainage, the plant is probably naturally seral, establishing at certain stages of wetland vegetation development and dying out at others, and various, as yet poorly understood, factors seem to sway the delicate balances of relatively rapidly changing wetland vegetation communities. There is some evidence of increase in its abundance following extensive flooding, such as that which occurred in Norfolk in 1952 (M. H. Rickard, personal communication). More recently, continued increase has been noted. *D. cristata* has been recorded to occur typically in the Norfolk Broads area on small islands of slightly raised ground with young birch scrub and hummocks of *Sphagnum* bog moss, amongst dying remains of old birch trees, in habitats which may mark a transition zone between reed swamp and old fen carr. Jermy *et al.* (1978) attribute some of its local increase to its encouragement following the cessation of fen-cutting. This has allowed tussocky growth of Purple Moorgrass (*Molinia caerulea*), Tufted Sedge (*Carex elata*) and various bogmosses (*Sphagnum* spp.) to form, suitable for renewed establishment of Fen Buckler-fern. In addition, incompletely understood but more widespread vegetational changes involving increase in base-poor communities have further created conditions beneficial for its survival. Continuing research into the ecological status of this interesting species seems much needed.

Dryopteris dilatata
(Hoffm.) A. Gray

Broad Buckler-fern

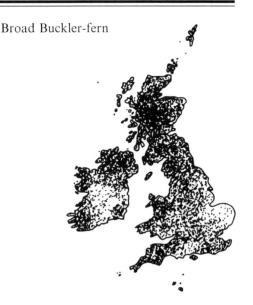

(*D. austriaca* (Jacq.) Woynar)

Preliminary recognition A medium to large, sometimes very large, vigorous and swarthy fern, with broad, arching, deep-green, bipinnate fronds, arising in clumps. The scales at the stipe bases have a deep-brown central stripe, or are nearly entirely deep brown.

Occurrence Very abundant in the British Isles through a wide range of altitude, and one of the most common ferns of acidic woodlands of all types, often present in large numbers.

Identification Plants are very variable in size. Juvenile plants with fronds 23–30 cm long can be common, whilst adult plants may be equally common, with fronds up to about a metre in length, or much more. In juvenile plants, the blades have a *broadly triangular outline, whilst those on mature plants have a broad lanceolate or ovate outline.* All but the smallest are bipinnately cut. The fronds are of slightly leathery texture, *of deep-green or sometimes blue-green colour,* and arise in tufts, from short ascending or decumbent stocky rhizomes. Most large fronds adopt a distinctive arching habit, and the edges of their pinnules characteristically *all turn slightly downwards* (except in specimens growing in the deepest shade), *giving each pinnule surface a convex appearance.* All the divisions of the frond usually appear less widely spaced than do those of *D. expansa* (q.v.), the whole thus seeming less lace like, whilst the pinnule segments are more squarish or oblong, and less obviously toothed.

The stipes usually occupy about 1/3 or less the length of the frond (much less in the largest fronds), and have *abundant, large, upward-pointing scales, which densely overlap each other at the base of the stipe, and in old, well-established plants, these are very large and of crisp texture. In the majority of plants, these scales have pale edges, with a conspicuous central dark-brown lengthwise stripe, sometimes sufficiently broad to make the whole of each scale, and hence the whole base of the stipe, appear a deep shining blackish brown.*

All but the smallest plants are normally very freely fertile, producing large sori in abundance

Jan.	Feb.	Mar.	Apr.	May	Jun.	Jly	Aug.	Sep.	Oct.	Nov.	Dec.

Dryopteris dilatata, habit of plant in life.

over most or all of the underside of the frond, and these, when ripe, release sooty black spores in great quantity.

Variation A highly variable species in size, frond outline and degree of dissection, depending much on specimen age and differences in habitat. Juvenile fronds and those arising from occasional, long-creeping, offset rhizomes (see under 'Field notes'), frequently have only sparse, pale, more concolorous stipe scales, each with only a few darker, centrally located cells at the extreme base.

Possible confusion Chiefly likely only with the other Buckler Ferns, although *D. dilatata* is by far the commonest of this group. The deep-green colour of the fronds, the downward-curving pinnule margins, the large arching fronds, and the abundant dark-centred scales at the base of the stipe of most large fronds should distinguish it clearly from all other species of *Dryopteris*, as well as from *Athyrium filix-femina*, which is often common in the same environment. The abundant good spores separate it from *D.* ˣ*deweveri* (q.v.), its hybrid with *D. carthusiana*, as well as from *D.* ˣ*ambroseae* (q.v.), its hybrid with *D. expansa*.

Technical confirmation Plants are tetraploids, with $n = 82$, $2n = 164$ chromosomes. Their spores appear darker than those of *D. expansa*, with a more densely tuberculate surface. The bases of the stipe scales are usually cordate – i.e. hollowed and lobed either side in a heart-shaped manner.

Field notes *Dryopteris dilatata* is a mainly European and western Asian species which is widespread at middle latitudes of the continent, with related forms at similar latitudes of the North American continent. Broad Buckler-fern occurs almost throughout Britain and Ireland, through a wide range of altitude from near sea-level to around 1130 m (3700 ft). At its higher-altitude stations in the mountainous districts of North Wales, northern England and Scotland, it adopts a rather stunted habit. At lower altitudes, however, *D. dilatata* usually becomes a very much larger and more abundant plant. Here it occurs in habitats from quite dense woodland to open rocky places, although it thrives well only where reasonably sheltered. It reaches its most extensive development on cool moist acidic woodland slopes of valleys in hilly districts, where it is frequently the dominant fern.

Dryopteris dilatata, frond samples from semi-juvenile populations that are very frequent in acidic woodland habitats and on logs throughout Britain and Ireland: *a*, Kintyre; *b*, Moray; *c*, Westerness; *d*, West Cornwall.

Dryopteris dilatata, frond samples of adult individuals characteristically forming arching clumps in acidic woodland floor habitats: *a–c*, Kintyre; *b*, Isle of Arran; *d*, Renfrewshire.

Dryopteris dilatata, dorsal view of pinnules in life showing downcurved margins (cf. those of *D. aemula*, q.v.).

It occurs extensively in association with Oak (*Quercus*), Oak–Birch (*Betula*) and mixed deciduous woodlands, and, where these occur on deep humus-rich soils, this fern may form nearly continuous waist-high, or nearly shoulder-high, swards, with long fronds arising from massive old stocks.

In old, undisturbed woodland, in areas of high humidity and frequent rainfall, plants may spread from terrestrial habitats to occupy crevices in rock-faces and pockets where humus accumulates on tops of boulders. Where such woodland spreads into sheltered gorges near running water, this fern frequently becomes epiphytic on rough-barked trees, especially where humus collects at the bases of their branches. Elsewhere it also occurs in old moist pine and birch woodland, and in established modern spruce plantations. In the latter it is often the dominant fern along the flanks of ridges, streams, ravines and drainage channels within them, wherever more light penetrates the dense canopy onto moist ground.

In its wide range of habitats, *D. dilatata* can adapt to a wide range of conditions of humidity, light and substrate. It occurs in moderately to highly acidic, well-drained soils, derived from a very wide range of acidic rock types, including sandstones, slates and granites, and the grits and shales associated with coal-measure outcrops, and

Dryopteris dilatata, expanding crozier showing dark centred scales: West Cornwall, May.

also over more basic rocks, including limestones, where these are capped by acidic humus layers. In old woodlands, where thin coverings of raw humus (such as that of semi-decayed pine needles) accumulate over hard clay soils or hidden boulders, an interesting, probably environmentally induced, form of *D. dilatata* often occurs, in which fronds remain small and arise from small crowns along slender, long-creeping, pale-scaled rhizomes, quite unlike those of typical plants.

In mixed deciduous woodland, Broad Buckler-fern thrives in ground which is permanently moist but not water-logged. It is frequently associated with Hard Fern (*Blechnum spicant*), which often replaces it in wetter spots with greater groundwater

movement, and with Male Fern (*Dryopteris filix-mas*), usually in lighter, somewhat more exposed conditions and in drier ground. Lady Fern (*Athyrium filix-femina*) is also often frequent with it, but spreads also into patches of wetter ground. Other common associates of *D. dilatata* in mixed woodlands include Bilberry (*Vaccinium myrtillus*), Greater Woodrush (*Luzula sylvatica*), Honeysuckle (*Lonicera periclymenum*), Wood-sorrel (*Oxalis acetosella*), Creeping Soft-grass (*Holcus mollis*) and Bramble (*Rubus fruticosus* agg.). In wet lowland, acidic woodland, it can occur with Narrow Buckler-fern (*D. carthusiana*), but in such vegetation *D. dilatata* is usually confined to better-drained hummocks, especially around rocks or tree boles, or on

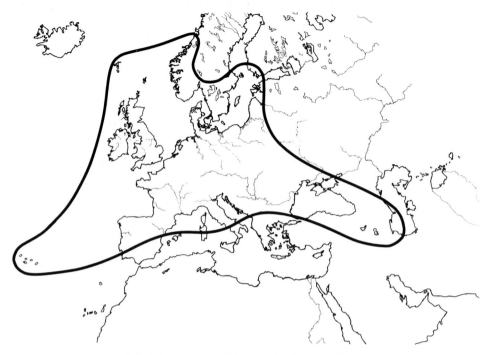

Sub-Atlantic range of *Dryopteris dilatata* in Europe.

half-submerged, moss-clad, fallen trunks and rotting stumps (see also the 'Field notes' under *D. car-thusiana*). On less-acidic soils, especially in milder areas of Britain, where very acidic humus layers are lacking, *D. dilatata* becomes extensively replaced by Soft Shield-fern (*Polystichum setiferum*) and Hart's-tongue Fern (*Phyllitis scolopendrium*), whilst in moist, acidic, rocky ground around the western oceanic fringe of Britain and Ireland, it becomes replaced locally by Hay-scented Buckler-fern (*Dryopteris aemula*). In lighter, more open wood-land, on sandy sites everywhere, which are more prone to summer droughts, *D. dilatata* is usually replaced by Bracken (*Pteridium aquilinum*).

Dryopteris dilatata is a rapidly growing species and a vigorous fern to the point of becoming weedy, especially in the cooler north of Britain. It often appears, together with Male Fern (*D. filix-mas*) and young plants of Bracken, in old damp brickwork of city basement stairways, and at bases of old walls and on brick canal bridges, as well as in damp sites around areas of industrial dereliction. Small plants also frequently establish as weeds in moist greenhouses, and in pots of other plants.

Dryopteris expansa (C. Presl) Fr.-Jk. & Jermy

Northern Buckler-fern

(*D. assimilis* S. Walker)

Preliminary recognition A small to moderately large fern with broad, finely dissected fronds, looking like a more delicately cut, slightly stiffer-fronded, yellower-green form of *D. dilatata*, but usually not so large. The scales at the stipe bases are usually a uniform tan or ginger-brown colour.

Occurrence Plants occur rather sparsely in widely scattered stations throughout the mountainous parts of Wales, northern England and Scotland, but occur most frequently in the Scottish Highlands. They grow mainly in sheltered niches in mountain scree, but in some localities in the west (especially western Scotland), they also occur at low altitude in wet, acidic woodlands. This species has not so far been confirmed in Ireland.

Identification Plants from sheltered woodland may have fronds 70–80 cm in length (occasionally to 1 m) whilst those from mountain boulder scree are much smaller, frequently 10–25 cm or less when mature, the size varying with degree of exposure.

Most have fronds with *a rather broad, triangular-ovate outline*, which becomes more nearly triangular in smaller specimens, although in some populations, the outline is more often narrowly triangular. Fronds on all plants are *finely and deeply divided, the divisions typically being well spaced (not overlapping) and the pinnules thin and flat, giving the whole frond a characteristically lace-like appearance.* The pinnae in the middle of the frond have pinnule segments which are more oval and more obviously toothed than those of *D. dilatata*. The pinnae are also particularly broad across their bases, and *many fronds have a lower pinnule adjacent to the stipe which is distinctly larger than the rest, especially on the basal pinnae*.

In most specimens, fronds arise in crowns from erect or decumbent short stocky rhizomes, and are held fairly rigidly at a strongly ascending angle, thus arching only slightly. The lamina is, however, quite soft in texture, thin and pellucid during early summer, and of *pale yellow-green colour*. The stipes, which occupy about half the length of the frond, are pale green, except near their extreme bases where they become brown, and carry *large scattered chaffy scales. The scales mostly vary in colour from a uniform pale tan brown to distinct ginger*, but some specimens may have darker brown central stripes (but seldom as dark or distinct as those of *D. dilatata*). Sori usually occur throughout the length of mature fronds, although the sori and indusia are characteristically small.

Jan. Feb. Mar.	Apr.	May	Jun.	Jly	Aug.	Sep.	Oct. Nov. Dec.

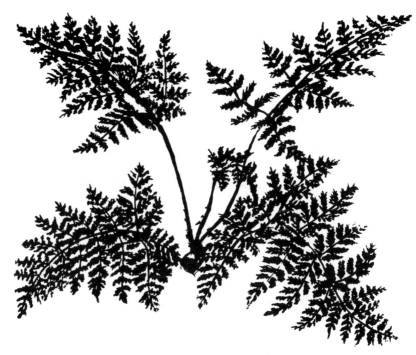

Dryopteris expansa, habit of plant in life.

Fronds usually flush much later in season than those of *D. carthusiana*, which may grow with it at low altitudes, and, unlike *D. carthusiana*, those of *D. expansa* often remain green for much of the winter.

Variation *Dryopteris expansa* is a variable species, making identification sometimes difficult, and technical confirmation may well be necessary. As well as variation in size, habit and scale colour, there is much detailed variation in frond outline shape and pinna cutting. Some plants have more overlapping pinnae than is usual, and some do not possess the enlarged lowermost pinnules.

Possible confusion Plants are most likely to be confused with *D. dilatata* or *D. carthusiana*, and their hybrids. The usually broader, much more finely dissected, more winter-hardy fronds, arising, for the most part, in distinct tufts (cf. usually much narrower, more coarsely cut, winter-deciduous fronds arising, singly or in small groups) help to separate *D. expansa* from *D. carthusiana*. The generally smaller fronds with stipes up to half their length, distinctly flat pinnules, more oval, more obviously toothed, pinnule segments, often concolorous and gingery scales and yellow-green frond colour (cf. larger fronds with stipes about 1/3 of their length, downcurving-edged pinnules, more squarish or oblong, less obviously toothed, pinnule segments, normally dark-striped scales and blue-green frond colour) help to separate most plants of *D. expansa* from those of *D. dilatata*. The broad-based pinnae further separate *D. expansa* from both other species. The presence of only good spores separates *D. expansa* from *D.* ×*ambroseae* and *D.* ×*sarvelae*, its hybrids with *D. dilatata* (q.v.) and *D. carthusiana* (q.v.) respectively as well as from the unrelated *D.* ×*deweveri* (*D. dilatata* × *carthusiana*), which can also look similar. There is a resemblance in the fineness of frond dissection also to plants of

0 _____ 10

cm

Dryopteris expansa, mature frond examples; *a*, Isle of Mull, Inner Hebrides; *b*, Easterness.

Athyrium filix-femina, the separation from which the triangular rather than ovate frond outline and the kidney-shaped sori (cf. usually J-shaped in *Athyrium filix-femina*) provide firm guides.

Dryopteris expansa, frequent upland juvenile frond phases: *a*, Orkney; *b*, *c*, Banff; *d*, *e*, Easterness; *f*, Orkney.

Technical confirmation Plants are diploids, with $n = 41$, $2n = 82$ chromosomes. The spores are 45–50 μm in length and have a surface sculpturing pattern of low, blunt spines, which are much less dense than in *D. dilatata* or *D. carthusiana*, and the spores are paler in colour.

Field notes *Dryopteris expansa* is a European and western Asian species which is widespread at predominantly northerly latitudes of the continent, and which has an amphi-Atlantic distribution to essentially similarly latitudes of the North American continent. In Europe, its centre of distribution is throughout Fennoscandia, with smaller, outlying centres of occurrence mainly in Iceland, northern Britain and the Alps and central European mountains.

In Britain and Ireland, Northern Buckler-fern has a distinctly northern and largely montane range. It is a highly desiccation-sensitive species, enjoying cool humid climatic conditions, permanently moist ground, a high degree of shelter and relatively good illumination. Such conditions occur both in thin lowland, wet woodland (chiefly in the north of Britain) and in damp sheltered rocky fissures of mountain scree.

Lowland plants are typically the largest, and occur at sites which are slightly flushed and have a low, thin canopy. In such habitats, species of *Juncus* (rushes) and *Spaghnum* (bog mosses) often form a well-developed hummock and hollow structure over a gently sloping woodland floor, and plants of *D. expansa* are often associated also with *D. carthusiana* in wetter hollows and *D. dilatata* on better-drained hummocks. *Dryopteris expansa* appears to be a relatively poorly aggressive species, and its ecology in wet woodlands may well be rather finely balanced: increased wetness probably favouring *D. carthusiana* dominance, increased drainage and denser tree canopy producing a swing to *D. dilatata. Dryopteris expansa* may also come into competition with the unrelated but relatively vigorous hybrid in *D. ×deweveri* (*D. dilatata × carthusiana*, q.v.), and in flushing appreciably later in the growing season than all these competitors, *D. expansa* appears to be at a competitive disadvantage in lowland sites. However, *D. expansa* appears to thrive under conditions of better illumination than do either *D. dilatata* or *D. ×deweveri*, but as conditions of better illumination at low altitude are often also more exposed and more prone to summer drought and high summer temperatures, *D. expansa* is effectively limited at low altitude to woodlands which are cool and wet.

Dryopteris expansa often succeeds well on mountains, perhaps because in such habitats it enjoys freedom from the competition of *D. carthusiana* and *×deweveri*, whilst *D. dilatata* is also of reduced vigour. Its greater tolerance of lower mean temperatures, and its later flushing are doubtless of advantage at such altitudes, where conditions of illumination are also more generally favourable. On mountains, *D. expansa* succeeds particularly in moist, sheltered pockets in large semi-stable block-scree slopes, where there is only a slow rate of boulder accumulation. Plants of varying sizes occur with their rhizomes anchored deeply between interlocked boulders, rooting into moister ground below and from the crevices of which often only the tips of long fronds typically emerge. In such screes, *D. expansa* may become the dominant plant. It occurs chiefly in screes of mica schist, especially those with a northerly and easterly aspect. Its dominance in such habitats is probably seral, becoming replaced under conditions of high scree stability by vigorous flowering plants, especially by more blanketing Alpine Lady's-mantle (*Alchemilla alpina*). It is entirely absent from scree of small boulder size (less than about 0.5–1.0 m), where it is usually replaced by Mountain Male-fern (*Dryopteris oreades*, q.v.). It is also absent from screes composed of entirely acidic rock, where it is replaced by Parsley Fern (*Cryptogramma crispa*). In many of its Scottish mountain habitats, it is associated with a number of other mountain ferns, including Holly Fern (*Polystichum lonchitis*), Brittle Bladder-fern (*Cystopteris fragilis*), Green Spleenwort (*Asplenium viride*), and occasionally too with Alpine Lady-fern (*Athyrium distentifolium*) and Alpine Bladder-fern (*Cystopteris montana*).

In mountain habitats, the emergent tips of its fronds seem very often to be taken by grazing Red Deer, although they appear not to be touched by Mountain Hares, around the entrances to the burrows of which it sometimes establishes.

Dryopteris filix-mas (L.) Schott   Common Male-fern

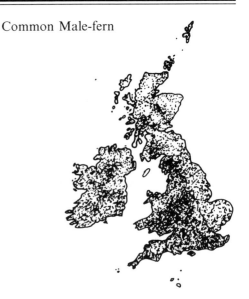

Preliminary recognition A large fern with robust lance-shaped mid-green fronds, arising usually in single shuttlecock-like clusters, from the crown of a large, upright rhizome, their stout stipes clothed with pale-brown scales.

Occurrence A common and ubiquitous species, widespread throughout the whole of Britain and Ireland and frequent in most places at low to moderate latitudes, in woodlands, hedgerows, thickets, streambanks and in rocky open sites, especially on most lighter, better-drained soils. It is present even in most industrial areas, and often naturalised in gardens and shrubberies.

Identification Plants have numerous, semi-evergreen, pinnate-pinnatifid fronds, up to 120 cm in length and 40 cm wide. The fronds arise in an *ascending-arching fashion, in fairly dense, shuttlecock-like clusters*, from the crowns of stout ascending or erect rhizomes. The rhizomes are seldom branched, so that crowns usually arise singly, only rarely two or three together. The fronds are of ovate-lanceolate outline, nearly 2 × pinnately divided, widest at about the middle of the frond or just below it, and tapering above. Below, the outline is eventually truncated by the pinnae stopping short of the base of the frond, with the lowermost pinna pair smaller than those above, but still of considerable size and *never very short*. The stipes are about 1/4–1/3 of the length of the frond, are often dull straw-brown in colour and are clothed with *mixed wide and narrow, slightly shining, pale buff-brown to straw-coloured, chaffy scales, in moderate density, but never so numerous as those of* D. affinis *(q.v.) or hybrids with it. Above the stipe, the scales thin out rather rapidly along the green rachis, to become sparse in the upper half of the frond.*

The fronds bear about 20–35 or more distinct, seldom-overlapping, mostly straight pinnae along either side. Each pinna is *quite broad at its base* (often more than 1/5 as broad as long) and narrows outwards to a gradually tapering tip. Throughout much of the frond, the pinnae are *quite deeply pinnatifidly divided* (cut almost to the pinna midribs), with the resulting segments themselves undivided, and attached to the pinna midribs by the whole of their bases, although there is much variation in this, and the inner ones are sometimes more nearly stalked. The segments usually *touch one another at their bases, but narrow slightly upwards so that their tips appear well spaced and distinctly round topped* (not abruptly truncated as in *D. affinis*, q.v.). The segments are more or less toothed along their edges and usually around the tip as well, into a few acute teeth which incurve towards the tip of each segment. The segments are notably flat,

Jan. Feb. Mar.	Apr.	May	Jun.	Jly	Aug.	Sep.	Oct. Nov. Dec.

Dryopteris filix-mas, habit of plant in life.

or slightly down curved along their outer edges, but usually *not strongly convex as in* D. affinis *nor concave as in* D. oreades *(q.v.)*.

The fronds are of moderately firm, but not tough, texture, and are of pale mid-green colour when flushing, turning to a slightly shining deep mid-green when fully expanded. At the junctions of pinna midribs and rachis, the fronds are green, *without any local dark colouration* (cf. the dark marking at this point in *D. affinis*).

Fertile fronds are similar to the vegetative ones, and bear two lines of conspicuous, kidney-shaped sori along the underside of each segment throughout most of the upper half of each frond. The indusium is about 1.5 m in diameter, hence larger than that of *D. oreades* (q.v.). The young indusium is at first a translucent pale-green colour, but *later becomes a leaden grey, and finally rust-brown*, before shrivelling at maturity to reveal the numerous black sori. In its young state, its cross-sectional shape is particularly characteristic: each indusium is *like an inverted soup plate in shape, with flattened edges spreading out all round*, instead of doughnut shaped and tucked under the developing sporangia as in *D. affinis* (q.v.).

Fronds arise early in spring, and persist well into the following winter. When the unfurling croziers are about half-expanded, *the tips usually completely unwind and hang laxly downwards forming a shape like that of a shepherd's crook*, contrasting with those of *D. oreades* which remain more tightly circinate. In winter, fronds are semi-persistent (often completely persistent in mild areas), and by spring are usually lying horizontally on the ground, still partly green and partly olive-brown and decaying.

Variation Plants show considerable variation between populations and sometimes between individuals, in size, length, width and spacing of pinnae, and shape of ultimate segments. Extreme forms with very broad pinnae with segments themselves cut into serrated

Dryopteris filix-mas: a, Merionethshire; *b, c,* Perthshire; *d,* Ross-shire.

Dryopteris filix-mas, expanding crozier: May.

lobes occasionally occur (var. *inscisa* Moore), and can be locally frequent in some populations.

Possible confusion Plants differ from *D. oreades* (= *D. abbreviata* particularly in their seldom-branched, short, thick rhizomes, less erect frond habit of broader outline, less coriaceous texture with longer stipe, less greyish scales, more open crozier on flushing, larger sori, flatter segments, and usually much larger size. They differ from *D. affinis* especially in their much less golden-scaled, less densely scaly stipes, less yellow-green frond colour, absence of dark colouration at the pinna–rachis junction, less convex pinna segments with more rounded ends, earlier spring-flushing habit and usually somewhat larger ultimate size. They also differ from all hybrids between *D. filix-mas* and *D. affinis* (*D.* *ˣcomplexa*, q.v.) in degree of development of each of these features.

The lack of very short lower pinna, proceeding to near the base of the frond, distinguishes *D.*

Dryopteris filix-mas, young frond stages (after Orth, 1938).

filix-mas from *Oreopteris limbosperma* (q.v.), which shares an otherwise similar frond outline and degree of frond division.

Technical confirmation Plants are tetraploids, with $n = 82$, $2n = 164$ chromosomes. Spore size is about 32–45 μm (cf. 36–40 μm in *D. oreades* and 40–70 μm in the *D. affinis* agg.). *Dryopteris filix-mas* is the allotetraploid derivative of *D. oreades* × *D. caucasica*, which are both diploids (see figure on p. 209). (*D. caucasica* is absent from the flora of Britain and Ireland. It is a large plant of graceful habit occurring in high-rainfall areas of south-west Asia – the Black Sea coast of Turkey, the Caucasus Mountains and the Caspian Forest area of Iran.)

Field notes *Dryopteris filix-mas* is a predominantly European and western Asian species that is widespread at mainly middle latitudes of the continent. What are probably related forms occur in Asia and in North America. In Europe its range extends more thinly northwards to reach extreme northern Scandinavia, and similarly southwards to southern Spain and extreme north-west Africa, and eastwards to Turkey and the Caucasus.

In Britain and Ireland, Common Male-fern is widespread and one of the most abundant species of woodland fern in Britain and Ireland, and in lowland deciduous woodlands is usually rivalled only in frequency by Woodland Lady-fern (*Athyrium filix-femina*) and Broad Buckler-fern (*Dryopteris dilatata*).

It is a large, vigorous and generally long-lived species, occurring in a very wide range of habitats in lowland woods, glades, rides, streambanks and copses, from where it ascends to upland valleys and glens to around 610 m (2000 ft) (higher in south-west Ireland). It occurs in a very wide range of generally light, well-drained lowland soils, whilst at higher altitudes and in areas of heavier rainfall, becomes increasingly confined to better-drained knolls, hummocks and steep sides of upland valley woodland. It frequently survives in lowland roadside hedgerow communities, especially where the main hedgerow plants are Hawthorn (*Crataegus monogyna*) and Hazel (*Corylus avellana*). In such lightly-shaded situations, its large size enables it to compete well with many smaller flowering plants, especially in shallow soils over rocky substrates. Common woodland and hedgerow associates include Dog's Mercury (*Mercurialis perennis*), Bramble (*Rubus fruticosus* agg.), Wild Strawberry (*Fragaria vesca*), Common Dog-violet (*Viola riviniana*), Bugle (*Ajuga reptans*), Ivy (*Hedera helix*), Lesser Celandine (*Ranunculus ficaria*), Ground Ivy (*Glechoma hederacea*), Primrose (*Primula vulgaris*), Wood Avens (*Geum urbanum*), Wood Anemone (*Anemone nemorosa*), Red Campion (*Silene dioica*), Wild Garlic (*Allium ursinum*), Bluebell (*Hyacinthoides non-scriptus*), Field Woodrush (*Luzula campestris*), Slender False-brome (*Brachypodium sylvaticum*), Red and Sheep's Fescues (*Festuca rubra* and *F. ovina*), Wood Melick (*Melica uniflora*), Sweet Vernal-grass (*Anthoxanthum odoratum*), and woodland mosses such as *Thuidium tamariscinum, Hypnum cupressiforme, Dicranum scoparium, Pseudoscleropodium purum, Rhytidiadelphus squarrosus* and *R. triquetrus.* In woodlands, Woodland Lady-fern often occurs with Common Male-fern, usually becoming the more abundant of the two in moister soils alongside streams, as does Broad Buckler-fern in more acidic ones, whilst at the other extreme, or relatively dry, sandy and rocky banks, *D. filix-mas* may occur with Western Polypody (*Polypodium interjectum*) and Black Spleenwort (*Asplenium adiantum-nigrum*), or, more rarely, with Lanceolate Spleenwort (*A. billotii*). Other more unusual species with which *D. filix-mas* may occur in old, undisturbed woodland can include Twayblade (*Listera ovata*) and Herb Paris (*Paris quadrifolia*), whilst in woodlands where Golden-scaled Male-fern (*Dryopteris affinis*) also grows, the hybrid *D. ˣcomplexa* (q.v.) may sometimes be present.

Apart from its often naturalised state in gardens and shrubberies, *D. filix-mas* occurs quite commonly in larger cities and in industrial areas, where it seems tolerant of considerable atmospheric pollution. In such areas, it can establish well on derelict land, on mortar rubble, amongst old brickwork, on shaded aspects of railway-cutting retaining walls and tunnel mouths, bridges and along canal-bank hedgerows, especially where Hawthorn and Bramble provide modest shelter.

In most of its sites, *D. filix-mas* seems to occur in soils of slight acidic to more or less circumneutral character, extending occasionally into somewhat basic ones. It seems to thrive generally, however, in less acidic sites than do most members of the Golden-scaled Male-fern group, which tend gradually to replace it in more open conditions and in western areas of higher rainfall. In upland screes, it becomes generally replaced by Mountain Male-fern (*D. oreades*), although where the lower margins of screes penetrate into remnant woodland, the two may occur together. In cool, acidic, lowland woodlands, especially in the north, *D. filix-mas* is extensively replaced by Broad Buckler-fern, whilst on more basic soils, especially on basic clays in woodland valley slopes in western and south-western districts, it can become replaced locally by dense growths of Soft Shield-fern (*Polystichum setiferum*), and Hart's Tongue (*Phyllitis scolopendrium*).

Originally, *D. filix-mas* was probably a fern of Oak (*Quercus*) woodland and mixed deciduous forest in lowland Britain, especially over more sandy soils. With gradual removal of forest vegetation through Roman and Medieval times, the species has very probably greatly diminished, giving way instead to Bracken (*Pteridium aquilinum*) in many of its former sites. Much of the present genetic variation of *D. filix-mas* can perhaps be attributed to the large size of such former populations. Although still frequent today, the species is undoubtedly steadily diminishing in abundance in natural areas with erosion of natural and semi-natural woodland area and continuing removal of hedgerow and copse vegetation.

Dryopteris oreades Fomin

Mountain Male-fern

(*D. abbreviata* auct.)

Preliminary recognition A usually medium-sized fern with rather stiff, strongly upright, lance-shaped fronds, like a smaller, more compact-fronded Common Male-fern, with crisp-textured pinnae.

Occurrence Plants are restricted to mountainous districts of Britain and Ireland, with their chief stations in North Wales, the English Lake District, southern Scotland and the central and western Scottish Highlands, where they occur at moderate elevations in mountain scree and on rocky ledges.

Identification Plants have numerous, *winter-deciduous*, pinnate-pinnatifid fronds, mostly 40–50 cm high, but occasionally up to 80 cm or more. The fronds arise in a strongly ascending fashion, in fairly close, irregular, shuttlecock-like clusters, from much-branched ascending or spreading rhizomes, *which branch freely, eventually to build up large clumps with many nearby crowns.* The fronds are of ovate-lanceolate outline, pinnate-pinnatifidly divided, are widest about the middle of the frond, and taper above and slightly below this point. The stipes are of variable length, but usually about 1/4 or less the length of the frond, and are moderately densely covered *with dull, chaffy, very pale grey-brown scales, which thin out to become sparse by the middle of the stipes.*

The fronds bear about 12–25 or more, closely spaced pinnae along either side. Each pinna is fairly broad at the base (often about 1/4 as broad as long), tapering outwards, and deeply pinnatifidly divided (out almost to the pinna midribs) into *rather oblong, but round-tipped, segments*, attached to the pinna-midribs by the whole of their bases. *The segments are usually touching at their bases, appear fairly crowded along the pinnae, but taper slightly outwards to more-separated, distinctly rounded tips.* The segment margins are somewhat crenately lobed and auricled, with a few broad, blunt serrations, and the *segments curl distinctly upwards around most of their edges (especially the lowermost ones) giving both the segments and pinnae a distinctly concave appearance.* Most of the pinnae (especially on the lower part of the frond) are also inclined nearly horizontally, and the combination of these steeply inclined pinnae, and the upturning margins of the segments, usually *gives the whole frond a distinctly crisped or crinkled appearance*, characteristic of this species and distinctive even from a distance.

The fronds are of more rigid and stiffer texture than those of *D. filix-mas*, of greyish

Jan. Feb. Mar.	Apr.	May	Jun.	Jly	Aug.	Sep.	Oct. Nov. Dec.

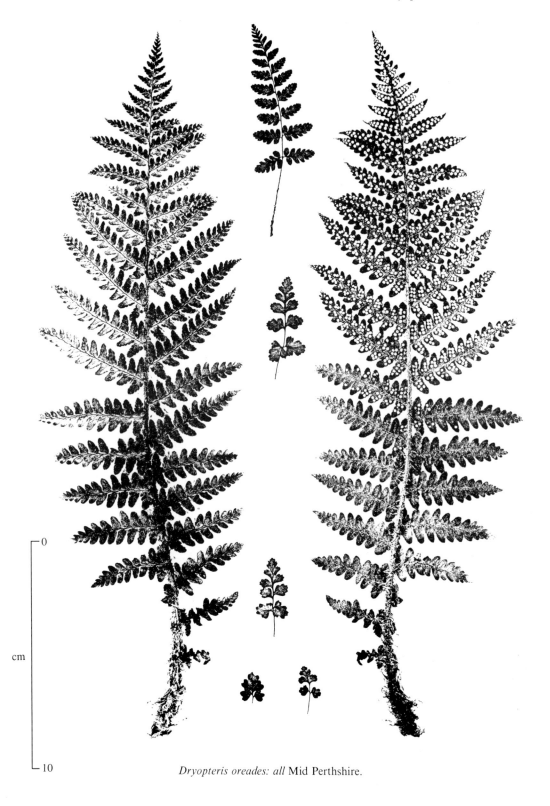

Dryopteris oreades: all Mid Perthshire.

Dryopteris oreades, habit of plant in life.

mid-green colour and have a *notably dull, non-glossy upper surface*. The junctions of pinnae-midribs and rachis are the same green colour as the rest of the frond and *totally lack the concentration of local dark pigment at this point* seen in fronds of the *D. affinis* agg. (q.v.). When lightly bruised, the fronds of *D. oreades* smell slightly fragrant.

Fertile fronds are similar to the vegetative ones, and have rather few sori confined usually to the bases of the pinnae (absent from their outer tips), but even fairly small fronds can be fertile. *The sori are small* (usually not much more than about 1 mm diameter), and the *indusium is usually biconvex or turban shaped*, like two discs one upon the other, *and the edges tuck under and embrace the developing sporangia* in a similar manner to those of *D. affinis* (q.v.). They never have spreading flattened margins as in *D. filix-mas*. The indusium is green when young, becoming tawny-grey-brown towards maturity.

Fronds begin to flush very early in spring and are usually fully expanded by the beginning of June (earlier than *D. filix-mas* and much earlier than those of the *D. affinis* agg.), when the differences between all the Male-ferns are very distinctive. The croziers, when expanding, are of a *dull greyish-buff colour* and remain *tightly circinately coiled* in a watch-spring-like manner, until completely unfurled, giving expanding fronds a characteristic appearance in the field, distinctive from all other Male-ferns.

Fronds die down in autumn, become brittle and detached, but decay only slowly, turning to a rich russet brown colour in winter. Such dead fronds often accumulate and persist well into the following season, giving a characteristic colouration in early spring to patches of hillsides where the species occurs in large numbers.

Possible confusion Plants are distinguished from *D. filix-mas* especially by their smaller size when fertile, fewer smaller sori, and tucked-under indusial margin. They are distinguished from all members of the *D. affinis* agg. particularly by their dull, more grey-green frond colour,

Dryopteris oreades, typical small and somewhat stunted fronds from populations in exposed sites: *a*, *b*, *d–f*, Caernarvonshire: *c*, Cardiganshire.

0 _____ 10

cm

Dryopteris oreades, relatively large frond from scree plant in sheltered site: Merionethshire.

paler, much less-abundant scales to the stipe, round-tipped segments and lack of black patch at pinna junctions. In addition, the crisped frond texture, early flushing season, tightly coiled crozier throughout frond expansion, much-branched rhizome, plus the lack of black patch, also distinguish *D. oreades* from all members of the *D.* *complexa* agg.

Technical confirmation Plants are diploids, with $n = 41$, $2n = 82$ chromosomes. The indusia and blade surface (especially when young) have numerous minute globose glands. Spores are 36–40 μm in size (cf. 32–45 μm in *D. filix-mas*, 40–70 μm in the *D. affinis* agg.).

Dryopteris oreades: Cumbria, June.

Field notes *Dryopteris oreades* shows a generally arctic-alpine distribution, with scattered stations at mainly middle and latitudes of the continent stretching south to northern Spain and eastwards to at least central Europe.

Within Britain and Ireland, *Dryopteris oreades* is a widespread and exclusively montane plant, occurring usually between 240 and 610 m (800 and 2000 ft), where it grows as scattered groups on well-drained rocky ledges or as extensive, nearly pure stands on well-drained open slopes of steep mountain scree. Its habitats are all in regions of high and frequent rainfall, with prolonged cloud cover. It often grows particularly prolifically on scree of sandstones and slates in North Wales, in the English Lake District and parts of western Scotland, and scree of mica schists in the Breadal-bane Range of the Scottish Highlands. They are usually associated with the least stable areas of such scree slopes. Hence they occur most often on the looser, upper parts of long talus fans, where the rock size is small and there is continual addition of new material, maintaining the openness of the site and minimising both grazing and competition. In such habitats, the plants are effective pioneers,

providing at least some initial stabilisation, and, under such circumstances, eventually build up massive clumps of much-branched rhizomes each with many crowns of fronds. In such situations, individual plants probably live to a particularly great age. Plants seem to diminish in numbers when scree stabilisation promotes eventual success of other competing vegetation, although they may persist for a considerable time. In its Breadalbane Range habitats, it seems to become extensively replaced by Northern Buckler-fern (*Dryopteris expansa*) once the average boulder size increases above about 10–20 cm diameter. In the most acidic scree, and at higher altitudes, it usually becomes replaced by Parsley Fern (*Cryptogramma crispa*).

As with Parsley Fern, the bulky rhizome masses often become encumbered at the end of the season with a covering of old, dark-brown fronds, which may persist until new fronds flush in the following spring. It seems possible that the old, dark-coloured fronds may themselves form a heat-absorbing blanket, assisting with the normal thermal balance of the plant. A very similar adaptation seems present in comparable mountain habitats in *Athyrium distentifolium* (q.v.).

Dryopteris submontana (Fras.-Jenk. & Jermy) Fras.-Jenk.

Limestone
Buckler-fern

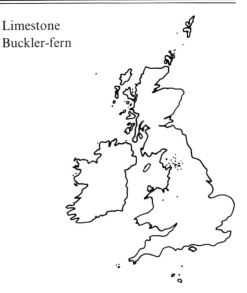

(D. villarii (Bellardi) Woynar)

Preliminary recognition A medium-sized, finely divided, rather stiff, upright-fronded Buckler-fern, with a distinctly greyish-green, mealy surfaced appearance.

Occurrence Confined mainly to a limited area of limestone country on the northern Pennines on the Lancashire–Westmorland–North Yorkshire borders, with rare outlying stations in North Wales and the north-west Midlands.

Identification Plants have rather *stiff and rigid fronds, of slightly leathery texture and distinctly mealy (farinose) appearance.* They are mostly about 20–60 cm high, bipinnately divided and of triangular-lanceolate outline, with the lowest pinnae pair not notably shorter than those above and often the longest. The fronds stand *fairly strongly erect, and have quite long, dull pale-brown, rather stout* and slightly succulent stipes, which are often as long as the blade length, especially in specimens growing in deep fissures. The fronds arise as small groups in rather irregular shuttlecocks from the crowns of decumbent or ascending rhizomes, which branch freely in old specimens to give many nearby crowns. The older parts of such rhizomes become densely clothed with the erect bases of old, dead fronds, and the younger parts, as well as the bases of the stipes, have numerous, *chaffy, long-pointed, shining bright pale-brown scales.*

The pinnae are slenderly tapering and often slightly ascending on the frond, with their *pinnules fairly widely spaced.* The pinnules are irregularly notched into rather acute teeth, but these are not spinulose. *The frond is of dull greyish-green colour*, with the mealy bloom produced by a dense covering of minute glands all over both surfaces of the blade.

Most fronds on mature plants bear rather large sori beneath, throughout their upper halves. *When lightly bruised, fronds have a balsam-like fragrance.* The fronds die down rapidly in autumn, with the first frosts.

Variation Not a very variable species.

Possible confusion The dull, mealy surfaced fronds readily distinguish *D. submontana* from all other Buckler-ferns.

Jan. Feb. Mar.	Apr.	May	Jun.	Jly	Aug.	Sep.	Oct. Nov. Dec.

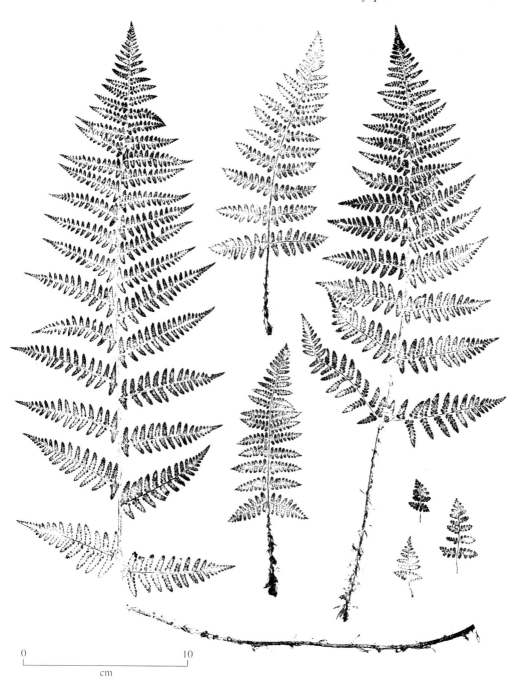

Dryopteris submontana: all North Yorkshire.

Technical confirmation Plants are tetraploids, with $n = 82$, $2n = 164$ chromosomes. The indusia have glandular margins, visible under a hand lens.

Field notes *Dryopteris submontana* is an outlying and insular part of a widespread European complex whose other members occur more widely in the more mountainous regions of the northern Mediterranean from Spain eastwards at least to northern Greece.

In Britain, Limestone Buckler-fern is a probably long-lived and usually gregarious, strongly calcicolous fern, growing in exposures of limestone rock, mainly in coarse limestone screes, sheltered cliff ledges and within deep, narrow broken-sided fissures (grykes) of limestone pavements. It is a locally frequent plant over outcrops, of the Great Scar Limestone in northern England, where it occurs from near sea-level to about 490 m (1600 ft) in the Pennines, above which height suitable habitats are absent. At their higher-altitude sites, plants tolerate frequent winter frost and snow.

Significant studies on its ecology have been made by Gilbert (1966, 1970). The plant requires a moist, well-drained calcareous soil in an open or only very lightly shaded situation. It thrives where there is a micro-relief providing some protection from grazing and from strong winds, and normally requires a layer of at least 15–20 cm of reasonably sheltered air above its rhizome to survive. It nevertheless seems to require also reasonably free air movement, and avoids stagnant situations. Such exacting conditions restrict the plant mainly to limestone block scree where there is a fairly large average boulder size (above about 60 cm average diameter), to sheltered pockets of limestone cliffs (especially summer-cooler ones with an aspect between north and south-east) and to limestone pavement grykes which are deeper than about 30–60 cm, but wide enough for reasonably good light penetration. In such fissures, it usually grows with only the tips of its fronds emerging at the surface of the limestone pavement. Occasional plants, however, grow on the surface where there is a low growth of scrubby woody surrounding vegetation (usually Yew (*Taxus*), Juniper (*Juniperus*) or Hawthorn (*Crataegus*)) to give the necessary local shelter and protection.

In the bottom of such grykes, it grows in local pockets of moist, highly calcareous soils, where mineral grains are intimately mixed with black, well-decomposed humus. Such strongly alkaline soil with a high proportion of calcium carbonate is usually highly fertile and has good water-retentive properties. *Dryopteris submontana* may dominate on such soils in the bottom of most of the deeper grykes, giving way usually to Limestone Oak-fern (*Gymnocarpium robertianum*, q.v.) in areas where

the grykes become shallower. *Dryopteris submontana* is accompanied in gryke habitats by a characteristic flora of other markedly calcicolous species, growing either deeply within the grykes or, in the case of most of the smaller ones, in pockets at various depths on the sides, or around the fissure mouths. These include Herb Robert (*Geranium robertianum*). Wood-sorrel (*Oxalis acetosella*), Common Violet (*Viola riviniana*), Lily-of-the-Valley (*Convallaria majalis*), Wood Anemone (*Anemone nemorosa*), Dog's Mercury (*Mercurialis perennis*), Wall Lettuce (*Mycelis muralis*), Blue Sesleria (*Sesleria albicans*), Red and Sheep's Fescue grasses (*Festuca rubra* and *F. ovina*), Hart's Tongue (*Phyllitis scolopendrium*), Brittle Bladder-fern (*Cystopteris fragilis*), Limestone Oak-fern (*Gymnocarpium robertianum*), Green Spleenwort (*Asplenium viride*), Common Maidenhair Spleenwort (*A. trichomanes* subsp. *quadrivalens*), Wall Rue (*A. ruta-muraria*) and dense growths of mosses, especially *Ctenidium molluscum*. Common Maidenhair Spleenwort and Wall Rue grow mostly in mossy cushions around the tops of the grykes, and often become dominant in shallower fissures. Hart's Tongue usually replaces Limestone Buckler-fern in the deepest, narrowest grykes where there is less light penetration and more stagnant air conditions. In more open grykes with deeper soils it seems eventually excluded when *Mercurialis* growth becomes dense.

The heavy dependence of Limestone Buckler-fern, for such a large fern, on conditions combining appropriate soils, shelter, and light, as well as perhaps particular mineral intolerances, is possibly one of the main factors that restrict it to its small and concentrated main area. Intolerance of even the relatively light shade of ashwoods over limestone further restricts the plant to open situations. Proctor (1972) suggests that it probably is a post glacial survivor in limestone cliff sites that rose above the forest cover during warmer post-glacial stages.

Today, many of the larger plants may be of considerable antiquity. The plants are slow growing, and some with fairly densely branched rhizome stocks have been estimated to be about 150 years old. They seem, however, highly grazing-sensitive, with fronds readily taken by sheep, probably making populations vulnerable to pressure from grazing, should this become heavy.

Dryopteris ˣambroseae
Fraser-Jenkins & Jermy

Gibby's Hybrid
Buckler-fern

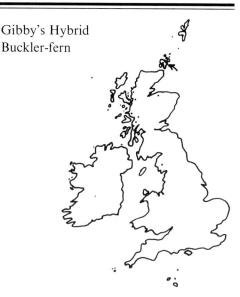

(*D. dilatata* × *D. expansa*)

Preliminary recognition A large fern, notably intermediate in general appearance between *Dryopteris dilatata* and *D. expansa*, but often looking most like a slightly more finely cut and yellower-green version of the former.

Occurrence Currently known from a small number of widely scattered localities, chiefly in Cumbria, Inverness and Argyll, growing as scattered individuals amongst mixed populations of both parents. It may well yet be found elsewhere.

Identification Plants usually have ascending-arching fronds, to about 60 cm or more in length. They are more-or-less intermediate in general appearance between those of the two parents, and arise from short, stout, erect or ascending rhizomes.

Plants differ from *D. expansa* (q.v.) mainly in having a frond of slightly tougher, sometimes bluer-green colour, and scales which are *less gingery and more frequently have a darker central area, which often becomes a dark median stripe, as in D. dilatata.* They differ from *D. dilatata* in having a rather more triangular-outlined frond, which is overall *more finely cut and appears more delicate and lace-like* because of the wider gaps between the bases of the ultimate segments, and because it has the flatter, less downcurving pinnule margins of *D. expansa.* Fronds usually emerge later in spring and die down earlier in autumn than do those of *D. dilatata* growing in the same environment.

Plants are, however, often difficult to separate with certainty from either, rather variable, parent, without recourse to the spores, *which are always entirely abortive.* In the field, often the whole sporangia appear sparse and abortive too.

Variation Not fully assessed, but likely to be high because of the considerable variability of both parents.

Possible confusion The abortive spores distinguish *D.* ˣ*ambroseae* from *D. dilatata* and *D. expansa*, but not from *D.* ˣ*deweveri* (q.v.) which shares the common *D. dilatata* parent and also has abortive spores. *Dryopteris* ˣ*ambroseae* has a broader, more triangular, outline to the frond than has *D.* ˣ*deweveri*, has a less erect frond habit, and inherits the short ascending rhizome of both its parents (cf. the stronger tendency for the rhizome to be prostrate or creep in *D.* ˣ*deweveri*).

Technical confirmation Plants are triploids, with $2n = 123$ chromosomes, showing approximately 41 pairs and 41 single chromosomes at meiosis (Walker, 1955; Gibby & Walker, 1977).

Field notes Scattered individuals of this hybrid have been found in widely spread localities in the Lake District and Scotland (especially in the western Highlands), in sites where both parents grow together in some quantity. Mostly these have been found in low-altitude, wet, acidic woodland, with a

Dryopteris [×]*ambroseae*: *a–a*, Cumbria; *b–b*, Westerness.

Dryopteris ˣ*ambroseae: a*, West Argyll; *b*, Cumbria.

low canopy (see 'Field notes' under *D. expansa*), but the hybrid is probably present, as are both parents, through a wide range of altitude, and it has been recorded from amongst mountain rocks and boulders at about 915 m (3000 ft) (Jermy & Walker, 1975). Jermy & Walker also record that under experimental conditions, this hybrid can be readily synthesised. It seems very probable that, in the field, it is frequently overlooked as a form of one or other parent, and is probably under-recorded.

Dryopteris ˟brathaica
Fr.-Jk. & Reichstein

Brathay Fern

(*D. carthusiana* × *D. filix-mas*)

Preliminary recognition A medium-sized, erect-fronded Buckler-fern, looking like a more robust, narrower-fronded, less divided form of *D. carthusiana*, with stouter, more scaly fronds, arising from an upright, not creeping, rhizome.

Occurrence Known once in damp woodland in the vicinity of Lake Windermere in the Lake District, but could yet be found again anywhere within the widespread British and Irish ranges of the two parent species.

Identification Plants are vigorous hybrids, of a form more or less intermediate between those of their rather different-looking parents. They have erect, oblong-lanceolate outlined fronds, to 60 cm or more in height, which are almost tripinnately divided.

Plants differ from *D. carthusiana* in having the upper half of the frond with *much more bluntly rounded segments*, which are joined to the pinna midribs usually by the whole width of their bases (adnate), much stouter stipe and rachis, which has more numerous, broadly ovate, pale-brown scales, *ascending or erect (not prostrate) rhizome*, and large sori with partly convex indusia, similar in form to those of *D. filix-mas*. Plants differ from *D. filix-mas* in having more distantly spaced lower pinnae, a more divided overall frond, which is *almost tripinnate, especially in its lower half*, where the segments are more stalked than in *D. filix-mas* and more acute, the margins serrated, more spine-tipped teeth (approaching the shape of those of *D. carthusiana*, q.v.), and sori occurring over almost the whole of the back of mature fronds.

The fronds shed abundant, but totally abortive, spores. When flushing in spring, the expanding fronds remain fairly tightly circinately coiled as in *D. carthusiana*, seldom forming the more laxly open shepherd's-crook shape of *D. filix-mas* (q.v.).

Variation Insufficient material is known to assess this.

Possible confusion The strong intermediacy of plants between their rather different-looking parents, makes them fairly easily distinguishable from both. They differ from *D.* ˟*uliginosa*, which shares the common *D. carthusiana* parent, in having fronds which are all alike (not at all dimorphic) and in its more upright, more robust and scaly rhizome. They differ from *D.* ˟*deweveri*, which also has *D. carthusiana* in its parentage, by the same rhizome characters as

Dryopteris ˣ*brathaica: a*, Westmorland (Windermere), wild source; *b*, frond ex horticulture, from material introduced from the same wild source in 1854 and cytologically authenticated in 1973.

well as by the narrower, less-divided frond. Plants are much more likely to be confused with *D.* ˣ*remota* (q.v.), and long were, although *D.* ˣ*remota* (probably *D. affinis* × *D. expansa*) and *D.* ˣ*brathaica* (*D. carthusiana* × *D. filix-mas*) have neither parent in common. *Dryopteris* ˣ*brathaica* always has entirely abortive spores (cf. some good spores in *D.* ˣ*remota*, q.v.), although technical confirmation is required to separate these two hybrids with certainty.

Technical confirmation Plants are sterile tetraploid hybrids, with $2n = c.$ 164 chromosomes (Manton, 1950), showing only a little pairing at meiosis (cf. an apogamous triploid hybrid with $n = 123$, $2n = 123$ chromosomes in *D.* ˣ*remota*, q.v.).

Field notes Despite the general frequency of its parents in both Britain and Ireland, *D.* ˣ*brathaica* has been found only once (in 1867) in low-lying, acidic damp, hummocky woodland by Lake Windermere (Brathay Woods). The hybrid occurred in woodland in which both parents grow in numbers nearby (as well as *D. dilatata* and *D.* ˣ*deweveri*). Following its original discovery, material was introduced into cultivation, and has been maintained and propagated vegetatively through the intervening century, and still survives in some botanic gardens.

Dryopteris ˣcomplexa Rothm.   Hybrid Male-fern

(*D.* ˣ*tavellii* Rothm.)
(*D. filix-mas* × *D. affinis* agg.)

Preliminary recognition A large to very large, vigorous and handsome fern, with tall, arching-tipped, stiff-based, lance-shaped fronds with densely scaly stipes, looking like a particularly strongly growing form of *D. affinis*, but with more leafy, less truncated pinna segments.

Occurrence A probably under-recognised and under-recorded fern, likely to occur wherever the parent species are found in proximity, especially in wetter, western climates, along streamsides and in mountain valley woodland.

Identification *The vigour and sheer size of individuals*, with fronds to 1.6 m or more in length, is the hybrid's most outstanding feature, enabling plants to be suspected even at a distance. They most resemble members of *D. affinis* in having *relatively bright, usually glossy, somewhat yellow-green-coloured fronds, with stiff, stout stipes covered in dense, chaffy, fairly golden-coloured scales*, which ascend along the rachis, thinning gradually. The pinna bases usually have *prominent dark spots* (sometimes only pale grey, sometimes as dark as in *D. affinis* and even more extensive) at the junction with the rachis. The fronds are, however, of much less leathery texture and usually less glossy surface than those of *D. affinis*, with much *less truncated, larger, more cut and lobed pinna segments with serrated margins*, the resulting much broader pinnae thus resembling a very luxuriant version of *D. filix-mas* in frond dissection. Plants usually form very large and regular shuttlecock-like baskets, from single-crowned, massive, stocky, erect, scaly rhizomes. From these, the fronds arise rather stiffly and steeply in their basal halves, but

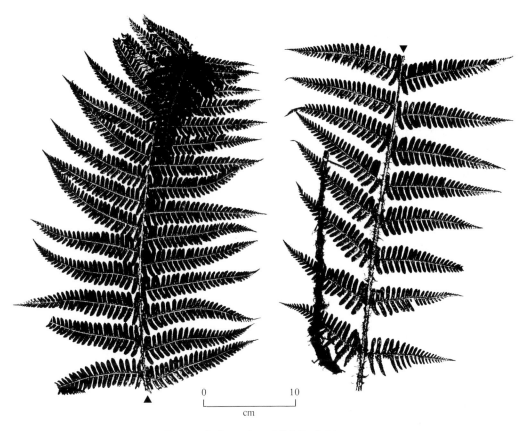

Dryopteris ×*complexa*: Mid Perthshire.

because of their length, usually arch gracefully over near their tips. Most fronds bear abundant sori, and the indusium margin varies from the doughnut-shaped, tucked-under type of *D. affinis* (q.v.) to the soup-plate-edged, flatly spreading type of *D. filix-mas* (q.v.), even on a single segment.

Variation Little studied – possibly greater than generally appreciated through the variability of parents on both sides, the possible involvement of all subspecies of *D. affinis* (q.v.) in the parentage, and the genetic isolation of resulting apogamously breeding individuals from one another. The possibility of back-cross hybrids to *D. filix-mas* (see 'Technical confirmation') may further blur the taxonomic outlines of *D.* ×*complexa* as a whole.

Possible confusion The more leafy frond, and presence of the black spot are the best field characters for distinguishing most hybrids from *D. affinis* and *D. filix-mas* respectively, whilst the much larger size is often distinctive from both. Doubtful material may need technical confirmation, especially to separate them from *D. affinis* subsp. *robusta* (q.v.).

Technical confirmation Plants are either apogamous tetraploids, with $n = 164$, $2n = 164$ chromosomes, resulting from the cross *D. filix-mas* × *D. affinis* subsp. *affinis*, or are apogamous pentaploids, resulting from the cross *D. filix-mas* × *D. affinis* subsp. *borreri* or *D. filix-mas* × *D. affinis* subsp. *cambrensis* or *D. filix-mas* × *D. affinis* var. *robusta*. Any of these combinations are likely to inherit the apogamous life-cycle of the respective *D. affinis* parent, and thus themselves produce at least a proportion of good spores, which are likely to be of quite large size and mixed with a proportion of bad ones. Until the differences between these different possible hybrids have been more fully studied, it is possible to differentiate only between the tetraploid and pentaploid crosses with a chromosome count.

Field notes As all types of *D.* ×*complexa* hybrids breed true, once formed, their numbers can multiply within a suitable habitat from a single original crossing, enabling plants to become more common locally as separate individuals than a sterile hybrid otherwise could. Plants thus occur sporadically as large individuals or can sometimes form small groups of several nearby plants. Habitats are usually in well-drained spots on slopes in damp, acidic, fairly sheltered western and upland woodland, especially along river slopes, on the tops of steep banks of stream gullies and along ditches and tracksides through forest. They most often occur where original vegetation has been disturbed, or partially or wholly modified by coniferous plantings (especially when Norway Spruce and Douglas Fir have been used and the plantings matured). The formation and subsequent spread of the hybrid may well be most often associated with the preliminary disturbance of such ground, construction of drainage ditches alongside tracks, and the shelter provided by developing plantation forest.

Because of their apogamous mode of reproduction, genetic isolation, true-breeding properties and morphological distinctness, such plants could, with justification, be regarded as a species in their own right. However their continual sporadic reformation anew wherever the parents meet in suitable situations, suggests that it would be least confusing to retain them, as here, under hybrid status, whilst nevertheless noting their potentially true-breeding properties.

Dryopteris ×deweveri
(Jansen) Jansen & Wachter

Hybrid Narrow
Buckler-fern

(*D. carthusiana* × *D. dilatata*)

Preliminary recognition A medium sized fern, with bipinnately divided, fairly erect fronds, looking like a bluer-green, slightly more coarsely cut, much more vigorous form of *D. carthusiana*.

Occurrence In numerous widely scattered stations throughout the whole of Britain and possibly Ireland, in slightly damp woodlands, where the two parents grow together.

Identification Plants have fronds up to 120 cm or more in height, which are intermediate in width and cutting between those of *D. carthusiana* and *D. dilatata*, but tend to look much like a larger more robust version of the former.

They differ from *D. dilatata* mainly in having fronds of *much more erect habit*, arising in more irregular clusters than in regular shuttlecocks (more like the habit of *D. carthusiana*), *more decumbent, often semi-prostrate or entirely creeping rhizome, more* D. carthusiana-*like pattern of frond cutting*, and less winter hardy, much earlier-dying, fronds. They differ from

Dryopteris [×]*deweveri: a, b,* Argyll (Loch Awe); *c,* (Loch Vennachar).

D. carthusiana mainly in their *much larger frond size*, less stiff, more flexible frond, slightly coarser frond cutting, giving it a less delicate and more solid-looking texture, deeper-green frond colour, proportionately less distantly spaced pinnae, rather shorter length of stipe, more robust rachis which is more abundantly scale-clad, and scales which either have a concolorous dark orange-brown colour or have a diffuse dark lengthwise marking (though usually not such a dark or distinct stripe as in *D. dilatata*).

Plants also show *considerable hybrid vigour*, and their rhizomes may creep and branch extensively to build up colonies many metres in diameter. These are normally very much larger than well-established patches of *D. carthusiana* growing in the same environment. Spores of the hybrid are mostly abortive.

Variation Plants show some variation in size and vigour, as well as in colouring of scales.

Possible confusion Most plants are fairly clearly distinguishable from either parent by the above characters. They differ from *D.* [×]*uliginosa* (q.v.), which shares the *D. carthusiana* parent, mainly in the much broader frond outline and darker, sometimes dark-striped scales, and from *D.* [×]*ambroseae* (q.v.), which shares the common *D. dilatata* parent, in having a narrower frond outline, more erect frond habit and more prostrate rhizome.

Technical confirmation Plants are tetraploid hybrids, with $2n = 164$ chromosomes. They show approximately 41 pairs and 82 single chromosomes at meiosis. The indusia are glandular.

Field notes Hybrid Narrow Buckler-fern is not uncommon in a wide range of slightly damp, fairly acidic woodland habitats, both natural and plantation. Where both parents grow together in quantity, scattered clumps of the hybrid are likely to be found, but plants also quite frequently occur in the absence of one or both parents. If one is absent, it is usually *D. carthusiana*, as the hybrid seems much more tolerant of drier ground conditions than this parent, and often outlives it in formerly wet woodland that has suffered from drainage. Indeed, the occurrence of extensive *D.* [×]*deweveri* without *D. carthusiana* seems often symptomatic of such a woodland history. With continued drainage, such sites usually give way then to *D. dilatata* dominance and eventually to Bracken (*Pteridium aquilinum*).

Dryopteris [×]*deweveri* can thus often be found in coniferous plantation forest, near ditches and gullies and damp edges of tracks, either surviving from indigenous plants, or becoming formed in areas of habitat disturbance from nearby parents during clearance and commencement of drainage operations. It then continues to gain ground, as establishment of coniferous trees further dries out the ground conditions, until it too becomes weakened and dies out through desiccation or whenever the dense plantation canopy closes.

Dryopteris ˣmantoniae Fras.-Jenk. & Corley

Hybrid Mountain
Male-fern

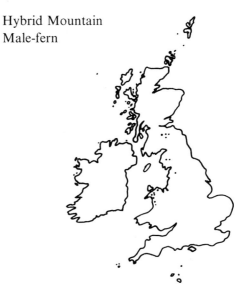

(*D. filix-mas* × *D. oreades*)

Preliminary recognition A large-sized fern with lance-shaped fronds arising in irregular, shuttlecock-like clusters, very similar in appearance to Common Male-fern, but with slightly more crisped-segmented fronds arising from much more freely branched rhizomes.

Occurrence Known from a small number of scattered, mostly montane, stations in North Wales, the Lake District, the Moffat Hills of Southern Scotland and scattered localities in the Scottish Highlands, mainly on screes and in scrub at moderate altitudes.

Identification Plants have fronds 70–90 cm or more in length, and look much like *D. filix-mas*, but have fronds with pinnae and segments which appear *slightly more concave when viewed from above* (including having the lower one or more segments on the lower side at the base of each pinna upturned). They also arise from a *much more branched rhizome mass*, as in *D. oreades*. Fertile fronds have indusia which vary, even in a single pinna, from the flat, soup plate-edged type of *D. filix-mas* (q.v.) to the doughnut-shaped, tucked-under type of *D. oreades* (q.v.). They inherit the slightly earlier flushing season and laxer crozier of *D. filix-mas*, but the *greyer-scaled appearance at the flushing stage* of *D. oreades* (q.v.).

Variation Little known.

Possible confusion Easily overlooked as a form of *D. filix-mas*, but the above features, plus its usual occurrence together with both parents, provide the best field guides.

Technical confirmation Plants are triploid hybrids, with $n = 123$ chromosomes showing $c.$ 41 pairs and $c.$ 41 single chromosomes at meiosis. They are highly sterile, with only entirely abortive spores.

Field notes Plants occur in scrubland on lower mountain slopes or in screes of slate or schist or other rock in mountain districts at moderate altitude. They usually occur as scattered, but vegetatively prolific, individuals, in the presence of both parents, usually at slightly lower elevations and in more sheltered pockets than are typical of the habitats of *D. oreades* alone.

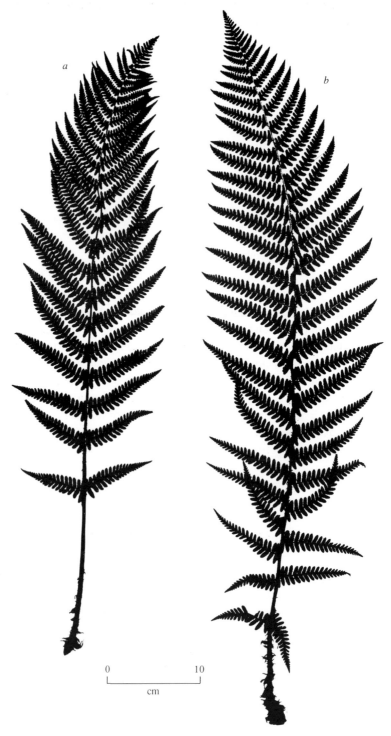

Dryopteris [×]*mantoniae: a, b,* Caernarvonshire.

Dryopteris [×]pseudoabbreviata Jermy   Mull Fern

(*D. aemula* × *D. oreades*)

Preliminary recognition A small fern, with distinctly crisped pinnules, looking like a smaller, more dissected, more long-stiped, more triangular-bladed form of Mountain Male-fern.

Occurrence Known only from the island of Mull, western Scotland, in oak-birch scrub on a steep, west-facing hillside.

Identification Plants approach *D. aemula* in size, with fronds about 12–15 cm in length. They have overall ovate-triangular outlined blades, with the lower 3–4 pinna pairs of approximately similar length, but shortening above to give the frond a tapering apex. Fronds have fairly long, thick stipes, the lower part of which inherits the purplish tinge of the *D. aemula* parent, covered with pale, rather limp, concolorous, pale, rather greyish-tan scales. The *dissection of the frond is intermediate between that of the parents*, with at least the basal pairs of pinnae in the lower part of the frond pronounced and stalked. Fronds arise in irregular shuttlecocks from rhizomes which are branched, suberect, and bear multiple crowns. All the segments have upturning margins, giving fronds a distinctly crisped appearance, and the *surface is glandular and hay scented*, as in *D. aemula*.

Possible confusion The intermediacy of appearance between the two very different looking parents is striking, and helps to distinguish plants fairly clearly from both.

Variation Unknown, but from the constancy of appearance of both parents, perhaps not likely to be great.

Technical confirmation Plants are diploid hybrids, with *n* = 82 chromosomes, and have only entirely abortive spores. Their ploidy level and lack of apogamy confirms that *D. oreades*, which is a normal diploid, is involved in the parentage, and not *D. affinis* (which also grows with it) but is either an apogamous diploid or apogamous triploid.

Field notes In its single known locality, in Oak–Birch (*Quercus-Betula*) scrub, on a steep west facing hillside on the island of Mull, western Scotland. *D.* [×]*pseudoabbreviata* was found with both parents and with *D. affinis* agg.

Despite both parents preferring high-rainfall climates, their altitudinal separation throughout most of the limited range of each in both Britain and Ireland (*D. aemula* always near sea-level, *D. oreades* in mostly mountain habitats) helps to ensure that opportunity for hybridisation is geographically limited to a few extreme west-coast sites.

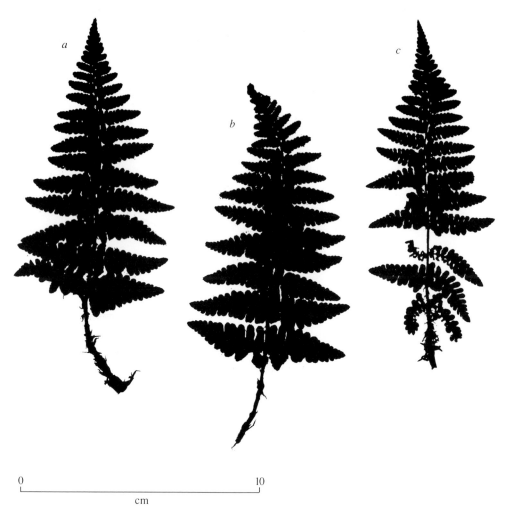

0   10
cm

Dryopteris ×*pseudoabbreviata: a, b*, Isle of Mull; *c*, Hebrides (Mid Ebudes).

Dryopteris ˣremota (A. Br.) Druce

Distant-leaved
Buckler-fern

(*D. affinis* subsp. *affinis* × *D. expansa*)

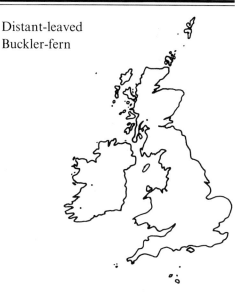

Preliminary recognition A medium-sized Buckler-fern with regular 'shuttlecocks' of ascending, slightly arching, finely divided fronds, looking like a much narrower-bladed form of *D. expansa*, with much more densely golden-scaled stipes.

Occurrence Of isolated occurrence, with one (original) station on the shore of Loch Lomond and two others in western Ireland. It could yet be found again anywhere within the ranges of the two parents, and possibly beyond them (see 'Field notes').

Identification Plants have fairly strongly ascending fronds of ovate-lance-shaped outline, with pinnae all of about equal length throughout the lower half of the fronds, except for the lowermost pair, which are usually shorter than those above. The fronds approach those of *D. expansa* (q.v.) in fineness of division, but have stipes which are densely clothed near the base, with quite large, golden-brown scales, and arise from robust stocky, usually erect scaly rhizomes.

   Plants differ from *D. expansa* in their narrower frond, larger number of much narrower, more gradually tapering, more closely spaced pinnae, densely scale-clad lower stipe, and sori with highly convex indusia, like those of *D. affinis* (q.v.). Plants differ from *D. affinis* in their thinner, less leathery, more flexible texture, very much finer, rather lace-like pattern of frond-cutting, distinctively longer stipes, and scales thinning out rapidly along the rachis. The scales also often have a dark patch near their centre, not seen in *D. affinis*.

   Plants produce abundant spores, at least some of which appear to be not totally abortive.

Variation Considerable variation seems likely to be inherited from the *D. affinis* side of the parentage, but too few plants are known to assess this.

Possible confusion The above features should distinguish *D.* ˣ*remota* fairly clearly from both parents. This hybrid does, however, look very similar to *D.* ˣ*brathaica* (although neither parent is common to both hybrids), and technical confirmation is essential to distinguish between them with certainty.

Technical confirmation Plants are apogamous triploid hybrids, with $n = 123$, $2n = 123$ chromosomes, producing at least some good spores. They are similar, in this respect, to plants known from continental Europe (cf. a sterile tetraploid, with $2n = c.$ 164 chromosomes in *D.* ˣ*brathaica*, q.v.).

Field notes *Dryopteris* ˣ*remota* was originally found by W. B. Boyd on 9 August 1894, a little way from the west shore of Loch Lomond, Dunbartonshire, between Tarbet and Ardlui, presumably growing in the damp, acidic, deciduous rocky woodlands that abound in this area. Slightly later, it was also found in two localities in western Ireland, but has not been seen anywhere in Britain or Ireland more

0

cm

10

Dryopteris [×]*remota:* Dunbartonshire.

recently. From the Loch Lomond locality, it was introduced into cultivation in Britain and maintained by such growers as Stansfield, who referred to it as *Lastrea 'boydii'*. Living material of both the Scottish and Irish plants was studied by Manton (1950), who first distinguished them cytologically from *D.* ˣ*brathaica*, and showed them to compare with the specimens of similar morphology known in continental Europe.

Dryopteris expansa, and particularly, *D. affinis*, occur extensively within the general vicinity of the Loch Lomond site, but only *D. affinis*, not as yet *D.*

expansa, is known in Ireland. However, in continental Europe, *D.* ˣ*remota* is known to occur as small groups of plants sometimes in the absence of both parents (Reichstein, 1965, and personal communication), perhaps the result of establishment by spores from its apogamous breeding system. The plant could yet be refound virtually anywhere the two parent species occur in quantity together, especially perhaps, in Scotland, or even, as in continental Europe, outside the known range of *D. expansa.*

Dryopteris ˣsarvelae Fraser-Jenkins & Jermy

Kintyre Buckler-fern

(*D. carthusiana* × *D. expansa*)

Preliminary recognition A large fern of appearance generally intermediate between those of the parents, and combining the fine frond dissection of *D. expansa* with the more spine-toothed margin of *D. carthusiana.*

Occurrence Known with certainty from a single low-altitude locality in the west of Scotland (Kintyre), growing with both parents. It is, however, likely to be easily overlooked as a form of one or other variable parent, and could well occur elsewhere in north-western districts, where the two parents grow closely together.

Identification Plants are large, with fronds to 75 cm. The fronds inherit the general broadness of blade outline and the fine, lace-like dissection of the *D. expansa* parent, but do not have the lower pinna pair so prominently developed, as in this parent. They resemble *D. carthusiana* (q.v.) in the rather spinulose margins to the pinnules, but the fronds are more arching and the pinnae usually less widely spaced on the frond than in this parent. The stipe is also usually somewhat shorter and stouter than that of *D. carthusiana*, but arises from a relatively thick, short rhizome which is set more horizontally than that of *D. expansa*. At the base of the stipe, the scales are nearly concolorous or have only an extremely faint, small, dark central patch, but are nearer to the rather gingery colour of *D. expansa*. The lamina of the frond has a fairly densely glandular surface, and the spores are entirely abortive.

Variant Not yet assessed, but because of the wide variation shown by each parent, may well be expected to be considerable.

Possible confusion The abortive spores distinguish *D.* ˣ*sarvelae* from *D. carthusiana* and *D. expansa*, with which it is most likely to be confused, but not from *D.* ˣ*ambroseae* (which also shares the *D. expansa* parent) or *D.* ˣ*deweveri* (which also shares the *D. carthusiana* parent).

Dryopteris [×]*sarvelae*, fronds of mature and cytologically verified material, Scotland: *a, c–d*, Kintyre; *b*, Westerness.

Dryopteris ×*sarvelae*, semi-juvenile frond specimens, Scotland: *a*, Westerness; *b*, Kintyre.

Dryopteris ×*sarvelae* has a more delicate lamina texture, with narrower fronds arising from a more horizontal rhizome than that of *D.* ×*ambroseae*, and fronds are more tufted on a shorter rhizome and more arching in habit than those of *D.* ×*deweveri*. In the field, fronds of *D.* ×*sarvelae* are paler green in texture and might be expected to die down earlier than those of either *D.* ×*deweveri* or *D.* ×*ambroseae*.

Technical confirmation Plants are triploid hybrids, with $2n = 123$ chromosomes, all of which remain unpaired at meiosis (cf. the presence of equal numbers of bivalents and univalents in *D.* ×*ambroseae*, which is also a triploid, and *D.* ×*deweveri* which is tetraploid).

Field notes This hybrid is probably scarce, yet very likely overlooked in lowland wet woodland situations in the west of Scotland where the parents grow alongside. Its first native record was confirmed from Kintyre, western Scotland (Corley & Gibby, 1981).

Dryopteris ˣuliginosa
(Newm.) Kuntze ex Druce

Hybrid Fen
Buckler-fern

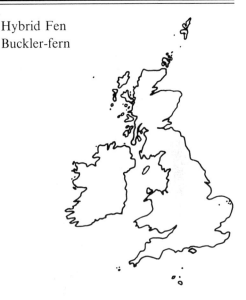

(*D. carthusiana* × *D. cristata*)

Preliminary recognition A medium to large-sized plant, looking like a much larger, narrow-fronded form of *D. carthusiana*, with less divided, more solid pinnae.

Occurrence A rare and local hybrid, probably now confined to Norfolk (see under 'Field notes').

Identification Plants are large and vigorous with fronds often 50–100 cm tall and about 10–15 cm wide, the taller, fertile ones of which stand stiffly erect. Their stipes are up to about half the length of the frond. They bear sparse, pale-brown, broadly ovate scales, and arise from stocky, prostrately creeping rhizomes, which branch freely to build up dense local colonies.

Fronds differ from those of *D. carthusiana* (q.v.) mainly in their usually much larger size, less delicate appearance, more leathery texture, narrower frond outline, and broader, less divided pinnae. They differ from *D. cristata* (q.v.) in their broader frond outline, more acutely tapering and more lobed pinnae, more pointed, more separated pinnules, with the basal ones usually distinct and pinnatifidly lobed, and with their edges more deeply toothed with distinct spinulose-tipped teeth (as in *D. carthusiana*). They also differ from *D. cristata* in having the lower pinna pair usually shorter than those above.

The spores are completely abortive.

Variation A rather constant hybrid in appearance.

Possible confusion The considerable individual distinctness of its parents from one another and the constancy and close intermediacy of the hybrid between them, make it sufficiently distinctive in the field that it has long been suspected to be of hybrid origin and of this parentage (e.g. Sowerby, 1866).

Technical confirmation Plants are tetraploid hybrids, with $2n = 164$ chromosomes, showing approximately 41 pairs and 82 single chromosomes at meiosis.

Field notes Hybrid Fen Buckler-fern is a large and vigorous plant, occurring mostly in fenland habitats similar to those of *D. cristata* (q.v.). It thrives especially in areas in which there is slight shade, such as Birch–Willow (*Betula–Salix*) carr developed over deep highly humified peat, and usually where

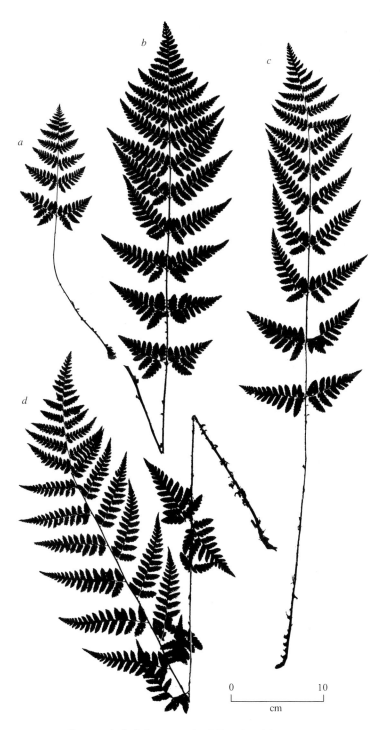

Dryopteris ×*uliginosa: a– d, all* East Norfolk.

both parents are also present in the vicinity in some numbers. The pH range of its sites has been reported to be 5.5–6.5 (Jermy & Walker, 1975), and associated ground flora can include Lesser Pond-sedge (*Carex acutiformis*), Purple Smallreed (*Calamagrostis canescens*), Common Reed (*Phragmites communis*) and various bogmosses (especially *Sphagnum fimbriatum*).

Like *D. cristata*, the range and abundance of *D. ×uliginosa* has declined substantially over the last century and, although probably now confined in Britain to Norfolk, was formerly known also in a number of stations in the north Midlands, and hence occurred in many of the places where *D. cristata* was formerly present. Many of the comments made about the decline of *D. cristata* probably also apply to this hybrid, although the hybrid could outlast *D. cristata* in sites of former occurrence of both.

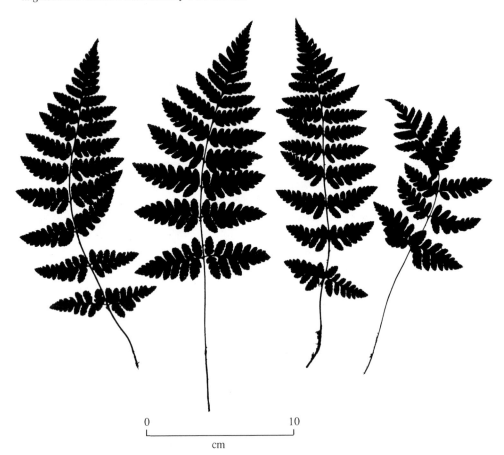

0   10
cm

Dryopteris ×uliginosa, typical juvenile fronds: ex horticulture.

Gymnocarpium dryopteris (L.) Newm.

Woodland Oak-fern

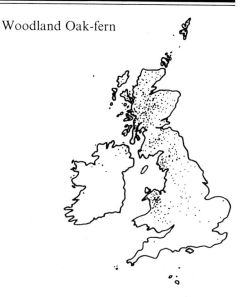

Preliminary recognition A small, colony-forming fern of delicate texture and graceful habit, with broadly triangular, vivid green, tripartite blades, supported on slender, upright, black stipes.

Occurrence A locally frequent plant of upland districts of Wales, northern England and Scotland, but scarce elsewhere. It grows mainly in moist, shady soil banks in upland valley woodland and along stream-courses, where it may form extensive colonies.

Identification Fronds arise individually at rather distant intervals from slender, freely branching, long-creeping, wiry, underground rhizomes, which eventually form a matted mass about 2–3 cm below the surface. The blades are borne on *very slender, brittle, polished, dark purplish-brown, stipes* about 12–20 cm or more high, and much longer than the length of the blade. *The stipes grow strongly erect, and at the top of each, the blade spreads horizontally.* The stipes at first have scattered, pale-brown scales, which, except for a few near the base, are later shed.

The blade is of *broadly triangular-deltoid outline*, 5–15 cm long, and slightly broader than long. The lower pair of pinnae, borne exactly oppositely at the point of junction of the horizontal blade on the vertical stem, are somewhat distant from the others, long-stalked, and particularly well and asymmetrically developed, with the largest side away from the apex of the frond. *Each nearly equals the whole of the remaining portion of the blade in size, so that the whole leafy portion of the frond appears tripartitely (ternately) divided.*

The lamina of a *rich, vivid green colour*, and of very thin, delicate and almost membranous texture, with a smooth non-glandular upper surface which is non-shining. The edges of each segment usually bend downwards, giving the whole frond a gracefully drooping appearance. The dull upper surface of the lamina readily repels water, which consequently forms into droplets and rolls off the sloping surfaces. Lightly bruised blades have no notable smell.

Fronds are either all similar, or the fertile ones have only slightly narrower segments than the vegetative, are borne on somewhat longer stipes and bear small, round sori beneath, in a line near the margin of the segments. There is no indusium, and the sori are sometimes so close that they nearly coalesce at maturity. When fully developed, they are black, before shedding their sooty black spores.

Jan. Feb. Mar.	Apr.	May	Jun.	Jly	Aug.	Sep.	Oct. Nov. Dec.

Gymnocarpium dryopteris: all Mid Perthshire.

Gymnocarpium dryopteris, habit of plant in life.

Plants flush early and rapidly in spring, with each of the three main divisions of the frond apparent at an early stage, *appearing like three small balls suspended on thin black wires*. In autumn the fronds die down completely with the onset of the earliest frosts.

Variation This is not a variable species, except in frond size, but even small fronds show the remarkably broad shape of the blade and the characteristic vivid green colour.

Possible confusion Unlikely to be confused with any other fern except *Gymnocarpium robertianum* (q.v.), in limestone districts, but the more tripartite frond, polished dark stipe, non-glandular, non-mealy surface, vivid green colour, and lack of smell when bruised, distinguish *G. dryopteris*. The contrast in colour is apparent even at a sporeling stage.

Technical confirmation Plants are tetraploids, with $n = 80$ chromosomes.

Field notes *Gymnocarpium dryopteris* is a widespread species in Europe and western Asia, present through most middle and fairly high-latitude districts and lower montane areas, and the same or other closely related species disjunctly in scattered stations at mainly mid to high latitudes discontinuously in northern, central and eastern Asia to Kamchatka, and in mountainous regions of North America. Throughout this range, it is a fern predominantly associated with north temperate deciduous or occasionally coniferous forests, and the overall range of this species and that of *Phegopteris connectilis* (q.v.) and its near allies show remarkable similarities.

Within Britain, Woodland Oak-fern occurs mostly in areas where the climate is cold in winter and cool and moist in summer, with a high humidity and frequent precipitation. Within such areas, it grows especially on deep moist, but open and light-textured mineral soil banks and moist, brown forest loams with surface humus layers, which are slightly flushed, and retain moisture in summer. It thrives best where such habitats occur in sheltered ravines along mountain streamsides, where lightly shaded by aspect and by light tree overgrowth. It seems tolerant of a moderate range of pH conditions, occurring in soils over mainly acidic rocks, but is perhaps more common on ground which has a reasonable base content.

Woodland Oak-fern grows very frequently in

Gymnocarpium dryopteris, young frond stages (after Orth, 1938).

the company of Beech Fern, whose ecology and geographic range in Britain and Ireland it closely parallels. The two species have a considerable ecological overlap (and sometimes occur as inter-grown colonies), but on the whole, usually form separate nearby patches, the *Gymnocarpium* on slightly drier slopes than the *Phegopteris.*

Woodland Oak-fern is a frequent plant of upland Oak–Birch (*Quercus–Betula*) woodland, descending to nearly sea-level in the west of Scotland. Like Beech Fern, it grows with a small number of other herbaceous species of vernal spring growth habit, especially with Wood-sorrel (*Oxalis acetosella*), Common Violet (*Viola riviniana*), and Wood Anemone (*Anemone nemorosa*). It also occurs with similar woodland species in moist rocky pockets amongst mossy screes on mountains, where it is often accompanied by Moschatel (*Adoxa mos-chatellina*), Opposite-leaved Golden-saxifrage (*Chrysosplenium oppositifolium*) and Alpine Lady's-mantle (*Alchemilla alpina*).

Gymnocarpium robertianum (Hoffm.) Newman

Limestone Oak-fern

Preliminary recognition A small to medium-sized fern, of similar general appearance to Woodland Oak-fern, but with larger, more robust fronds and narrower, less tripartite-looking blades, which are always of firmer, but of duller and more mealy texture.

Occurrence A local plant, chiefly confined to low or moderate altitudes in England and Wales, with a few widely scattered outlying stations elsewhere. It is virtually confined, as a wild plant, to fissures of limestone outcrops, with its main headquarters in the Cotswold hills, the Peak District and North Yorkshire Dales.

Identification Fronds arise individually, in rather small numbers, from slender tortuous, occasionally branching, creeping, dark-brown rhizomes. The blades are borne on stipes which are *longer, thicker, firmer and more robust* than those of *G. dryopteris* (q.v.), 10–25 cm in length, and are of dull green colour with a glandular surface. They at first bear fairly numerous pale scales, which are later shed. The stipes grow strongly erect, and the blade spreads horizontally, although the bend between stipe and rachis is less marked and less abrupt than in *G. dryopteris*.

The blade is of triangular-deltoid outline, 8–20 cm in length and *narrower than in G. dryopteris*. It is also less finely cut and has its ultimate segments rather more distantly spaced. The lower pair of pinnae are only slightly distant from the others, and are only about half as long as the rest of the blade. They are thus *less developed* than those of *G. dryopteris*, and the blade consequently has a *less tripartite appearance*.

The lamina is of a fairly firm texture, and is of dull greyish-green colour, with minute scattered glands covering the whole of the surface. This gives the fronds *a characteristically dull, dusty and mealy appearance*, with a particular bloom to the midrib and stipe regions where the glands are most concentrated. The edges of the segments bend down much less than in *G. dryopteris*, and *when a portion of frond is bruised lightly between finger and thumb, it gives off a distinctive, rather sweet, apple-like smell*, not found in *G. dryopteris*.

The fronds are mostly all similar, the fertile ones bearing numerous small rounded sori beneath, in a line near to the margin of each segment.

Plants flush early and rapidly in spring, but *lack the distinctive three-balled appearance* at the crozier stage of *G. dryopteris*. The glandular surface is, however, particularly conspicuous

Jan.	Feb.	Mar.	Apr.	May	Jun.	Jly	Aug.	Sep.	Oct.	Nov.	Dec.

Gymnocarpium robertianum, habit of plant in life.

during frond emergence, especially when viewed against the light. In autumn, the fronds die down with the onset of the first frosts.

Variation Limestone Oak-fern is not a very variable species.

Possible confusion Likely to be confused only with *G. dryopteris*, but besides the differences in sites of the two species, the narrower, scarcely tripartite frond, duller more greyish-green colour and mealy surface, paler stipe, and distinctive smell, always distinguish even juvenile specimens of this species.

Technical confirmation Plants are tetraploids, with $n = c.$ 80 chromosomes.

Field notes *Gymnocarpium robertianum* is reasonably widespread through mostly middle to lower-latitude districts and lower montane areas of Europe, and the same or other closely related species disjunctly in scattered stations at mainly mid latitudes discontinuously across central and eastern Asia to Kamchatka, and in mid to high latitudes and mountainous regions of North America. In Europe, plants spread more sparsely southward to northern Spain and northwards to northern Scandinavia. Overall, the plant is centred in somewhat drier (and sometimes more continental) climates than is Woodland Oak-fern (*G. dryopteris*) (q.v.).

Within Britain, Limestone Oak-fern thrives in more southern, sunnier and drier climates than does Woodland Oak-fern. It also seems tolerant of much greater exposure and direct sunlight, and its mealy surface may possibly help in reflecting incoming bright light.

It is confined almost exclusively to calcium-rich substrates, and its natural habitats all occur in broken ground on limestone rock of various types. Plants usually grow half-hidden in cracks and fissures of rocks with only their blades emerging at the surface.

Limestone Oak-fern is often abundant in limestone scree, in ledges and crevices of limestone cliffs and scars, and in the shallower fissures (grykes) of limestone pavements. It thrives on rock with sunny exposure. It may form extensive colonies amongst grasses and low bushes of Hawthorn (*Crataegus*), Juniper (*Juniperus*) and Yew (*Taxus*), and where there is a light woodland cover, such as that of Ash (*Fraxinus*), over rocky ground. But it is intolerant of dense woodland shade and stagnant air conditions.

Most of its stations are at fairly low altitude (*c.* 75–275 m, 250–900 ft), although scattered plants occasionally occur higher up.

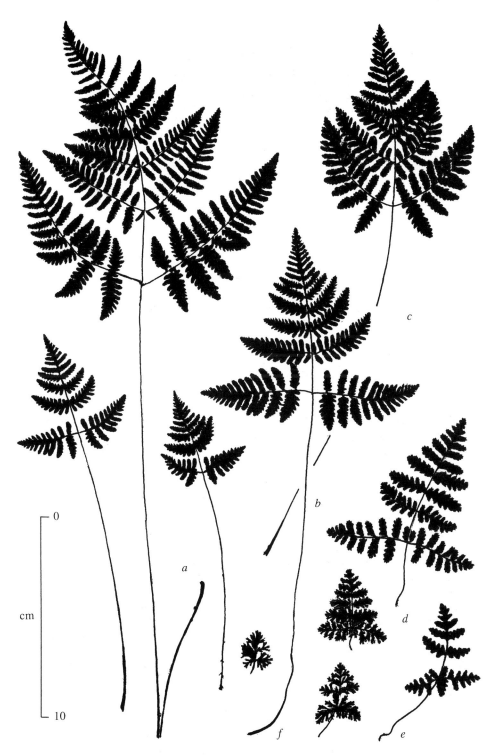

Gymnocarpium robertianum: a–c, North Yorkshire; *d–f*, Mid Perthshire.

In the limestone pavement habitats of north-west England, Limestone Oak-fern frequently occurs nearby Limestone Buckler-fern (*Dryopteris submontana*), but is usually confined to shallower fissures of around 30–40 cm depth, whilst Limestone Buckler-fern is restricted mainly to fissures that are deeper than this.

In its limestone habitats, Limestone Oak-fern associates with a great number of lime-loving flowering plants and other ferns. These include Herb Robert (*Geranium robertianum*), Wood-sorrel (*Oxalis acetosella*), Common Dog-violet (*Viola riviniana*), Wood Anemone (*Anemone nemorosa*), Lily-of-the-Valley (*Convallaria majalis*), Dog's Mercury (*Mercurialis perennis*), Red and Sheep's Fescue (*Festuca rubra* and *F. ovina*), Blue Seslaria (*Seslaria albicans*), Common Maidenhair, Green and Wall Rue Spleenworts (*Asplenium trichomanes* subsp. *quadrivalens, A. viride* and *A. ruta-muraria*), Hart's Tongue (*Phyllitis scolopendrium*) and Brittle Bladder-fern (*Cystopteris fragilis*), although the latter two ferns normally become more luxuriant in deeper grykes or in more shaded conditions.

Limestone Oak-fern seems to become slowly excluded if dense growths of Dog's Mercury develop, whilst the species also seems susceptible to grazing pressure (especially by sheep but possibly also by deer and rabbits). Away from its native habitats, it has been reported to have established in urban areas and on disused railway platforms or sidings in the east of England (Jermy *et al.*, 1978), presumably succeeding in niches formed in old mortar rubble.

Hymenophyllum tunbrigense (L.) Sm.

Tunbridge Filmy-fern

Preliminary recognition A very small fern, with numerous, membranous, translucent pale-green glistening fronds, mostly nearly as wide as long, each rather flattened end tipping over another in masses, like the half-ruffled plumage of a bird.

Occurrence A western, Atlantic species, locally frequent at low altitude in western Scotland, north-west Wales and south-west Ireland, and present only in scattered stations elsewhere, usually on vertical surfaces of damp, shaded, rocks or tree boles, in very humid, sheltered, moist conditions.

Identification Plants form mat-like masses with very numerous, thin, translucent overlapping fronds, about 2–5 cm in length, occasionally much more, and of rather broadly ovate-oblong outline, often nearly as wide as long. The blades have an *overall flattened appearance* (not with markedly downswept sides as in *H. wilsonii*, q.v.), and are bi- or tripinnatifidly divided into a rather small number of ultimate segments (*more numerous, however, than in* H. wilsonii). They are borne on slender, rather wiry stipes usually about 1/3 or more of the length of the lamina, and are frequently held at an angle to it. The fronds arise singly from slender, dark, thread-like, surface-creeping, wiry rhizomes, which spread and branch locally to build up much-interlaced, dense masses.

Each pinna is divided in an irregularly forking manner into usually more than five, flat, spreading, parallel-sided, linear-oblong segments, each with a spinosely serrulate margin. In the upper half of the frond, the segments are predominantly forward-pointing towards the

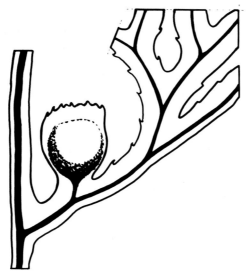

Hymenophyllum tunbrigense, detail of sorus – note toothed indusial margin visible with a hand lens.

frond apex, but *at least some of them on the lower part of the frond also emerge from the basal side of the pinnae.* The bases of all pinnae unite into a wing down either side of the rachis, which also extends a short distance down the stipe.

Fertile fronds are similar to the vegetative ones, and usually occur mixed amongst them. They bear sporangia within small, somewhat purse-like, two-valved indusia. These occur usually singly on the upper margin of each pinna, terminally on the segment nearest the rachis. The outer edges of the valves are flattened together in the plane of the frond, and are *coarsely and irregularly cut into a few, small, sharp teeth*, readily visible under a × 10 hand lens (cf. entire on *H. wilsonii*, q.v.). A useful mnemonic for remembering this is that *tunbrigense* is 'toothed', *wilsonii* is 'whole'. The diagram above shows detail of a sorus, its general shape and toothed margin.

Fronds are of a rather bright, translucent green colour, but are somewhat paler than those of *H. wilsonii*. The fronds are wintergreen and, once fully expanded, last at least a year, and probably most often for two or three years, before finally browning and withering. They seem, however, to lack the potential for indefinite apical grown of *H. wilsonii* (q.v.).

Variation Luxuriance of plants varies greatly with habitat. Degree of fertility seems also environmentally influenced, with plants in the deepest shade often vegetatively the most luxuriant, as well as the least fertile.

Possible confusion Plants are most likely to be confused with *H. wilsonii*, but the generally broader, shorter, flatter fronds lacking indefinite terminal growth, more numerous, more crowded segments, paler-green colour and always-toothed indusial margins distinguish *H. tunbrigense.*

Technical confirmation Plants have $n = 13$ chromosomes, with some as large as those of *H. wilsonii*, but others quite small (Manton, 1950: 273). The spores are about 40–48 μm diameter (cf. 62–74 μm in *H. wilsonii*), whilst the vegetative cells of the frond are said to contain relatively small numbers (about 30–40) of fairly large chloroplasts (cf. 60–80 in *H. wilsonii*) (Richards & Evans, 1972). The vegetative cells are usually also not much longer than broad (cf. several times longer in *H. wilsonii*) and, in the ultimate segments, the conspicuous veins stop short of the apices (cf. usually extending to the very tip in *H. wilsonii*) (H.McHaffie, personal communication). These latter features are of particular value in separating small and non-fertile specimens of the two species.

Field notes *Hymenophyllum tunbrigense* is chiefly of western European range, occurring from the Canary Islands, Madeira and Azores northward chiefly to Brittany, Ireland and Britain, with a small number of much more scattered inland sites in western mainland Europe (Jalas & Suominen, 1972). What are probably related members of the genus occur today in mostly southern hemisphere sites, including South America and Madagascar. The fragmentary nature of this range indicates the species to be relictual and probably of mid- to late-Tertiary origins, the most extensive northern hemisphere areas of *H. tunbrigense* today being those within Britain and Ireland.

Throughout its range in these islands, *H. tunbrigense* is a gregarious fern, usually occurring as numerous, nearby, discontinuous patches in suitable habitats. It is less widespread and generally less abundant than *H. wilsonii*, especially in the northern part of its range, although in south-west Ireland and south-west England, it can be locally the more common, and is the only filmy fern to occur in the Weald of Kent.

It typically grows amongst moist, shaded, rocky outcrops, in well-wooded, sheltered, stream valleys and in deep rocky stream gorges. It usually thrives on freely drained, steep or vertical rock surfaces, with the fronds mostly in a hanging position. Frequently it forms cushion-like carpets on and around large boulders along courses of steep, tumbling stream ravines, or festoons the boles of trees beside cascading water-courses. It is most frequently epiphytic in south-west Scotland and south-west Ireland and, when it grows epiphytically alongside *H. wilsonii*, *H. tunbrigense* usually occurs only on the lower parts of the trees, in moister, more humid conditions.

Hymenophyllum tunbrigense is generally more abundant at lower altitudes and seems to be confined to sites below about 90 m (300 ft) in Scotland, *c*. 380 m (1250 ft) in Wales, and *c*. 412 m (1350 ft) in Devon. Above these altitudes, it becomes replaced entirely by populations of *H. wilsonii*.

Plants grow on a wide range of rocks of circum-neutral to acidic type, but mainly on hard rocks such as sandstones, quartzites and granites, which provide a firm, non-crumbling surface. The water-holding capacity of sandstones during dry periods, compared with that of the harder metamorphic and volcanic rocks, may account for its frequency on this type of rock, especially in areas where the climate is otherwise marginal. Its overall geographic range follows closely climates with good cloud cover and frequent light rain. Where these factors are especially high, this fern can spread into more open situations, occasionally occurring in deep hollows of block scree.

Plants seem, however, particularly intolerant of desiccation, both at the sporophyte and gametophyte stage. They are confined mainly to niches where the air is constantly cool and moist, shaded from summer sun, and sheltered from frost and desiccating winter winds.

Although less common generally than *H. wilsonii* (q.v.), cushions of *H. tunbrigense* may be nearly as extensive and luxuriant in the sites in which they occur, and especially in Sessile Oak (*Quercus petraea*) woodlands at low altitude near to rocky western coasts, in sites in which Atlantic bryophytes also typically abound. They thrive on aspects generally untouched by the sun, and beneath an insulating tree canopy. In such sites, developed over centuries, and forming a part of a mosaic amongst hummocky mosses and liverwort patches, *H. tunbrigense* may form large but inconspicuous silver-green patches over lower tree trunks, fallen oak branches and especially over large rounded forest floor boulders on lower valley sides and in the splash zones of cool, permanently tumbling streams.

Under such conditions, cushions of *H. tunbrigense* expand at a rate of probably about 1–2 cm a year, and can eventually build up extensive masses over long periods. They spread initially over bare rock faces, but eventually accumulate a thin layer of raw humus beneath. They survive in such habitats largely because of their tolerance of conditions of continual very high humidity, very low light, and presumably low nutrient levels, which minimise vascular plant competition. There may, however,

Hymenophyllum tunbrigense: a–d, West Argyll; *e*, North Devon; *f*, Killarney.

be numerous hepatic and moss competitors, which include *Ctenidium molluscum, Leucobryum glaucum, Dicranum majus, Thuidium tamariscinum, Diplophyllum albicans, Mnium hornum, Lepidozia reptans, Isopterygium elegans, Plagiothecium undulatum, Hypnum cupressiforme, Bazzania trilobata, Mylia taylori, Ditrichum heteromallum* and *Isothecium myosuroides*, and rare Atlantic liverworts such as *Jamesoniella autumnalis, Cephaloziella pearsonii, Porella pinnata* and *Lepidozia pinnata*.

The stages of the life-cycle of species of *Hymenophyllum* are also rather bryophyte like. Their prothalli are perennial (and do not die after production of a sporophyte), and instead of growing as a single heart-shaped structure, as do most ferns, filmy fern gametophytes branch repeatedly, those of species of *Hymenophyllum* developing into small clusters of thin, thalloid tissue resembling mats of small liverworts, such as the liverwort genus *Riccardia* (Rumsey *et al.*, 1990). The perennial nature of their gametophytes is probably one factor in limiting their range closely to micro-habitats in which adequate moisture and humidity is available on a year-round basis.

In such permanently wet conditions, plants appear to require exceptionally free drainage, and seem highly intolerant of stagnant conditions. Their occurrence so frequently on steep or vertical surfaces contributes much to their free drainage. But cushions frequently also receive considerable throughflow of surface moisture from rocks above after heavy rainfall. It may well be that additional nutrients are beneficially brought to the plant in this way. At such times, the hanging fronds can frequently be seen to be running with surface water, which is dispelled by dripping freely from the frond and segment tips. As with *H. wilsonii*, the partly asymmetric structure of the frond, with, when hanging, most segments pointing steeply downwards, plus the winged rachis and the jagged, finely serrulate margins to the segments, probably help significantly in constantly shedding excess moisture in such permanently wet environments.

Hymenophyllum wilsonii
Hook

Wilson's Filmy-fern

(*H. unilaterale* Bory, *H. peltatum* Desvaux)

Preliminary recognition A small fern, with numerous, membranous, highly translucent, deep-green fronds, each longer than wide and swept down along either side of their midribs, arching over one another in descending masses, like the ridged tiles of a pantiled roof.

Occurrence A western, Atlantic species widespread and frequent in western Scotland, the Lake District, north Wales and south-west Ireland, and in scattered, mainly western, stations elsewhere. It usually grows in wet, shady, mossy mountain glens, generally near streams, in mossy cushions on vertical surfaces of damp rock or on boles of old trees.

Identification Plants form mat-like masses, with numerous thin, translucent, arching, overlapping fronds, each about 3–10 cm or more in length, of oblong outline at first, *eventually becoming longer and more lanceolate with age and often much longer than wide*. The blades have an overall ruffled appearance, as their *tips and edges are frequently held in a downswept manner*, whilst the indusia (see below) frequently *stand out upwards out of the plane of the frond*. The fronds are usually bipinnatifidly divided, with their ultimate segments less numerous than those of *H. tunbrigense*, with the blades held on fairly short, wiry stipes. The fronds arise singly

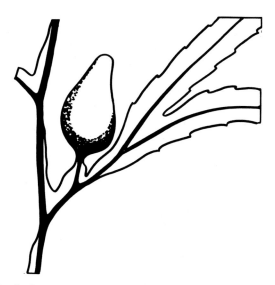

Hymenophyllum wilsonii, detail of sorus – note untoothed and attenuate indusial margin visible with a hand lens.

from dark, thread-like, surface-creeping, wiry rhizomes, which spread and branch locally to build up much-interlaced dense masses, or run far through mossy cushions.

The pinnae are divided in an *irregularly but predominantly unilaterally forking manner* into *a rather small number* (usually about 3–5), of parallel-sided, linear-oblong segments, each with spinosely serrulate margins. Throughout the frond, most segments arise from the upper side of most pinnae, towards the apex of the frond. The bases of all pinnae unite into a wing down either side of the rachis, which extends a short distance down the stipe.

Fertile fronds are similar to the vegetative ones, and usually occur mixed amongst them, often in large numbers. They bear sporangia within small, two-valved indusia, which are usually larger than those of *H. tunbrigense*, and are *more elongated in a rather flask-like manner.* Their flattened outer margins are *entire, conspicuously lacking the toothed serrations* of the latter species. Like *H. tunbrigense*, however, the indusia are solitary and borne on the upper margin of each pinna, terminally on the segment nearest to the rachis, but in *H. wilsonii*, are usually slightly stalked, and often stand out angled upwards above the plane of the rest of the frond. The diagram above shows detail of a sorus, its shape and entire margin. Fronds are of a *rather dark olive-green colour*, darker than those of *H. tunbrigense*, but in the field seem to have a much more translucent quality. This is probably because of their much more open habit of growth, arching more distinctly away from the rock surface and thus allowing more light to penetrate in behind them. The fronds also have a rather more rigid, crisper quality than those of *H. tunbrigense.* The fronds of *H. wilsonii* are wintergreen, and last, once fully expanded, at least a year. But in many populations, *individual fronds acquire the habit of indefinite growth*, unique amongst British or Irish ferns, *allowing them to continue growing by further annual increments, each as much as the first year's growth*, over a subsequent 3 to 5 years or more.* Such growth flushes eventually result in fronds with regularly alternating sequences of sterile and fertile growth, the fertile marking the end of each season's increment. On long fronds,

---

*I am grateful to J. W. Dyce, who first pointed out this phenomenon to me in the field.

Hymenophyllum wilsonii: a–e, Isle of Arran; *f–g*, West Argyll

flushes of several most recent seasons remain green, whilst the lamina of earlier seasons at the base of the frond is becoming brown and withering.

Variation Size of plants varies greatly with habitat.

Possible confusion Plants are most likely to be confused with *H. tunbrigense* (q.v.), but the generally narrower, longer fronds with less crowded, more unilaterally arranged segments, indefinite terminal growth habit, darker green colour and untoothed indusial margins, distinguish *H. wilsonii.*

Plants might also be confused with the large, leafy moss, *Mnium undulatum*, which often grows in habitats similar to those of the filmy ferns, and has similar, very thin and translucent leaves, but the divided pinnae as well as presence of indusia distinguish the fern.

Technical confirmation Plants have $n = 18$ chromosomes, which are all of unusually large size (Manton, 1950: 272). The spores are about $62-74\,\mu$m diameter (cf. $40-48\,\mu$m in *H. tunbrigense*), whilst the vegetative cells of the frond are said to contain large numbers (about 60–80) of very small chloroplasts (cf. 30–40 in *H. tunbrigense* (Richards & Evans, 1972)). The vegetative cells are usually also markedly longer than broad (cf. often not much longer than broad in *H. tunbrigense*) and in the ultimate segments, the conspicuous veins usually reach to the very tips (cf. stopping short of the apices in *H. tunbrigense*) (H. McHaffie, personal communication). These latter features are of particular value in separating small and non-fertile specimens of the two species.

Field notes *Hymenophyllum wilsonii* is a plant chiefly of extreme western European range, occurring from Madeira and the Azores in the North Atlantic northward to Brittany, Ireland, Britain, extreme south-west Norway and the fringes of Iceland (Jalas & Suominen, 1972). What are possibly further related forms occur highly discontinuously only in several other extremely oceanic regions of the world, including Japan, New Zealand, Tasmania, Kerguelen and Tristan da Cunha. The highly fragmentary nature of this range indicates the members of this alliance to be considerably relictual.

Within its North Atlantic area, *H. wilsonii* is geographically at its most extensive in Ireland and western Britain. It occurs in a large but highly discontinuous array of stations throughout much of Ireland (especially in western districts), Cornwall and Devon (rare in the former), west Wales, Cumbria and western Scotland (very widely) northward to the Orkneys and Shetlands. Throughout this range it is a gregarious fern, usually forming many nearby patches within deep ravines and river gorges, on vertical faces of rock by waterfalls and along cascading streamsides, on tree trunks in valley woodlands and in sheltered ravines, or amongst scree on mountainside. It is a more widely distributed species than *H. tunbrigense*, and much

the more common of the two in northerly areas, and is particularly widely spread in western Scotland. It also ascends to higher altitude, occurring from sea-level to about 1000 m (3300 ft) in south-west Ireland, but is most extensive between about 500 and 1000 ft (*c.* 150–300 m). Its most abundant habitats are in steep, sheltered and well-shaded stream ravines near western Atlantic coasts, where humidity is normally constantly high, where winter temperatures are mild, and where there is a long season of growth. In such sites, and especially on more acidic rocks, large patches of this delicate fern may occur, intermingling with very numerous bryophytes and locally with *H. tunbrigense*, beneath steep rocky overhangs.

Its ecology and range within Britain and Ireland share many features in common with *H. tunbrigense* (q.v.). *Hymenophyllum wilsonii* does, however, seem rather more drought tolerant, occurring more frequently epiphytically than *H. tunbrigense*, with specimens emerging into less deeply shaded and slightly more exposed conditions. But in many places in western Scotland, north Wales and western Ireland in particular, the two species grow alongside one another, with cushions of *H. wilsonii* also penetrating into very deep shade. Because of its greater ecological amplitude, *H. wilsonii* associates with an even wider range of bryophytes (including

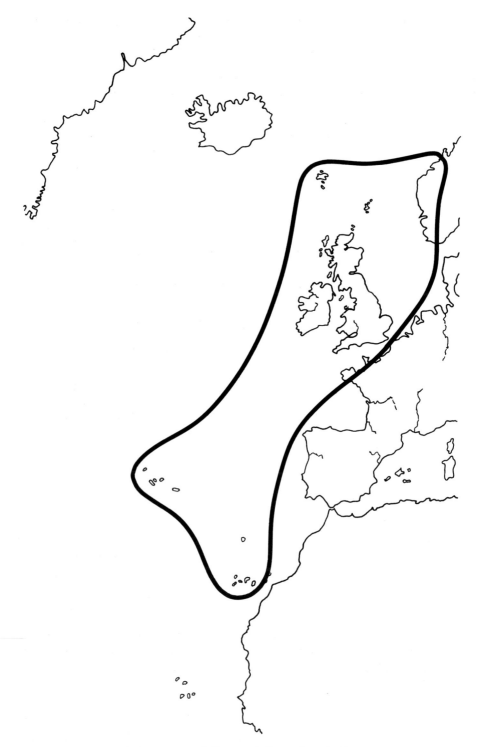

Atlantic range of *Hymenophyllum wilsonii* in Europe.

Rhacomitrium lanuginosum at higher altitudes) and lichens.

Individual fronds seem to tolerate considerable degrees of drying from time to time, returning from a fairly desiccated condition in much the same manner as do mosses. In epiphytic habitats, plants usually revive after droughts, but on smoother, vertical rock surfaces, prolonged droughts can cause whole cushions to shrink, peel and fall away.

Most plants grow on steep, vertical surfaces, in very freely draining situations. The long fronds of *H. wilsonii* arch much more away from the substrate than do those of *H. tunbrigense* (which hang more vertically down), but like the latter species, the fronds of *H. wilsonii* can often be seen to drip freely with water from the tips during and after frequent rain. Like those of *H. tunbrigense*, the fronds of *H. wilsonii* seem particularly appropriately constructed to remove and shed such excesses of surface water from the rhizomes, in which the winged upper stipe, arching pendulous-tipped frond, downswept pinnae, unilaterally divided downpointing segments on inverted fronds and serrulately jagged lamina edges probably all play a significant role.

Its method of indeterminate frond growth seems also of field significance. There is a new flush of growth in spring, but further growth flushes seem to occur under favourable weather conditions at any time of year. There is considerable resulting range in ripening times of sori, which, coupled with their graduated development on the frond, seems to produce an unusually long period of spore release. This rather opportunistic growth habit seems particularly appropriately adapted to the mild, oceanic west coast climates of Britain and Ireland, where the growing season is anyway long and spells of particularly mild temperatures can occur at any time of year.

Ophioglossum azoricum C. Presl

Small Adder's-tongue

(*O. vulgatum* L. subsp. *ambiguum* (Ciss, & Germ.) E. F. Warb.)

(*O. vulgatum* L. subsp. *polyphyllum* (A. Br.) E. F. Warb.)

Preliminary recognition A small, plantain-like fern, of gregarious habit, looking like a smaller version of *O. vulgatum*, but with narrower leafy segments, fewer sporangia, and shoots frequently arising in pairs and threes from a single stock.

Occurrence An Atlantic species of unusual distribution, occurring in dune slacks and short turf in exposed places always near the sea, in very widely scattered and discontinuous stations, almost always on islands or peninsulas.

Identification Plants are mostly 3–8 cm high, occasionally slightly more, with a leafy segment about 3.0–3.5 cm in length. Shoots frequently arise in twos and threes from a single underground rootstock, 3–4 cm below the ground surface, and plants bearing leaves but no fertile spike occur frequently, and may be the dominant form. Where fertile spikes are present,

Jan.	Feb.	Mar.	Apr.	May	Jun.	Jly	Aug.	Sep.	Oct.	Nov.	Dec.

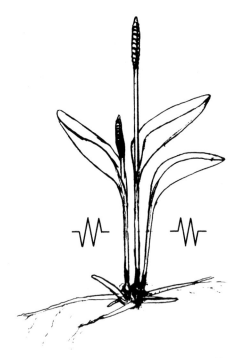

Ophioglossum azoricum, habit of plant in life.

these are small, about 8–20 mm long, and bear *about 6–14 sporangia* (commonly 8–10, but very exceptionally as many as 20) on either side. The boat-shaped, leafy segments are *usually narrower in outline than those of* O. vulgatum (q.v.), widest at or above the middle, *and held at a more spreading, less steeply ascending angle.* The leafy portion arises from much nearer the base of the plant than does that of *O. vulgatum, and tapers much more gradually towards its base.* In herbarium material, the base of the leaf can become folded and so slender on drying, that the leaf appears quite long-stalked.

Shoots begin growth about early May, become mature by about late June, and wither by late summer. Plants gradually build up colonies by root-budding, new plants from which can be leafy for about 3–4 years before bearing their first fertile spike.

Variation Specimens vary slightly in form and habit between stations. In any one colony, they also vary in size and luxuriance according to season and state of development.

Possible confusion The late spring and early summer growing-season clearly distinguishes *O. azoricum* from the winter-growing *O. lusitanicum* (q.v.). Plants of *O. azoricum* are most likely to be confused with small forms of *O. vulgatum*, which can, and frequently do, spread in stunted form to similar coastal habitats, the two species often occurring nearby. The narrower outlines, more-spreading habit, lower-set more tapering-based leaf, fewer sporangia and paired or triple-grouped plants are the best guides to field separation of *O. azoricum*.

Technical confirmation Quite difficult. No chromosome counts of British or Irish material have been made, but if conspecificity with plants from Iceland can be assumed, are highly polyploids with $n = c.$ 360, $2n = c.$ 720 chromosomes (Löve & Kapoor, 1967). British material has

Ophioglossum azoricum: *all* Orkney Islands (from nineteenth century herbarium material).

Ophioglossum azoricum, Atlantic-Mediterranean section of range.

somewhat larger spores (about 38–47 μm) compared with native *O. vulgatum* (*c.* 26–41 μm) or *O. lusitanicum* (*c.* 23–32 μm). The spore wall is only moderately tuberculate, and thus intermediate between the nearly smooth-walled spore of *O. lusitanicum* and the highly tuberculate spore of *O. vulgatum*. The meshes of the veins in the leafy portion of the plant usually include only a small number of free vein endings, which are less conspicuous than those of *O. vulgatum*.

The morphological intermediacy and possibly very high chromosome number of *O. azoricum* has led to the suggestion that it might be an old allopolyploid derived from crossing *O. lusitanicum* and *O. vulgatum*.

Field notes *Ophioglossum azoricum* is a European endemic species, with a highly disjunct and widely scattered distribution chiefly from Iceland through Britain and Ireland, France and the Iberian Peninsula to the Azores and Canary Islands, with a small number of other scattered stations in the west Mediterranean basin and very locally in central Europe (Jalas & Suominen, 1972).

Within Britain and Ireland, plants occur in low-lying, damp, sandy-peaty habitats, always close to the sea, sometimes in dense, short turfy vegetation on forelands or fixed dune pastures, sometimes in fairly bare damp ground of moist sand-dune slacks. They are always in very exposed situations. Where *Ophioglossum vulgatum* also occurs in nearby sites, those of *O. azoricum* seem to be in habitats which are less extensively flooded and ones which perhaps warm up more quickly and readily in spring.

The remote British and Irish sites for *O. azoricum* are scattered from the Channel Islands, Scillies and Lundy Island, to the North Wales, Cumberland, Northumberland (Lindisfarne) and Caithness (Dunnet Head) coasts, Orkneys, Shetlands, Fair Isle and St Kilda, and in Ireland in equally isolated stations from Great Blasket Island and the Dingle Peninsula of Kerry to the Connemara, Mayo and west Donegal coasts. The reasons for this very discontinuous and distinctly 'insular-peninsular' type of Atlantic distribution, and the way in which it has been achieved, are totally unknown, but might well imply the operation of some dispersal agent other than mere chance. The possibility of the involvement of seabirds of such habitats, known to migrate and eat herbage that might include ripe *Ophioglossum* spikes, might well be investigated.

Associated plant species probably vary from site to site, and more detailed study of the plant in its scattered stations is much needed. Bluebell (*Hyacinthoides non-scriptus*) sometimes occurs with it in turfy vegetation, whilst in damp sand-dune slacks, scattered plants of Lesser-Spearwort (*Ranunculus flammula*) and Marsh Pennywort (*Hydrocotyle vulgaris*) sometimes occur as the main vegetational dominants.

Plants seem more strongly calcicolous than those of *O. vulgatum*, and are probably less winter hardy. In addition to relative infrequency of severe prolonged frosts in its habitats and the benefit of full sun, the exposure of its maritime stations probably also helps to maintain surrounding herbage in a state of only slow summer growth. Moderate grazing pressure by rabbits as well as occasionally by sheep and cattle, probably further helps in maintaining the turfy stature of its communities, although with heavier grazing pressure, the *Ophioglossum* is itself readily eaten.

Colonies are vegetatively moderately vigorous and, once established, can build up clonal patches of considerable size by vegetative growth, and these are probably long lived.

Ophioglossum lusitanicum L.

Least Adder's-tongue

Preliminary recognition The smallest native fern, bearing minute, narrow, ovate leaves with occasional fertile spikes attached, visible above ground only in later winter and spring, and detectable only by close examination of turf on hands and knees in its few known localities.

Occurrence Very local and rare. Known only in a few cliff-top and rocky downland sites, in short, turfy coastal vegetation, in the Scilly Islands (St Agnes) and Channel Islands (south and west Guernsey), mostly in rather precipitous places. Although known from some of the Guernsey sites since the mid-nineteenth century, it has only more recently been found in others, and was not known in the Scillies until 1950. It would not be surprising if such an insignificant, winter-growing species, had been overlooked elsewhere.

Identification Plants have single or usually paired entire leaves, of which the emergent narrowly ovate parts visible in the turf, are scarcely 10–15 mm long, and *usually lie closely pressed to the ground, spreading oppositely apart, when in pairs.*

The leaves are of bright yellow-green colour, becoming gradually paler with age, and are satin-sheened and thinly fleshy, without any conspicuous veins. Occasional fronds also bear small, erect, fertile spikes, usually not more than 3–6 mm high and 1 mm broad, ending in a short, bare, flexible point, and bearing two rows of *about 3–5 sporangia* on either side.

The emergent portions of the leaves arise from minute, white, globose corms, 2–4 mm in diameter, 2–4 cm below the surface of the ground, from which arise long, slender, ascending white leaf bases, terminating in the green, boat-shaped portions of the leaves visible at the soil surface. From each corm, 2–4 rather thick, fleshy, white roots also spread horizontally through the ground, and give rise to new individuals by root-budding.

Possible confusion The only other fern with which it could be confused is *Ophioglossum azoricum*, but the very early growing season of *O. lusitanicum* (see 'Field notes'), as well as its smaller size and fewer sporangia, clearly distinguishes it. The leaves of *O. lusitanicum* are very similar in colour and texture to the emerging tips of leaves of Bluebell (*Hyacinthoides non-scriptus*), which accompanies it in turf at most of its sites. Paired leaves of the *Ophioglossum* lying flat on the turf superficially closely resemble the seedling leaves of a dicotyledonous plant.

Jan.	Feb.	Mar.	Apr.	May	Jun.	Jly	Aug.	Sep.	Oct.	Nov.	Dec.

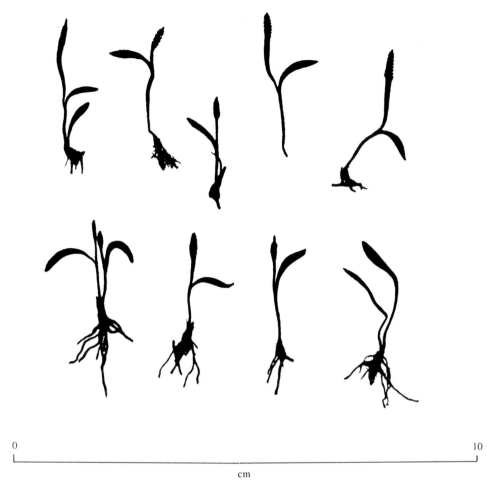

0   10
└───┘
cm

Ophioglossum lusitanicum: all Guernsey, Channel Islands. (Silhouettes prepared from nineteenth century herbarium specimens.)

Variation Not a variable species. Occasional plants in shelter of boulders or bushes were recorded last century to grow to a larger size and have a more upright habit than is usually seen today.

Technical confirmation Plants have approximately $n = 125$–130, $2n = 250$–260 chromosomes. The spores are smooth walled, and are smaller (about 23–32 μm) than those of either *O. azoricum* or *O. vulgatum*. The epidermal cells of the leaf have straight, not sinuous, walls, and the leaf venation, although very inconspicuous, lacks the free included veinlets of *O. vulgatum* (q.v.).

Field notes *Ophioglossum lusitanicum* is a mainly European species in range, occurring from extreme south-western Britain and the Channel Islands through Brittany to the Azores, Canary Islands, the Iberian Peninsula and the west Mediterranean basin, where it has its principal centre of distribution.

Least Adder's-tongue is thus a plant principally of Mediterranean climates, which occurs in Britain only in areas of mild climate and predominantly winter rainfall. Its localities are all unshaded and exposed, but are on south and south-westward aspects, which are sheltered by rocks from the

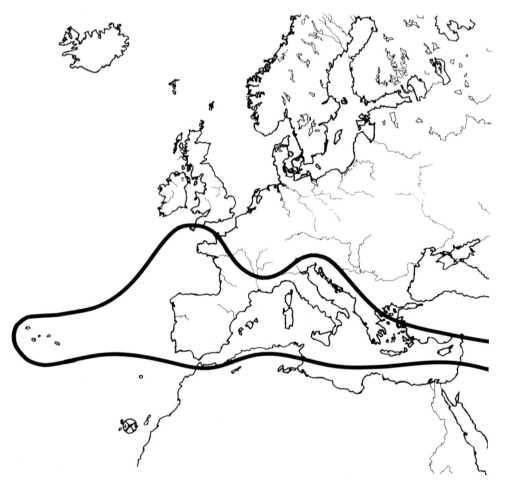

Ophioglossum lusitanicum, habit of plant in life.

Atlantic range of *Ophioglossum lusitanicum* in Europe with single outlying station in Tenerife.

Ophioglossum lusitanicum: St Agnes, Scilly Isles, March. Note 1p coin (20 mm) for scale (*below*).

north and east and warm rapidly in spring. Plants are perennials which grow after autumn rains, appear above ground in early November, mature by mid-January, and die down by late April. They thus complete the above-ground part of their life-cycle in the very short, winter-grazed turf, before becoming overtopped by taller, summer vegetation.

Least Adder's-tongue grows in patches of moist, mossy, sometimes slightly eroding, very short turf, mainly on steeply sloping shelves near the tops of sea cliffs and on lower rocky promontories. It occurs in small patches, mostly much less than a square metre in extent, but in which there may be several score individuals. It usually grows where there is an accumulation of about 5–8 cm depth of dark, humus-rich peaty, sandy soil over underlying sloping rock (granitic in the Channel Islands), and this is irrigated by gradual movement through it of seeping groundwater. Such conditions maintain patches of moist, spongy soil, in which the water table is a little below the surface, and are usually marked by an increase in the proportion of the moss components in the turf.

At one Guernsey locality, Land Quillwort, *Isoetes histrix* (q.v.), grows only a few metres away from the *Ophioglossum*. The habitats of the *Ophioglossum* seem to differ from those of the *Isoetes* in being slightly deeper soils that are less inundated by water in winter, and probably dry out rather less in summer. The Adder's-tongue habitats are also in less-eroded, denser, more vegetationally closed and floristically richer turf, than the Quillwort. In addition to the increase in moss (mostly *Polytrichum*), other plants associated with the Adder's-tongue are mainly species locally typical of short, rocky grasslands or turf vegetation, and in the Channel Island localities, include abundant Buck's-horn and Ribwort Plantain (*Plantago coronopus* and *P. lanceolata*), Autumnal Squill (*Scilla autum-*

nalis), 'Sand Crocus' (*Romulea columnae*), Lady's Bedstraw (*Galium verum*), Ox-eye Daisy (*Chrysanthemum leucanthemum*), Scarlet Pimpernel (*Anagallis arvensis*), Parsley Piert (*Aphanes arvensis*), scattered Bluebell (*Hyacinthoides non-scriptus*), and *Vulpia* grass, with English Stonecrop (*Sedum anglicum*), on surrounding shallower, drier soil pockets nearby. Most of the sites are surrounded by abundant Gorse (*Ulex*). In its Scilly Island site, Lousley (1971) has recorded similar associates: abundant *Sedum anglicum, Plantago coronopus*, and *Anagallis arvensis*, with stunted Heather (*Calluna vulgaris*) and seedling Gorse.

Along the southern cliffs of Guernsey today, spread of Gorse, especially over all the deeper soils, has probably been one of the main reasons for great diminution in abundance of the Adder's-tongue. McClintock (1975) notes that old prints of the area show how the cliffs were formerly much more open, the Gorse being constantly cut for fuel, as well as grazed by sheep and goats. At this time, records of the Adder's-tongue suggest it to have been much more plentiful and widespread and growing to a larger size. With changes in agricultural practices, the Gorse has greatly increased, reducing the areas of short turf. Today, it is mainly rabbits that graze the remaining turf patches short, but appear also to consume the leaves of the *Ophioglossum* avidly, only leaves with the most prostrate habit apparently surviving. Little is known about the reproduction of the *Ophioglossum*, but its occurrence in dense turf in patches suggests that these are probably each mostly clones, vegetatively propagated by root budding, and that it is this vegetative mechanism that has largely been responsible for the persistence of the surviving patches. It would be interesting to know if the creation of newly opened habitats by the removal of Gorse, would again allow this rare species to spread.

Ophioglossum vulgatum L.

Adder's-tongue

Preliminary recognition A small to moderate-sized plant, of rather plantain-like appearance, with a simple, leafy lobe attached in a spathe-like manner to one side of a tall, fleshy shaft, bearing a flattened fertile spike. Plants form spreading colonies, but can be difficult to find when growing amongst tall herbage.

Occurrence A low-altitude species, widespread throughout both Britain and Ireland especially on old, damp meadowland, but in a wide range of other moist lowland habitats. It is most frequent in calcareous soils of central and south-eastern England, thinning out both westwards and northwards, becoming infrequent in high-rainfall districts and where acidic soils predominate.

Identification Plants usually arise singly, more rarely in pairs from the same underground, tuberous rootstock. They are 8–30 cm or more tall, and although a few plants without a spike occur in most colonies, the great majority of specimens bear both a simple leafy segment and a fertile spike.

The leafy segment arises at an angle of about 45° or steeper, is usually about 5–15 cm in length (sometimes more), and is *of oblong-elliptic, or often rather spade-like outline*, often not much more than about twice as long as broad. Typically it has *its widest point in the basal half*. Above this, it tapers gently to a rounded or pointed apex, and below *narrows bluntly and rapidly* into the shaft-ensheathing base. The whole segment is at first a bright, sappy pea-green colour, but turns gradually paler as the season progresses. It has a fleshy, rather stiffly flexible texture, and is often slightly concave, with the outer margin tipped downwards. Its surface is totally devoid of hairs or scales, is minutely convolute and puckered, has a dull, waxy sheen, and has a distinctly non-slippery feel, like that of a rubber sole on a polished floor.

The fertile portion, which bears the leafy segment about half-way up, is a tall, erect, fleshy, slightly translucent green shaft, which ends in an erect, slightly tapering, linearly flattened spike, about 2.0–5.0 cm long, with two vertical, serially arranged rows, of rounded, thick-walled, partially immersed, bead-like sporangia. There are usually *about 10–40 sporangia in each row*, the number generally increasing proportionately to the size and luxuriance of the whole plant. A slender, green point (apiculum), 2–4 mm in length, caps the top of each spike. The length of the shaft bearing the spike increases as the spike matures. The spike arises about early May at the base of the leafy portion, develops and overtops it by mid-June, and is eventually held high above it by August, before plants die down again in late summer. The

Jan. Feb. Mar.	Apr.	May	Jun.	Jly	Aug.	Sep.	Oct. Nov. Dec.

0   10

cm

Ophioglossum vulgatum: *a*, *b*, Midlothian; *c*, Isle of Arran; *d*, Co. Cork; *e*, Northumberland; *f*, West Yorkshire.

sporangia are at first green, but become yellowish-green and then brownish with maturity, before each splits horizontally to release the spores and then remains gaping. The spores are fine, very numerous, yellow-coloured, and all of one size.

The aerial parts arise annually from the perennial underground portion. As with all other species of *Ophioglossum* and *Botrychium* (q.v.), plants lack the circinate vernation (uncoiling crozier) so characteristic of other ferns.

The rootstocks are swollen, vertical and tuberous, about 4–6 mm wide and 5–6 cm or more long (the length increasing with age of the plant), and occur at depths of about 5–15 cm below the surface of the soil. Each bears numerous, white fleshy roots, about 2–3 mm thick, most of which spread horizontally for appreciable distances. Some eventually give rise to new plants by root-budding. By such vegetative means, plants may eventually build up patches in which many score individuals may occur in an area of a few square metres.

Variation Plants show very considerable variation in size with luxuriance of the habitat, whilst there is considerable variation in details of leaf shape, especially between populations. Densities of populations seem considerably subject to annual climatic fluctuations.

Possible confusion The succulent texture and unusual appearance of large plants makes them unlikely to be confused with anything else. Smaller plants might be confused with *Ophioglossum azoricum* in rare coastal sites where the two are present. The broader-based, more ascending leaf and larger number of sporangia in the spike (10–40 in *O. vulgatum*, cf. usually about 8–14 in *O. azoricum*), plus the technical characters below and under *O. azoricum* (q.v.), should help to distinguish between the two.

Technical confirmation Plants are highly polyploid, with $n = c.$ 240–260 (perhaps 256), $2n = 480$–520 chromosomes, which are smaller and about twice as numerous as those of *O. lusitanicum* (Manton, 1950) but far fewer than those of *O. azoricum*. The mature spores have small, but distinct, blunt tubercles (cf. smooth-surfaced in *O. lusitanicum*, moderately tuberculate in *O. azoricum*), and are intermediate in size, about 26–41 μm, between those of *O. lusitanicum* (*c.* 23–32 μm) and *O. azoricum* (*c.* 38–47 μm). Epidermal cells of the leaf have sinuous, rather than smooth walls. The venation of the leaf, inconspicuous in fresh specimens but conspicuous when dried and held against the light, shows a network of veins forming elongate meshes at its base and centre. These meshes often show smaller, secondary meshes formed by finer anastomosing veins, from which numerous free veinlets end within the network (cf. no free veins in *O. lusitanicum*, only a few in *O. azoricum*).

Field notes *Ophioglossum vulgatum* is a northern boreal species, with a highly disjunct and widely scattered natural range mainly in Europe but also across northern Asia to Japan and in North America.

Within Britain and Ireland, it is the hardiest and most northerly of the species of *Ophioglossum*, and also the most widespread, although it is always a predominantly lowland plant. It occurs in a very wide variety of habitats, including sand-dune slacks, coastal machair, grassy links, damp ditches, coastal flushes, moist hollows in grassy pastures or heathy places, marl pits, and occasionally in moist bottoms of limestone grykes and in old damp quarry workings over most clays or limestones. But it is a species especially characteristic of old moist water-meadows pastures of the rural lowlands of central and southern England, where luxuriant stands sometimes still occur. In such sites, it thrives in moist grassy areas amongst luxuriant rich turfy pasture vegetation, especially in slight hollows which fill with rain. Here it grows on soft clay or loamy soils through which its roots can penetrate easily, where there is usually lime in the soil and presumably conditions conducive to the success of a rich mycorrhizal fungal soil flora.

Ophioglossum vulgatum: Mainland, Orkney Islands.

Prothalli of *O. vulgatum* are subterranean, lack chlorophyll, and have an essential association with endotrophic (internal) mycorrhizal fungi, which provide the slowly growing prothallus with nutrients which are said to be saprophytically obtained from decaying matter in the soil. It seems likely that some sort of association with other plants providing decaying organic matter, such as grasses, may be of importance. Manton (1950: 264) recorded prothalli in Cheshire at depths of 5 in. (about 12 cm) beneath the surface of grass, looking like 'little contorted worm-like objects' without vascular tissue. It seems likely, too, that an endotrophic mycorrhizal association within the fleshy roots of the sporophyte plant is also of importance in its success.

In earlier centuries, Adder's-tongue was clearly better known than today, and occurred in quantity. Reports suggest that it could be found covering acres of meadowland, and when it did so, gained the reputation of being highly injurious to the crop of grass. Plants were regarded as having great power for evil, hence perhaps the derivation of its name. It was prized as a remedial by old herbalists, as a reputed antidote to snake-bite. Today it is very much less common than formerly, having declined appreciably over the last 300 years, and even in the last 20 years, seems to have disappeared totally from many of its more peripheral (especially southern Scottish and Irish) stations. Although part of this may be due to subtle climatic changes, much of it is probably due to changes brought about by human beings. Plants seem avidly eaten by rabbits and sheep, and probably also by cattle, having a relatively palatable sweet taste.* The Adder's-tongue is also very sensitive to drying-out and to habitat disturbance of all kinds. Drainage of damper parts of old meadows and ploughing up of old pasture lands, which may have remained unchanged for centuries, have probably been some of the main factors in the plant's widespread reduction in recent years. But other factors may also be involved, and the plant's susceptibility to agricultural herbicides as well as to air pollution, might also stand investigation.

---

* However, a specimen experimentally eaten by the author caused swelling of the tongue, and this is not recommended!

Oreopteris limbosperma
(All.) Holub

Sweet Mountain-fern

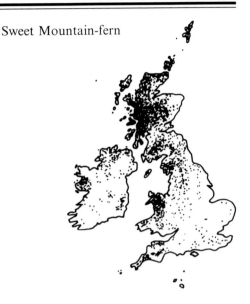

(*Thelypteris oreopteris* (Ehrh.) Slosson)

Preliminary recognition A moderate to large-sized fern, with ascending, pale bright green, pinnate-pinnatifid fronds, with an outline tapering steadily almost to the very base, arising in distinct shuttle-cock-like clusters.

Occurrence Widely spread and often frequent in most wetter, western upland parts of Wales, northern England and Scotland, in damp acidic ground, especially by mountain streams, but scarce and fairly local in most other parts of England and throughout Ireland.

Identification Plants have fronds mostly up to 90–120 cm high and to 20 cm broad, of lanceolate outline, broadest at or above the middle of the frond. They taper upwards to an acute apex and downwards very gradually *to almost the extreme frond base*. The stipe is consequently *very short* – usually much less than 1/4 of the frond length, and is slender and succulent. Fronds all stand at a *steeply ascending angle, arching gently backwards near the tip*, and form fairly distinct, shuttlecock-like groups, arising from the crowns of short, stocky, ascending rhizomes. Old plants eventually build up large clumps by rhizome branching, with often many nearby crowns.

The fronds are pinnately divided into numerous, parallel, spreading, pinnae. Throughout most of the frond, these are narrowly linearly, gradually tapering and pinnatifidly divided into numerous oblong, obtusely round-tipped segments. But the pinnae gradually reduce in length towards the base of the frond, progressively losing their pinnatifid character *to become eventually mere triangular auricles* with undulate margins. The segments of most of the pinnae are *usually flat, or turn slightly downwards around the margins*, especially in plants in exposed conditions. The rachis and stipe are rather sparsely covered with thin, concolorous scales, *which are silvery-white in bud and while the fronds are emerging*, but mature to a duller pale-buff colour.

The majority of large fronds bear sori, which are small and numerous, and lack obvious indusia, and are always more numerous on the upper parts of the fronds. They are each rounded and *form rows of spots around the margin of each segment*. These rows are green when young but become a very conspicuous *deep glossy black* when mature, before releasing their sooty black spores. The diagram opposite shows the characteristic arangement of sori, near to the margin of each segment.

Jan.	Feb.	Mar.	Apr.	May	Jun.	Jly	Aug.	Sep.	Oct.	Nov.	Dec.

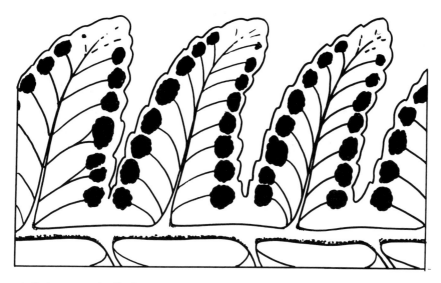

Oreopteris limbosperma, detail of undersurface of pinnules, showing nearly marginally set naked soral positions, turning black at spore maturity.

The fronds are of rather soft texture to touch, especially in shade, although they eventually become more leathery in very exposed situations. They are bright yellow green coloured when emerging and bright pale-green when mature. The lamina is sprinkled beneath with numerous, minute, unicellular glands, which *give off a scent of lemon when fronds are lightly brushed between the fingers*, especially when fronds are young and recently expanded.

When flushing in spring, fronds have a particularly characteristic appearance, distinctive from all other large ferns. The pinnae begin to *extend outwards sideways whilst still in the coiled crozier*. They are quite angular at this stage, and *stand out from the expanding croziers like jagged slender wings*, whilst on the croziers the white scales also stand out prominently. At the beginning of the season, even when fronds are still in bud, the silvery-white scales are also particularly conspicuous, and differ from those of all similar-sized *Dryopteris* and *Athyrium*. At the end of the season, fronds die rapidly with the first frosts, but in sheltered localities dead fronds may remain standing well into the winter months, becoming bleached to a pale yellow brown and eventually becoming parchment like, both in colour and texture.

Variation *Oreopteris limbosperma* is not a very variable species.

Possible confusion Plants are most likely to be confused with *Dryopteris filix-mas* (q.v.), which is of comparable size and has a similar frond form, but the more slender, less scaly stipe, with pinnae tapering to near the extreme frond base, more marginally set, non-kidney-shaped sori, and lemon smell to the frond distinguish *Oreopteris*. Fronds also resemble those of *Thelypteris palustris* (q.v.) in texture, but are easily distinguished by their gradually shortening lower pinnae, very short stipe and ascending, more massive rhizome.

Technical confirmation Plants are diploids, with $n = 34$, $2n = 68$ chromosomes.

Field notes *Oreopteris limbosperma* has a range which is predominantly and most abundantly north-west European, but occurring in Madeira and the Azores, and then more continuously from the Pyrenees northward to Britain and Ireland and western Norway (to just north of the Arctic Circle)

Oreopteris limbosperma: *a*, Inverness-shire; *b*, Argyll; *c*, *d*, Merionethshire.

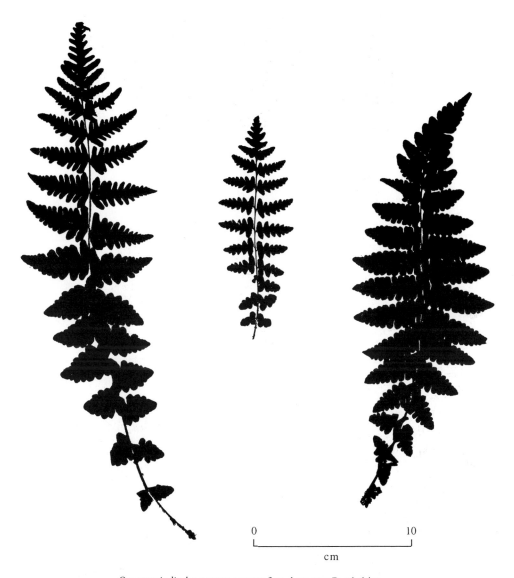

Oreopteris limbosperma, young frond stages: Perthshire.

and eastwards to central Europe and, much more discontinuously to the southern end of the Caspian Sea (Hultén, 1958, 1964; Jalas & Suominen, 1972). It is a member of an extremely small genus, whose only other two species include one (*O. qualpaertensis* (Christ) Holub, closely related to *O. limbosperma*) which ranges sparsely around the northern Pacific fringe, from Central Japan, Korea and Kamchatka north and eastward to the Aleutian Islands and western seaboard of British Columbia, and another (*O. elwesii* (Bak.) Holttum), endemic to Sikkim. In Britain (less so in Ireland – see below), *O. limbo-*

sperma increases in abundance strongly westwards and northwards, reaching its most frequent in western Scotland, where quite numerous plants occur nearly from sea-level to about 880 m (2900 ft). The very fragmentary nature of its overall generic range indicates *O. limbosperma* to be a geographically relictual species of a particularly small genus, which thus probably reaches its greatest abundance in western Scotland anywhere on a world scale.

In Britain and Ireland, Sweet Mountain-fern is also called Lemon-scented fern from the sweet and citrus-like smell of its fronds when lightly brushed,

Oreopteris limbosperma, Argyll: *above*, mid May; *below*, early May.

presumably a secretion from the glands copiously present over the lower frond surface, but of unknown ecological purpose. Its main climatic requirements in these islands appear to be of appropriately high rainfall and generally cool summer temperatures. Plants are clearly tolerant of full daylight in areas of high and frequent cloud cover and, in more upland areas, occur mostly in generally sheltered pockets where their relatively large fronds gain some protection from exposure to strong winds. In such upland sites, there may also be some gain in frost protection for the rhizomes from considerable depths of winter snow.

At all altitudes, *O. limbosperma* is a strongly calcifuge plant, which occurs mostly in damp, acidic, peaty situations over predominantly hard and acidic rocks, especially over mineral layers which are themselves impervious, and where the topography is sufficiently steep to cause strong surface run-off. In such sites, *O. limbosperma* is frequent over the Welsh and Cumbrian slates and over Scottish granites, slates and gneiss. In the highest rainfall areas, where run-off of acidic peaty water is most continuous, plants may also occur over harder, more base-yielding rock types such as schists, if these are largely overlain by acidic peaty layers. It is associated in such habitats with usually frequent *Luzula sylvatica*, *Blechnum spicant* and *Equisetum sylvaticum* in more upland areas, and sometimes *Hyacinthoides non-scriptus* in lower altitude sites, the main west European distributions of all of which, as well as those within these islands, show remarkable parallels. *Oreopteris* is, however, as with these other species, of much sparser occurrence in Ireland.

Within appropriate climatic regions over such rock types and within sheltered or semi-sheltered sites, *O. limbosperma* can be a fast growing and gregarious species. It thrives in sheltered sites such as on steep damp ledges overhanging streams, along the craggy flanks of sheltered upland gullies, and along streamlet banks in open acidic woodlands. It can be particularly frequent down the steeper aspects of moorland valleys, where very numerous individuals may form long, somewhat regimented,

meandering lines of bright-green fronded plants, crowding the damp banks and margins of a myriad of small rills and mountain stream courses and marking the many hidden lines where artificial drainage ditches or tumbling mountain torrents over steep acidic terrain have cut deep, narrow, hidden clefts to the underlying bedrock surface through thick layers of surface peat.

All these habitats are niches where trickling acidic peaty water provides the most constant surface run-off or subterranean seepage, creating cool and constantly moist but well-aerated edaphic conditions, which seem optimal for this plant. *Oreopteris limbosperma* is thus a particularly valuable indicator of specific edaphic niches throughout the hillsides, revealing where the groundwater is most continuous, acidic, clean, ample and always moving. Its much less frequent occurrence in Ireland is perhaps due largely to the less freely drained nature of many wetter lowland sites and the more widespread presence in higher ground of sedimentary rocks of much more base-yielding status, producing terrain with better downward drainage and, at their bases, seepage lines with a rather higher base content.

In autumn, fronds turn rapidly pale brown and eventually almost white following the first cold nights, and in sheltered low-altitude sites, such pale fronds can remain standing well into mid-November. The fronds have a remarkably high and rather uncontrolled rate of water loss through their large and leafy soft-textured blades. Its consequent high demand for shelter suggests that *O. limbosperma* was originally a woodland fern in these islands, enjoying the shelter of probably a light, open canopy, mainly Birch (*Betula*) and Rowan (*Sorbus*), where it was always mainly a species of the edges of streams and rivulets within them. Under the cloudy skies of western Scotland, the species has adapted remarkably well to woodland loss, although it is probably extensively reduced (and in some areas already largely absent) and currently still reducing under conditions of modern sustained high grazing pressures.

Osmunda regalis L.

Royal Fern

Preliminary recognition A large to very large fern, of majestic appearance, with tall, broad, coarsely divided fronds, many ending in a brown fertile tassel, like the inflorescence of a large rhubarb or a flowering dock.

Occurrence Of somewhat local occurrence, in a wide range of wet, acidic habitats, but especially around fens and boggy woodland. It occurs mainly near wetter, western coasts of England, Wales and Scotland, but is especially widespread in Ireland, where it is sometimes frequent in the extreme west.

Identification Plants have *tall, broad, mostly erect fronds*, usually at least 60–120 cm high, *but up to 3–4 m tall and 1 m broad in exceptional specimens*. The fronds have broadly ovate-oblong outlined blades with a triangular apex, and are borne on stipes that are usually about half the length of the frond and arise from the crowns of very stout, ascending or semi-erect rhizomes. Each crown usually bears only a small number of fronds, but, as rhizomes age, they branch to eventually build up *massive, heavy, woody, raised rhizome clumps*, with the old parts persistent and densely covered by many interwoven roots, and forming many nearby crowns. Such clumps give rise to fronds in large numbers.

All mature fronds are broad, regularly bipinnately divided throughout at least the lower 60–85% of the blade, with about 5–9 or more pairs of nearly opposite, often rather distant, pinnae, the very tips of which, as well as the more terminal parts of wholly vegetative fronds, are progressively less completely divided. The pinnae bear a series of often 8–10 or more opposite to subopposite, mostly shortly stalked, pinnules, each straight or slightly falcately curved. All ultimate segments of vegetative fronds have a similar leafy structure, with large, simple, broad, oblong segments tapering to a rounded or blunt point only near their apical portions, mostly 2–8 cm long and at least 10–15 mm broad. The fronds thus have an overall appearance of many, regular and usually widely spaced *large coarse ultimate segments*.

On mature plants, most of the fronds stand particularly erect, with the pinnae usually strongly rotated towards a more horizontal plane. On such fronds, the distal 3–10 or more pairs of pinnae have *highly reduced laminae, are strongly upsweeping, and bear densely clustered serial rows of large rounded sporangia borne along the stark pinnule midribs*, the whole forming a fertile tassel-like outgrowth to the top of the frond. All the sporangia mature and dehisce

Jan.	Feb.	Mar.	Apr.	May	Jun.	Jly	Aug.	Sep.	Oct.	Nov.	Dec.

Osmunda regalis, habit of plant in life.

almost simultaneously over a short period in early summer, to yield large quantities of short-lived, green spores.

When emerging in spring, the uncoiling croziers are usually a pale greenish-pink colour, are lightly bloomed and have an asparagus-like texture. They are at first covered with a fine, golden or cinnamon-coloured pubescence, which is soon lost as each frond matures. Expanded fronds have a pale pea-green colour, and leafy, somewhat leathery texture, with a surface that repels droplets of water without becoming itself wetted, like the surface of a fresh cabbage leaf. Although large, the fronds are rather cold-sensitive, and die down rapidly with the first autumn frosts.

Variation Not a very variable species within any one locality, except that of progressively larger fronds between individuals of differing ages. Between the leafy portion of the frond and the fertile tassel, there are often transitional regions with some pinnae partly leafy and some partly fertile. These half-way stages can be of much morphological interest. They can, however, occur on almost any plant, and there can be much variation in their appearance even between successive fronds of a single plant. More taxonomically significant, perhaps, is that there may be some overall variation in plant habit and form between specimens originating

Osmunda regalis, adult frond stages: *all* West Norfolk.

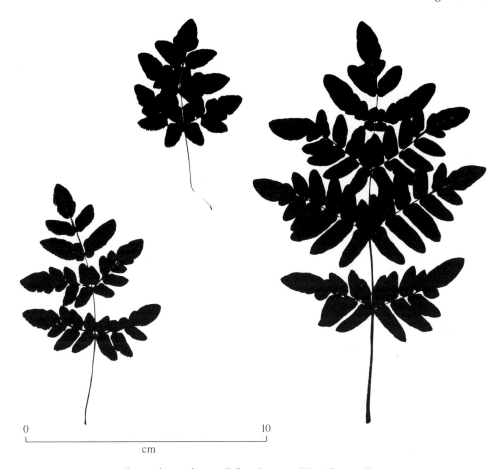

Osmunda regalis, small frond stages: West Cornwall.

from some extreme eastern and western localities in these islands, and this needs further investigation.

Possible confusion Its sheer size and coarse frond cutting, as well as the distinctive tassels of fertile terminal growth, ensure that *Osmunda* is unlikely to be confused with many other plants. Sometimes the large, pinnately divided leaves of the umbelliferous Hemlock Water-dropwort (*Oenanthe crocata*) in somewhat similar habitats can look similar to *Osmunda* from a distance.

Technical confirmation Plants have $n = 22$, $2n = 44$ chromosomes. the sporangial capsules are two valved, and totally lack an annulus ring. The gametophyte generation is typically of a very highly convoluted thallus with a notably upright-growing posture.

Field notes *Osmunda regalis* is a widespread but also variable species on a world scale, in the broadest taxonomic sense the aggregate species (amongst which probably several local species occur) is present highly discontinuously throughout Europe and Africa, to the Cape, India and the Himalayas, to Japan and in part of both North and South America. Within Europe, *O. regalis* is a predominantly western and southern species, occurring from Britain and Ireland and southern Sweden southwards to the Azores and Mediterranean to Turkey and the Caucasus.

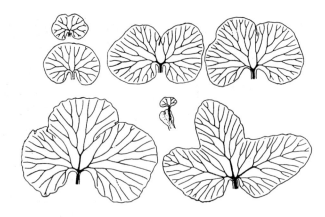

Osmunda regalis, young frond stages (after Orth, 1938).

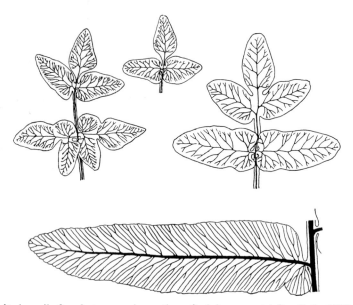

Osmunda regalis, juvenile frond stages and venation of adult segment (after Orth, 1938).

In Britain and Ireland it is also a predominantly southern and western species, but with other lowland locations too, especially in the fenlands of East Anglia. Royal Fern is a slow-growing, but probably very long-lived, markedly calcifuge species, thriving in predominantly wet, acidic, peaty conditions, especially in areas of mild western oceanic climates, with high and frequent rainfall. It is mainly confined to within about 30 m (100 ft) above sea-level, although scattered individuals ascend to nearly 305 m (1000 ft) in parts of Co. Kerry and Co. Waterford. It occurs in a very wide range of acidic, wetland habitats, including wet heaths, ditches, damp woodland, river banks, lake margins, ledges on sea-cliffs, and valley bogs and fens, often,

especially in the latter habitats, where these are developed over deep alluvial soils, and where there is some movement of water and mineral enrichment near the surface.

Plants can withstand considerable exposure, but in such situations are often stunted. They seem tolerant of exposure to salt-laden air. But plants reach their most luxuriant development, where they are sheltered in the bottoms of shallow valleys carrying woodland vegetation. Dense stands of individuals with head-high fronds can occur in such habitats, growing with their rhizomes just above the level of the prevailing water table.

Royal Fern shows a very considerable adaptability to different wet acidic habitats near extreme west

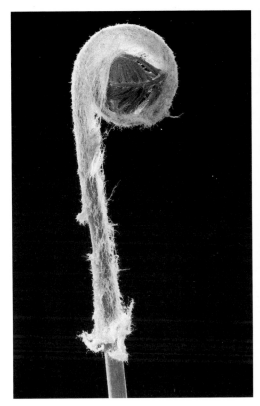

Osmunda regalis, expanding croziers showing copious indumentum: Argyll, May.

coasts. It is abundant around loughsides and edges of rivers draining them in Co. Kerry, and spreads extensively around margins of valley bogs and roadside drainage ditches, where there is considerable seepage of peaty water. In south Argyll, west Inverness-shire and in the Outer Hebrides, plants occur mainly around the stony margins of small lochs and peaty pools, extending away from standing water up some of the myriads of small, semi-sheltered streamlets and rills which drain surrounding moorlands, through deep, peaty bogs and trickling channels. In south Devon and east Cornwall, *Osmunda* sometimes grows abundantly along the edges of the courses of streams and rivers feeding peaty water from the upland granitic moorlands through densely wooded valleys. In such areas, extensive colonies of Royal Fern sometimes cling to the banks of rivers where there is additional local seepage, as well as colonising the margins of reservoirs, sometimes in considerable numbers. In west Cornwall, large numbers of plants occur locally on dripping edges of sea-cliffs, where there is more or less permanent abundant acidic seepage

and run-off from moorland areas inland. Here, dense colonies of small plants sometimes grow within a few feet of high tide level, many becoming precociously fertile. In south Dorset, Royal Fern occurs as extensive local patches in damper areas of wet heaths, and even away from the west coast influence, such as in fenland habitats adjacent to many of the east Norfolk Broads, *Osmunda* can occur in some numbers. Here it grows particularly on raised hummocks just above the general fen water level, often with established Birch (*Betula*) and dense growths of *Sphagnum* bogmosses.

Royal Fern is probably one of the longest lived of native ferns. Individual clumps have been known in cultivation for a century or more, and, judging by their still-small size, some of the more massive clumps occasionally surviving in the wild must be many centuries old. Drainage of wetland habitats has probably contributed to some of the decline in abundance of *Osmunda* in the last 100 years, but this was very greatly accentuated in Victorian times by extensive collection of plants, sometimes on a commercial scale (in certain areas, literally by the wagonload), mainly because of their former use as a horticultural medium for orchid culture. It was probably the natural gregariousness of the plants that enabled their collection on such a scale to be commercially undertaken, but against which their slow natural rate of growth is totally unable to compensate. Plants are now protected by law, and specimens should on no account be removed from the wild.

Phegopteris connectilis (Michx) Watt

Beech Fern

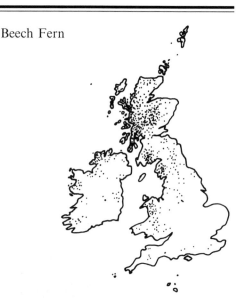

Preliminary recognition A small, pale-green fern with soft-textured fronds arising individually in colonies, each blade of triangular, tapering outline, all pointing regimentedly downslope.

Occurrence A locally frequent fern of upland districts of Wales, northern England and Scotland, but rare or absent elsewhere. It grows in moist, shady, soil banks in upland valley woodland, where it may form extensive colonies. It also spreads, in more stunted form, to moist patches on damp rocky slopes and cliff ledges in mountains to about 610 m (2000 ft).

Identification Fronds arise individually from slender, 1–2 mm wide, freely branching, long-creeping underground rhizomes. The blades are borne on *slender, very brittle, slightly translucent, pale creamish-green stipes, which are brown only at the extreme base.* The stipes are as long or longer than the length of the blade, are finely pubescent, and have a few, sparse, scattered, soft brown scales, which are soon lost. At the top of each erect stipe, the blade spreads horizontally (or even descends), with *the lowermost pinna pair* arising at this point, *reflexed backwards, away from the rest of the blade.* The blade is of triangular or triangular-ovate outline, usually 1.5 times as long as broad, mostly 8–20 cm but sometimes 30 cm or more in length, and *drawn into a narrowly tapering drip-tip apex.* The lowest (reflexed) pinna pair is usually attached just by the midribs, but the rest of the pinnae are attached by their entire bases (adnate), with the pinnae in the upper part of the frond becoming gradually combined into the frond apex. Each pinna is pinnatifidly cut into deep, round-ended lobes, and its apex is slenderly tapering. The lamina is of soft, flexible texture, and *minutely and finely pubescent over both surfaces.*

Fertile fronds usually have slightly more slender pinnae than those of purely vegetative ones, and are borne on rather longer stipes. Their sori are small, rounded or oval in shape, naked,

Jan. Feb. Mar.	Apr.	May	Jun.	Jly	Aug.	Sep.	Oct. Nov. Dec.

Phegopteris connectilis: all Mid Perthshire.

and are *widely spaced in a line around each segment near to the margin.* They are black at maturity and conspicuous in late July or early August, before shedding their sooty black spores, but, in the late season, the discharged sori are pale buff and indistinctive.

Plants flush early and rather rapidly in spring, and the croziers of expanding fronds droop backwards before fully expanding. At this stage, the dense, fine, silvery pubescence is particularly easily seen. The fronds die down completely in autumn with the onset of the earliest frosts.

Variation Not a very variable species, except in frond size. Small fronds are virtually scaled-down versions of larger ones, with fewer, less acutely pointed pinnae.

Possible confusion Its triangular shape, soft, pubescent texture, and reflexed lower pinnae, make its confusion with any other species unlikely.

Technical confirmation Plants are apogamous triploids, with $n = 90$, $2n = 90$ chromosomes.

Field notes *Phegopteris connectilis* is part of a small genus of ferns confined entirely to north temperate regions of both the Old and the New Worlds. *Phegopteris connectilis* occurs almost throughout the range of the genus, varying locally in cytotypes, with sexual diploid and tetraploid forms (presumably the parents of our apogamous triploid) found to be present more locally (e.g. within Japan) where their cytology has been researched. *Phegopteris connectilis* is a widespread species in Europe and western Asia, present through most middle and fairly high-latitude districts and lower montane areas, and the same or other closely related species disjunctly in scattered stations at mainly mid to high latitudes discontinuously in northern, central and eastern Asia to Kamchatka, and in mountainous regions of North America. Throughout much of this range, the species is associated with north temperate deciduous forests, occurring especially in more northern districts associated with shady upland streamsides, and the overall range of this species and that of *Gymnocarpium dryopteris* (q.v.) show remarkable similarities. The range of Beech Fern in Britain and Ireland seems to correspond mainly with areas of mountain climate, which are cold in winter and cool with a high humidity and frequent precipitation in summer.

Within such areas, it thrives especially on deep, moist, often only semi-stable, steeply sloping, unconsolidated, mineral soil banks, especially where these are gently irrigated by slight seepage, and there is good incorporation of organic matter into largely siliceous soils with an essentially sand structure, seldom far from outcropping rocks. It attains its most extensive development where such banks occur in sheltered ravines alongside streams,

especially where the summer humidity is augmented by light spray, which the unwettable surface of the fronds and descending pinnae and frond apices with drip tips, effectively shed. Plants seem to avoid flat and compacted ground, and are also intolerant of strong, direct sunshine, and hence occur mainly on northerly aspects, in shadowy areas of ravines, or on cloudy flanks of mountains and where there is light shade of a deciduous tree canopy. It seems, however, tolerant of a wide range of pH conditions, occurring in soils over mildly acidic rocks, but is perhaps more common in ground where there is a reasonable base content.

Beech Fern often grows in the company of Woodland Oak Fern (*Gymnocarpium dryopteris*, q.v.), whose ecology and geographic range in Britain and Ireland it very closely parallels. Of the two, Beech Fern seems more tolerant of slightly wetter ground conditions. Other species commonly associated with Beech Fern in upland Birch–Oak (*Betula–Quercus*) woodland habitats include several other creeping perennials of vernal growth, such as Wood-sorrel (*Oxalis acetosella*), Common Dog-violet (*Viola riviniana*), Wood Anemone (*Anemone nemorosa*) and occasionally, Shade Horsetail (*Equisetum pratense*, q.v.), as well as numerous mosses. The habitats of this fern are almost always high sheltered ones. Beech Fern grows relatively slowly, but, in moist and appropriately sheltered mainly purely natural sites, colonies gradually expand to fill usually the whole of each available steeply sloping habitat. It seems likely that the larger of such patches in northern valley woodlands of Britain may well be of considerable age and individual vegetative longevity. The plant clearly resents disturbance of its sheltering woodland canopy, and is clearly highly intolerant of exposure to sudden drought. The presence of this species, especially *as extensive and continuous canopies,* *may well thus be a good marker of long-undisturbed sites, especially those deep within the ravines of northern ancient broadleaf woodland.*

Phyllitis scolopendrium (L.) Newm.

Hart's Tongue Fern

Asplenium scolopendrium L.

Preliminary recognition A small- to medium-sized, wintergreen fern, with entire, undivided, strap-shaped leaf-like ascending fronds, arising in irregular 'shuttlecocks'.

Occurrence Widely spread throughout Britain and Ireland, becoming most common over lime-rich soils in oceanic, high-rainfall districts, especially in south-west England, South Wales and southern Ireland, and becoming progressively more infrequent in eastern and northern districts.

Identification Plants have numerous, *entire-bladed fronds*, mostly about 30–60 cm or more in length, which arise in somewhat irregular, shuttlecock-like tufts from the crowns of short, stocky, ascending rhizomes. The fronds are of *narrowly tongue-like, strap-shaped, linear-ovate outline*. They taper gradually above to an acute or obtuse apex, and below *to a base which is cordately lobed into two, often unequal and sometimes partly overlapping, heart-like auricles, immediately above which the outline of the frond is itself often slightly narrowed.* The entire blades are usually of a rather bright, grass-green colour, with a shining upper surface, and of distinctly firm and slightly leathery texture. They are borne on fairly short stipes, usually about 1/6 the length of the frond, but sometimes more, which are of *dark, somewhat purplish-brown colour*, slightly shaggy with soft, pale brown scales. The dark colour of the stipe usually persists into the lower part of the rachis, above which it gradually merges into a pale green colour, similar to that of the underside of the frond. Mature rhizomes usually have their older parts densely set with the persistent, erect bases of many previous years' stipes.

Most mature fronds bear sori throughout the upper half or more of the underside of the blade. The sori are numerous and regularly arranged into *parallel, long lines, ascending slightly diagonally from either side of the midrib*, ending and beginning about equidistantly from the midrib and the frond margin, and occupying much of the width of the leaf. Each line is made up of a closely set pair of twinned sori, lying parallel and closely together, and each opening

Jan. Feb. Mar.	Apr.	May	Jun.	Jly	Aug.	Sep.	Oct. Nov. Dec.

0 ——————————————————————— 10
cm

Phyllitis scolopendrium, juvenile and semi-juvenile frond stages: *all* West Cornwall.

towards the other, with a long, whitish indusium on either outer side. When mature, they bulge with numerous dark sporangia, and largely merge. The sori follow the course of veins, which can be seen elsewhere in the frond running in a closely parallel fashion diagonally outwards.

Fronds are wintergreen and persist for about a year. They are usually borne in a strongly ascending-arching manner, with the tips becoming more arching and curving over, but plants growing in steep situations may have their fronds mostly pendulously hanging. The margin of fronds on many plants, as they mature, takes on a slightly crinkled or undulate shape. Fronds flush mostly simultaneously in spring, and when flushing can be seen to have numerous scales *which are distinctly silver white at this stage*. The scales are soft and gradually darken with age, and eventually are mostly lost, except on the stipe. (These silver-white scales on flushing fronds are inherited as a dominant character in known crosses between *Phyllitis* and *Asplenium*, helping to distinguish the intergeneric ×*Asplenophyllitis* (q.v.) crosses at an early-flushing stage in spring from crosses involving only species of *Asplenium*.)

Phyllitis scolopendrium, adult frond stages: *all* West Cornwall.

Possible confusion The entire, shining fronds are very distinctive from those of any other fern in Britain or Ireland (even from its hybrids with *Asplenium*), and are unlikely to easily be confused with anything else.

Variation *Phyllitis scolopendrium* is a widely variable species over the whole of its British and

Phyllitis scolopendrium, habit of plant in life.

Irish range in many relatively minor details of frond form, which differ even between individuals in most populations. Much of this variation is genetically inherited. Individuals may vary in frond size, shape and length of lamina, breadth of frond, acuteness of apex, undulation of the margin and length of stipe. In addition, more monstrous forms can occur with fronds that have lacerated margins or forking frond tips. The majority of the more extreme forms are often sterile. Many have been introduced, selected and named in cultivation as fern varieties or 'cultivars'.

Technical confirmation Plants are diploids, with $n = 36$, $2n = 72$ chromosomes.

Field notes *Phyllitis scolopendrium* has a highly discontinuous natural world range. It occurs in both Europe (as a tetraploid) and in North America (as a much rarer diploid taxon), and what may be the same or a different but allied taxon occurs also very locally in both northern Japan and in Mexico. Only in Europe is there an extensive though seldom dense range, with scattered stations occurring from extreme south-western Norway to the Iberian Peninsula, Madeira and the Azores, eastwards with much more scattered stations to the Caspian Sea. Only in Britain and Ireland, and extreme western mainland Europe south to the Pyrenees, is it

at all abundant. In Britain and Ireland, Hart's Tongue Fern occurs in a wide range of mainly lowland habitats, seldom above about 185 m (600 ft) altitude. It is often widespread and abundant in western coastal districts but of much more local occurrence and restricted habitat range further east and north.

Its main habitats include rocky woodlands, streambanks, hedgebanks, lanebanks, damp, shaded stonework, and cliffsides, sides of steep ravines, grykes and fissures of limestone pavements, and around cave mouths and mining adits. It is a markedly calcicolous wintergreen species requiring

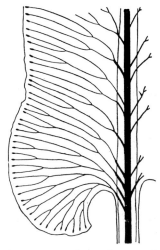

Phyllitis scolopendrium, venation pattern at base of mature frond (after Orth, 1938).

Phyllitis scolopendrium, young frond stages (after Orth, 1938).

habitats which are more or less permanently moist, have a high local humidity, and offer good shelter and shade. It thrives in soils with a high pH, often in mineral soils with very little humus, and can be especially abundant in mixed deciduous woodland valleys in western districts where there are exposures of basic shales, marls and red clays associated with the Old and New Red Standstone systems.

In sheltered, moist, well-drained sites, especially where the ground contains abundant rock fragments mixed with basic clay, and there is gradual seepage from adjacent cliffs through sloping soil surfaces, luxuriant lines and sometimes whole stands of pure *Phyllitis* may occur, containing individuals with fronds of exceptional size. Such individuals may have blades over 60 cm in length and 10 cm in breadth, plus a stipe of at least 25 cm, with over 60 lines of paired sori per side (thus over 120 per blade), with most such sori 3.5–4.0 cm in length. From such plants, spore production is clearly vigorous and abundant, and the individuals themselves appear to be long lived, at least over the course of many decades.

In such habitats measured in south Devon, west Gloucestershire, Anglesey, Cheshire and Kirkcudbrightshire, it occurs on clays with a pH range of about 7.5–8.6, associated particularly with Soft Shield-fern (*Polystichum setiferum*), and flowering plants such as Herb Robert (*Geranium robertianum*), Wood Anemone (*Anemone nemorosa*), Dog's Mer-

Phyllitis scolopendrium, West Cornwall: *above*, July; *below*, late June.

cury (*Mercurialis perennis*), Wild Garlic (*Allium ursinum*), Ivy (*Hedera helix*), Sweet Woodruff (*Galium odoratum*), Foxglove (*Digitalis purpurea*), Enchanter's Nightshade (*Circaea lutetiana*), Bramble (*Rubus fruticosus* agg.) and Slender False-brome (*Brachypodium sylvaticum*). In shaded rocky stream gorges, in northern England and Scotland, where steep lime-rich rocks are exposed, it often is

Mediterranean-Atlantic range of *Phyllitis scolopendrium* in Europe.

accompanied by Hard Shield-fern (*Polystichum aculeatum*), Brittle Bladder-fern (*Cystopteris fragilis*), Common Maidenhair Spleenwort (*Asplenium trichomanes* subsp. *quadrivalens*), and, sometimes, Green Spleenwort (*A. viride*), and the bryophytes *Ctenidium molluscum, Bryum* spp., and *Conocephalum conicum.* Many of the above species also accompany Hart's Tongue Fern in limestone pavement grykes in the northern Pennines, whilst in those of Co. Clare, Rusty-back Fern (*Ceterach officinarum*) and Maidenhair Fern (*Adiantum capillus-veneris*) may also grow nearby. In many limestone districts, plants descend around damp pothole mouths and sink-holes, and enter caves and mining adits wherever there is sufficient light to grow. In various limestone cave systems which have the benefit of more or less frequent artificial illumination for commercial tours (e.g. in the Cheddar Gorge systems of Somerset and the Torbay limestones of south Devon), isolated plants can often be seen clinging to moist rock surfaces in deep underground caverns around tungsten light bulbs. Such plants have almost certainly established from chance incoming airborne spores.

In other artificial habitats, plants occasionally establish, usually as scattered individuals, in the mortar of old brickwork, and, especially where this is shaded, there is a high permanent humidity and some seepage of local moisture. They grow especially well in such situations around many of the deeper nineteenth century railway constructions, such as along retaining walls and around deep tunnel mouths, as well as beneath canal bridges, whilst during earlier periods of historic time in England, in particular, Hart's Tongue Fern was a very common colonist of the insides of well shafts, where it received ample shade, shelter, moisture and humidity and, in folklore, became well known.

The presence of conspicuous clusters of plants of *P. scolopendrium* in areas in which the species is otherwise of generally infrequent occurrence has been used by archaeologists as an indicator of sites in which old masonry structures may lie hidden. Such occurrences are especially associated in valley bottom sites beside permanent flowing water with structures originally cemented by old lime mortar (Page, 1988*b*), where they are most frequently indicative of the former presence of old mill workings.

In former times and in many a rural setting within these islands, Hart's Tongue Fern must have been especially well known as an abundant colonist and regular denizen of the interior surfaces of a myriad of village wells throughout at least the more lowland parts of these islands. Not surprisingly for an evergreen plant of such unusual appearance, which appeared almost everywhere apparently spontaneously in such well sites, the plant became associated particularly with a rich and often bizarre folklore associated with various remedial and magical properties (Page, 1988*b*).

In experimental culture, the highly ornamented spores germinate relatively slowly and the subsequent development of the prothallus similarly takes place at a relatively slow rate. As with the sporophyte, plants of the gametophyte generation grow most vigorously on substrates with a neutral to basic pH. Young sporophytes also grow relatively slowly, typically giving rise to their first fertile fronds in about 2–5 years, and not reaching full size until 5–8 years or more. Thereafter, individual plants seem to be very long lived, surviving probably several decades in favourable sites, where they may eventually form large clumps with many dozens of fronds. Such slow but steady growth rates have clearly adapted the plant well to success in habitats on which conditions of shelter, edaphic moisture and surrounding humidity are relatively constant, and to exploiting sites in which the ambient light levels can be constantly low. Plants will, however, succeed well in fairly high light conditions, but under such circumstances competition usually confines them to niches in which their ability to gain and maintain a permanent hold within small rocky fissures gives them a permanent ecological advantage. On the mineral surface of many such sites, especially where it is steep and there is always some degree of erosion, plants clearly are often permanent pioneers. Further, in more luxuriant forest riverbank habitats, especially on damp clay slopes in south-western counties, the presence of populations containing very numerous large individuals is probably strongly indicative of long-term lack of disturbance of those sites.

The natural predominance of Hart's Tongue Fern in western climates, which are most influenced by the Gulf Stream, is probably associated with a requirement for a long growing season with adequate shelter from desiccating winds in winter, and high year-round humidity. It seems likely that under such conditions, much of this plant's photosynthetic activity takes place during the late autumn and winter months, when there is a light increase in many of its natural woodland habitats through shedding of the deciduous leafy canopies.

Pilularia globulifera L.

Pillwort

Preliminary recognition A small marginal or submerged aquatic plant with numerous, slender, yellow-green, cylindrical leaves, standing up like a short wavy grass when above water, and in quantity, sometimes forming a slightly curly, yellow-green lakeside turf.

Occurrence Very local and scarce throughout the whole of Britain and Ireland and absent from many areas, occurring mainly around damp pond margins and on shallow lakeside mud.

Identification Plants are perennial, with *numerous, yellow-green, ascending, filiform leaves*, about 3–8 cm or more in length and about 1 mm in diameter near the base. They arise at intervals of about 5–10 mm or more from the sides of pale greenish-cream coloured, far-creeping, cylindrical rhizomes, about 2 mm or less in thickness, which spread far over, or

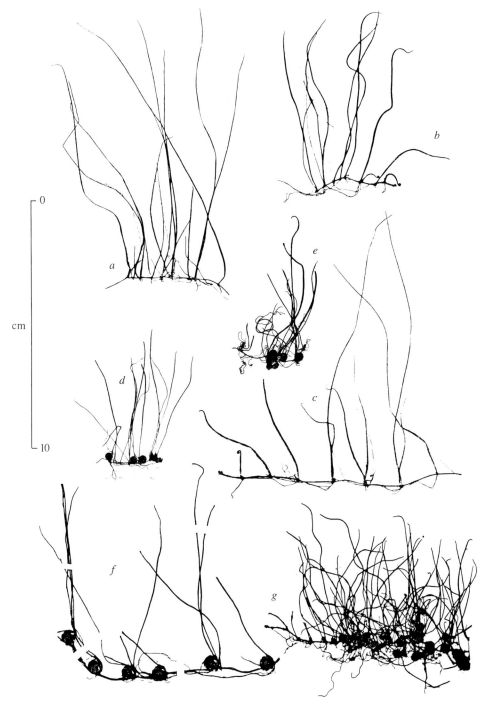

Pilularia globulifera: a–c, Hampshire; *d*, Sussex; *e–g*, Midlothian (courtesy Dr T. G. Walker). Note the masses of bean-shaped sporocarps in the last few specimens.

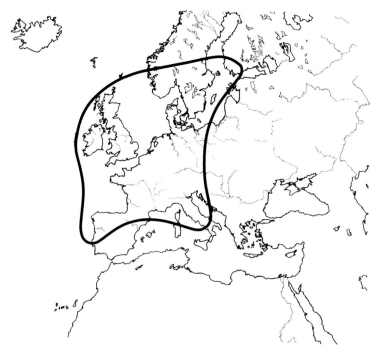

Sub-Atlantic range of *Pilularia globulifera* in Europe.

just beneath, the surfaces of damp silty mud. Long, slender, pale-brown roots arise from the rhizomes, and there may also be short spur-like side-branch rhizomes, each of which can give rise to further groups of about 2–10 leaves in a close-crowded succession. *All the leaves are solid, cylindrical, and slender*, and each tapers gradually throughout most of its length to a soft, acute tip. The leaves are yellow-green in colour and have a delicately translucent quality. When young they are circinately coiled in a watchspring-like manner and, when expanded, are slightly succulent and of a turgidly rigid yet flexible texture, with a surface that feels waxy and non-slippery to touch. Most stand fairly erect, *but have an irregularly flexuous and slightly curved habit*, in quantity looking like masses of coarse, slightly curly hair.

Established plants give rise to numerous, small, dark, hard, bean-shaped objects each of about 3 mm diameter, scattered along the rhizomes at the leaf-junctions, on very short stalks. These are the sporocarps, which are slightly hairy when young but hard and smooth when mature, and contain the sporangia in internal watertight compartments. These produce mega- or micro-spores, which, on eventual cracking of the sporocarps, exude in a jelly-like mass.

Variation Plants are extremely variable in size, and most of this variation seems environmentally induced.

Possible confusion Plants are most likely to be mistaken for forms of other linear-leaved plants, perhaps especially small rushes (*Juncus* spp.), but their solid, circinately coiled expanding leaves arising from creeping rhizomes, the absence of a blade, and the presence of dark, bean-shaped sporocarps when mature, distinguish *Pilularia*.

Technical confirmation Plants have $2n = 26$ chromosomes.

Field notes *Pilularia globulifera* is a European endemic pteridophyte species originally occurring (but now very much reduced – see below) throughout much of western mainly lowland Europe from southern Scandinavia to the Iberian Peninsula. Its stations penetrate remarkably little into central Europe.

In Britain and Ireland it has a widely scattered range throughout. Pillwort is a very local and scarce plant, which colonises soft, nearly level, muddy flats at the edges of low-lying freshwater lakes, occasionally in sufficient quantity to form a turf. It also grows on silty mud at the edges of ponds, slow-flowing river backwaters and river mouths, in wet sandy hollows in dunes and heaths, and sometimes invades muddy ditches and old claypit workings in shallow, wet situations. It occurs from sea-level to about 380 m (1250 ft) in altitude, and thrives where soft, semi-consolidated, silty mud is periodically flooded by freshwater conditions, especially where it can continually pioneer new, bare sites. It can survive periods of complete immersion, and sometimes occurs in water up to about a metre in depth. It grows mainly in habitats that are largely competition free, and especially where reedswamp is absent.

Occasionally associated with it, usually as scattered individual plants, or as scattered patches, are a range of marginal and submerged aquatic species, including several Water Mints (*Mentha* spp.), Pennywort (*Hydrocotyle vulgaris*), Shore-weed (*Littorella uniflora*), Bog Pondweed (*Potamogeton polygonifolius*), and Bladderworts (*Utricularia* spp.).

Pillwort usually grows in habitats that are circumneutral to rather acidic in character, which remain more or less permanently damp throughout the summer, and which warm rapidly whenever there is sunshine. Plants die back considerably in the winter months, but survive as subevergreen fragments to resume growth the following spring. Once established from such fragments, plants are able to make substantial vegetative growth in one season. Growth, however, can vary widely from year to year, even in the same locality, depending on climatic conditions and water levels, and their reappearance can thus be spasmodic. It seems possible that after colonies die back during years of exceptional droughts, sporocarps may persist and remain dormant in mud, perhaps for many years, and detailed observation and research are needed on this.

In addition to spore production, vegetative dispersal around lakes may also play a role in longshore spread, and occasional floating fragments have been seen after storms.

Pilularia globulifera is a European endemic species confined to a relatively small range in the western European lowlands. On the continent, it has declined sufficiently in number to be regarded already as an endangered species. In Britain and Ireland too, Pillwort was probably once a widespread plant, but seems to have decreased very substantially in recent years. Although the raising of water levels to form steep-banked reservoirs and drainage of other sites may have played a significant role in reducing the habitats available, other factors may well be involved, and its ecology seems in need of detailed and especially experimental research.

Polypodium cambricum L.

Southern Polypody

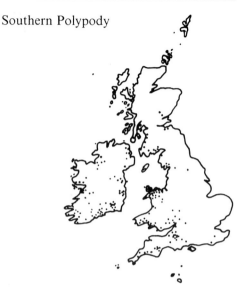

(*Polypodium australe* Fee)

Preliminary recognition A small to medium-sized wintergreen fern, with coarsely cut, herringbone-like fronds of thin textures and broadly oval or broadly triangular-deltoid outline, arising from scaly surface-creeping rhizomes.

Occurrence The least frequent of the three species of *Polypodium* in the British Isles, occurring mainly in mild, moist southern and western climates, especially in Wales and in Ireland, on high calcium-yielding rocky substrates (especially on limestones). Plants are occasionally epiphytic and are confined mostly to low altitude.

Identification *Polypodium cambricum* generally has fronds that are broader in relation to their length than have the other species or hybrids of *Polypodium*, and this broadness of outline provides a useful character for initially recognising it in the field. Small fronds, 5–15 cm in length, usually have a *broadly triangular-deltoid outline, like the shape of a wide gothic arch.* Larger fronds, 15–40 cm or more in length, become progressively more oval with size, *although remaining broad, with blades often half as wide as long.* All usually have *long, slender, tapering pinnae, which are mostly pointed at the tip and are frequently serrate along the margins* (see 'Variation'). Except in the most exposed situations, fronds are of thinner and more flexible texture, yellow-green colour and usually with a glossier upper surface than those of other species of *Polypodium*, the *yellow-green colour persisting even in deeply shaded situations*, where the other species become a dark mid-green. The pinnae of *P. cambricum* are also often slightly waved along their lengths. When growing on steep rock faces, their flexible texture makes large fronds arch steeply downwards, sometimes in densely overlapping masses, with the glossy surface of each enhanced by the thin and undulate texture of the pinnae.

In most plants, *the lowermost pinnae of each frond are usually angled forwards* ('inflexed'). On smaller fronds this may involve only the lowermost pair, whilst on larger, more luxuriant fronds, several of the lowermost pinna pairs may be affected, each angled progressively further forward with the lowermost pinnae on either side of the frond often standing perpendicularly forwards like the high-pinioned wings of an eagle. On larger fronds, these lower pinnae may also be widely spaced, so that the webbing of the blade between them may become drawn into a very broad U-shape.

Jan.	Feb.	Mar.	Apr.	May	Jun.	Jly	Aug.	Sep.	Oct.	Nov.	Dec.

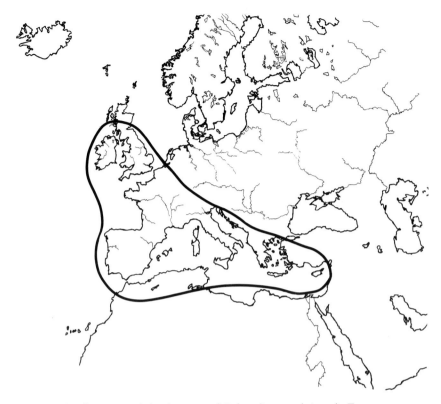

Mediterranean-Atlantic range of *Polypodium cambricum* in Europe.

The sori, when at a young (green) stage of development, *are frequently of oval, rather than circular, shape, especially those near to the base of each pinna.* When fully mature, the sori usually become a bright, more amber-yellow than those of other species (which instead become more yellow-brown or reddish-brown). Most fronds bear only a few sori, usually confined to the upper 1/3 or less on fronds in exposed conditions, and if present at all, to only a few of the upper pinnae on fronds in vegetatively luxuriant conditions.

Variation Frond size, frond form and fertility can be much modified by the environment. Outline shape and degree of pinna serration can vary widely from frond to frond even from the same rhizome, whilst average states of these features vary from population to population, indicating underlying genetic differences. More extreme forms with extensively serrated pinnae sometimes occur and have been referred to var. *semilacerum* Link, var. *serratum* (Wild) Milde, subsp. *serrulatum* Archangeli and var. *cambricum* (L.) Lightfoot. The last is sterile. Var. *semilacerum* is often particularly large, is fully fertile, and its strongly southern and western range in Britain and Ireland is particularly notable.

Possible confusion The broad frond outlines and calcicole habitats help to separate most plants of *P. cambricum* from *P. vulgare*, whilst their narrower, frequently serrated pinnae, yellower-green colour, fewer sori, and thinner, more flexible, softer, glossier frond texture help to distinguish *P. cambricum* from *P. interjectum*. Technical confirmation is, however,

Polypodium cambricum: a, Anglesey; *b,* Cheddar, Somerset; *c–f,* South Devon; *g,* Denbighshire; *h,* Ayrshire; silouhettes showing distribution of sori.

Polypodium cambricum; *all* East Cornwall, showing variation of frond outline.

recommended for all doubtful material. The good spores distinguish *P. cambricum* from *P. ×font-queri* and *P. ×shivasiae*, of which *P. cambricum* is the common parent.

Technical confirmation Plants are diploids, with $n = 37$, $2n = 74$ chromosomes. The spores are all good. Long, branched multicellular paraphyses, 500–1400 µm (0.5–1.4 mm) in length, occur within each sorus (Shivas, 1961a,b; Roberts, 1970). These paraphyses are distinct from the minute, short (40–80 µm long), glandular hairs consisting of only 3–4 cells, which occur on the lower surface of the frond of all species.

The combination of good spores and the presence of these paraphyses provide the best single feature, other than chromosome number, for technical confirmation. The number of indurated (thickened) cells of the annulus varies much more widely than had been previously realised, and has the range of 4–19, with a mean in different plants varying from 5 to 10 (cf. range 4–13, mean varying from 7 to 9 in *P. interjectum* (the 6n species), and range 7–17, mean varying from 10 to 14 in *P. vulgare* (the 4n species)). Because of the extensive overlap in ranges, the number of indurated cells on its own cannot be used to separate the diploid from the other two species, although it remains a reliable character for separating *P. interjectum* from *P. vulgare*, and in the hybrids especially for separating *P. ×shivasiae* from *P. ×font-queri* and *P. ×mantoniae* (q.v.). However, the indurated cells differ in shape from the other two species: they are shorter (21–26 µm) and broader (81–100 µm) than those of *P. interjectum* (28–35 µm × 76–86 µm) and much narrower than those of *P. vulgare* (22–28 µm × 60–80 µm). This gives the annulus of *P. cambricum* a distinctive contracted appearance, compared with that of the other two species. Usually 3–4 large basal cells separate the end of the annulus from the sporangium stalk (cf. 2–3 in *P. interjectum*, almost constantly 1 in *P. vulgare*), and these cells are much wider than the annulus.

In addition, the annulus is of a fairly constantly bright to pale yellow colour (cf. much more variable in colour in *P. interjectum*, and constantly dark orange-brown in *P. vulgare*). The rhizome scales at the base of the frond are long, slender, lanceolate and 5–16 mm in length (cf. 3.5–11 mm in *P. interjectum*, under 6 mm in *P. vulgare*) (Shivas, 1961a,b, 1962; Benoit, 1966; Roberts, 1965, 1970, 1980). (All sporangial and numerical data updated by kind courtesy of R. H. Roberts.)

Field notes *Polypodium cambricum* is entirely west and southern European in world range, with a sparse and somewhat disjunct distribution from Britain and Ireland to Portugal and North Africa and eastwards along both shores of the Mediterranean to Greece and at least as far east as Turkey. In Britain and Ireland, Southern Polypody grows in base-rich rocky situations, usually on steep slopes, and particularly on ones of southerly aspect in the northerly parts of its range. Most of the sites are within a few hundred metres of sea-level and within the 4°C winter minimum isotherm. Many are coastal, the plants growing around rocky mouths of stream gorges, on hinterland cliffs, on raised-beach terraces and on the walls of old quarries. On appropriate substrates near western and southern coasts in particular, it may become locally frequent, thriving under conditions of year-round high humidity, moderate light rainfall and a long growing season. Plants become most luxuriant where sheltered cliffs or rocky slopes are lightly shaded, and there is slight ground moisture seepage. In humid, western climates, Southern Polypody can also occur on walls, especially where these are constructed of natural limestone rock, and there are large populations in natural habitats nearby. Further inland, it becomes a species most of gorges in limestone areas. It ascends to higher altitudes only occasionally, and in drier, more exposed conditions, plants usually remain stunted.

Polypodium cambricum is a much more strongly calcicolous species than either *P. interjectum* or *P. vulgare*, occurring mostly on Carboniferous, Cretaceous or Devonian limestones, but also sometimes

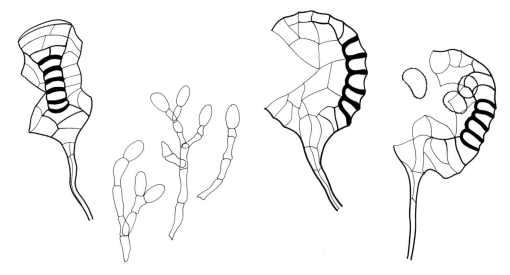

Polypodium cambricum, microscopic detail showing the low indurated cell number of the annulus, and the separate branched paraphyses (see the text).

on other lime-yielding rocks such as calciferous Old Red Sandstones, conglomerates, chlorite schists, basic agglomerates and basalts, and occasionally on ultrabasic rocks. Near coasts, plants also sometimes occur epiphytically on Oak (*Quercus*), Ash (*Fraxinus*), Elder (*Sambucus*), or Hawthorn (*Crataegus*), over base-rich soils in coastal woodland, or occur in fissures of acidic rocks where there is continual accumulation of sufficient wind-blown calcareous sand to maintain locally basic conditions. Hughes (1969) found the pH of soils at Welsh sites to vary from 6.4 to 6.9.

Associated species in drier, more-exposed sites frequently include Bloody Cranesbill (*Geranium sanguineum*), Common Rock-rose (*Helianthemum chamaecistus*), Meadow Oat Grass (*Helictotrichon pratense*), Wall Rue (*Asplenium ruta-muraria*), and Rusty-back Fern (*Ceterach officinarum*). In darker, moister sites it can be more aggressively dominant, and its few associates here usually include Ivy (*Hedera helix*), Common Maidenhair Spleenwort (*Asplenium trichomanes* subsp. *quadrivalens*), Hart's Tongue Fern (*Phyllitis scolopendrium*) and Soft Shield Fern (*Polystichum setiferum*).

Fronds emerge later in the season than do those of other *Polypodium* species and are less cold hardy – their pinna tips often suffering from winter damage by February. Sori mature from late autumn until the following spring, and this predominantly winter growing season seems characteristic of a species of warmer climates reaching its most northerly limits along the milder fringes of Britain and Ireland.

Polypodium interjectum Shivas   Western Polypody

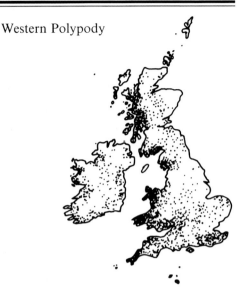

Preliminary recognition A small to medium-sized wintergreen fern, with coarsely cut herringbone-like fronds, of fairly thick texture and long, oval-outlined fronds, arising from scaly, surface-creeping rhizomes.

Occurrence Widespread throughout Britain and Ireland, becoming most abundant and luxuriant in humid, high rainfall areas of southern and western districts, on rocks, walls and hedgebanks at low altitude, and sometimes epiphytic in sheltered situations.

Identification Fronds show considerable variation in size, though are less variable in form than are those of *P. cambricum*. Fronds of mature plants, 15–60 cm or more in length, have a *long, distinctly oval outline shape*. The longest pinnae are situated near the middle of the blade, the pinnae shortening gradually both above and below this point. Overall, the pinnae are shorter in relation to frond length than in *P. cambricum*, and although they are sometimes tapering, many fronds have pinnae with bluntly rounded tips throughout. The pinnae usually have very shallowly serrate margins. The fronds are of *moderately thick, and slightly fleshy and rigid texture* and, when growing in shade, they have a mid-green, non-shining, upper surface. They are borne on rather thick, rigid stipes, and because of their weight and length, well-grown fronds usually adopt an arching habit in contrast to the more ascending one of *P. vulgare*. The lowermost pinna pair, and sometimes the next few pairs above it, are inflexed partially forward, but seldom so markedly so as in *P. cambricum* (q.v.).

The sori vary from oval to more circular in shape with the most oval ones usually occurring near the pinna–rachis junction, and are of a yellow-brown colour when mature. Plants are much more freely fertile than those of *P. cambricum*, and in exposed situations, they may bear sori throughout much of the upper 1/2 to 2/3 of the frond. Such sori are frequently slightly sunken into the frond undersurface so that the position is marked on the upper surface by a series of distinct low protuberances.

Variation Plants are widely influenced in size by the environment. In exposed situations fronds become much more coriaceous than those in shade, and of a much yellower-green colour. Occasional plants in sheltered habitats produce fronds with deeply serrate pinnae, as in some

Jan.	Feb.	Mar.	Apr.	May	Jun.	Jly	Aug.	Sep.	Oct.	Nov.	Dec.

Sub-Atlantic range of *Polypodium interjectum* in Europe.

forms of *P. cambricum*, and in some plants, such serrations are more developed on the lower than the upper pinna margins.

Possible confusion Particularly luxuriant (especially more serrated) specimens could be confused with *P. cambricum*, more stunted specimens are likely to be confused with *P. vulgare*, which they can closely resemble, and technical confirmation may be necessary. In its overall form, however, the narrower frond outline, more rounded pinna tips, more numerous, more yellow-brown sori, thicker frond texture, less yellow-green colour and less glossy surface when growing in equivalent, shaded habitats, separate *P. interjectum* from *P. cambricum*. The larger frond size with wider outline, more arching habit and more oval sori help to separate *P. interjectum* from *P. vulgare*. The presence of good spores, produced in quantity, separates it from all hybrids, including *P.* ×*mantoniae* and *P.* ×*shivasiae*, of which *P. interjectum* is the common parent, and only the former of which is at all frequent.

Technical confirmation Plants are hexaploids, with $n = 111$, $2n = 222$ chromosomes. The spores are all good, but their sori lack the large, branched paraphyses seen in *P. cambricum* (q.v.).

The combination of good spores and lack of paraphyses separate it from *P. cambricum*. The number of indurated cells of the annulus ranges from 4 to 13, with a mean in different plants varying from 7 to 9 (Shivas, 1961*b*; Roberts, 1970) (cf. range 4–19, mean varying from 5 to 10 in *P. cambricum*, range 7–17, mean varying from 10 to 14 in *P. vulgare*). Usually 2–3 basal cells separate the annulus from the sporangium stalk (cf. 3–4 in *P. cambricum*, usually 1 in *P.*

Polypodium interjectum: all West Cornwall.

vulgare). These features thus clearly differentiate between *P. interjectum* and *P. vulgare*. The indurated cells are 28–35 μm in length and 76–86 μm in breadth (cf. 21–26 × 91–100 μm in *P. cambricum*, 22–28 μm × 60–80 μm in *P. vulgare*), and are thus slightly longer and narrower than those of *P. cambricum*, compared with which the whole annulus appears narrower and more elongate.

In addition, the annulus colour is very variable, and may be a pale-buff colour similar to the sporangium wall, a pale yellow, a golden brown or a bright yellow. This contrasts with the fairly constantly bright to pale yellow annulus of *P. australe* and the constantly dark orange-brown one of *P. vulgare*. The rhizome scales are rather abruptly narrowed above a somewhat dilated base, and are generally broader and shorter (3.5–11 mm) than those of *P. cambricum*.

(All sporangial and numerical data updated by kind courtesy of R. H. Roberts.)

Field notes *Polypodium interjectum* is a species of essentially European range, where it is predominant in the west from Scotland and extreme south-west Norway southward to Portugal, and from this region more discontinuously and progressively more sparsely eastward through the west Mediterranean basin to southern central Europe, Turkey and northern Iran. Because *P. interjectum* is the allohexaploid derivative of *P. cambricum* crossed with *P. vulgare* (Shivas, 1961*a,b*), plants are in many ways morphologically intermediate in appearance between these two parents, and also follow a general intermediacy in climatic and edaphic distribution. As a hexaploid, plants are also normally larger and generally more vigorous than either parent species, and tend to form larger, more extensive and more often conspicuous colonies. They occur mostly in mildly basic conditions, overlapping to some extent with the habitats of the other two species, but are usually absent from the more extremely acidic conditions tolerated by *P. vulgare*. In Wales, in predominantly acidic country, *P. interjectum* becomes confined largely to wall mortar, where it may be locally common, whilst only in generally more base-rich regions does it become frequent in other habitats such as roadside banks (Benoit, 1966). Its additional vigour enables *P. interjectum* effectively to colonise an additionally wide range of edaphically intermediate habitats, including ones in which either one of the other two species may occur and can occasionally also be frequent.

Plants are generally least frequent in central districts and in the more continental climates of the east, but become much more abundant in many areas of the west, especially along coastal strips, where there is year-round high humidity. Under such conditions, *P. interjectum* shows a relatively adaptable ecology. It can be particularly luxuriant in sheltered stream valleys, and frequently becomes epiphytic in the forks and along the branches of large, rough-barked trees, especially Oak (*Quercus*). In such situations, as well as on rocky banks and cliffs, its rhizomes may creep extensively, clinging by masses of fine, surface roots, plants sometimes reaching branches over 12 m (40 ft) from the ground. Colonies may be particularly abundant covering boulders in thin-canopied, coastal woodlands, on a wide range of rock types, including acidic ones where these are influenced by salt-laden air and wind-blown calcareous sand. On nearby walls, plants are often stunted in height although rhizomatously extensive and probably long-lived, and on old rural walls originally capped by grass turfs, colonies of *P. interjectum* sometimes extend for as far as such walls stretch. Sometimes plants occur also as a component of mature sand dune systems (Jermy *et al.*, 1978), and in south-west Scotland, in exposed turfy coastal vegetation, many seem to mark sites of nests of the pale brown ant *Lavius flavus*.

In wall mortar habitats, associated plant species include mostly other wall ferns: Black Spleenwort (*Asplenium adiantum-nigrum*), Common Maidenhair Spleenwort (*A. trichomanes* subsp. *quadrivalens*), Wall Rue (*A. ruta-muraria*), and occasionally Rusty-back Fern (*Ceterach officinarum*). In roadside banks, frequent associates include Ivy (*Hedera helix*), Black Spleenwort (*Asplenium adiantum-nigrum*) and a number of grasses.

Fronds of *P. interjectum* flush earlier than do those of *P. cambricum* but later than those of *P. vulgare*. They remain green throughout winter, and appear more winter-hardy than those of *P. cambricum*, seldom suffering equivalently from cold-scorch. They thrive in conditions of good drainage

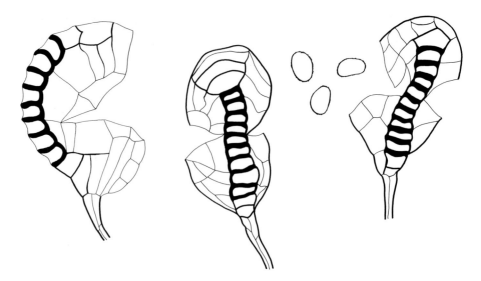

Polypodium interjectum, microscopic detail showing the intermediate indurated cell number of the annulus (see the text).

and moderate soil moisture, especially where shaded by light deciduous tree canopies, providing some shade in summer but allowing good winter and spring-time illumination.

Even more than the other two species of *Polypodium*, *P. interjectum* is the single native fern most closely reminiscent in kind, through its general verdure and extent, of the great array of epiphytic and rupestral fern luxuriance so characteristic of moist mountain flanks in the forest of the tropics (Page, 1979a). Appropriately, throughout almost the whole of west Wales and the English West Country, and in much of the counties of Merioneth, Caernarvonshire, Cardigan, Pembrokeshire, Devon and Cornwall in particular, *P. interjectum* becomes the dominant fern of very many steep rocky slopes, especially on their more sheltered and gently shaded aspects. In such sites sometimes extensive colonies can occur, including along the rocky bases of the north-facing aspects of many deeper railway cuttings, where verdant growths of fronds flushing in mid-July draw vivid attention to their widespread and often luxuriant presence, clearly visible from the window of even a speedily passing train.

Observations made by the author suggest that in different localities in Britain and Ireland, various different forms of *P. interjectum* exist, each of them often with a morphological structure that can very specifically reflect the different morphologies also to be seen in the local forms of plants of *P. cambricum*. Such correspondence between local morphological variation in a diploid species with that also to be seen in the local allohexaploid derivative of it is of great evolutionary interest in pteridophytes. It certainly would appear to suggest that, in several independent localities in these islands, in the case of *P. interjectum*, the same hexaploid has formed anew, each time incorporating the particular variants of the genomes of the local diploid species. These observations provide intriguing possible evidence of likely multiple origins, at different times and perhaps in many different places, of the same genomic complement of polyploid fern and hence of the same polyploid taxon arising separately and independently persisting locally.

Polypodium vulgare L.

Common Polypody

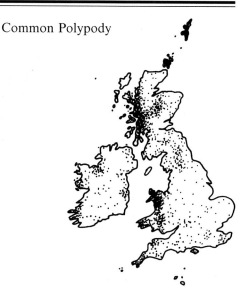

Preliminary recognition A small-sized wintergreen fern with coarsely cut herringbone-like fronds, of thick texture and narrow, often quite parallel-sided outline, arising from scaly, surface-creeping rhizomes.

Occurrence The most widespread and northerly of the three closely allied species of *Polypodium*, occurring frequently throughout much of Britain and Ireland, mainly on acidic rock outcrops, sometimes in exposed situations.

Identification Except in size, this is the least variable of the species of *Polypodium*. Fronds of young or stunted specimens, under 10 cm in length, are usually of triangular outline. Those of mature plants, 10–25 cm in length, become progressively *more narrowly ovate or linear strap-shaped with size, often with the lower 1/3 to 1/2 or more of the blade approximately parallel sided*. All pinnae are short, usually with entire or scarcely serrate margins, and bluntly rounded ends. Fronds are *of thick and rigid texture*, and the pinnae are frequently somewhat twisted, especially the lower ones, with their ends turned into a more nearly horizontal plane. In exposed situations, fronds usually adopt a particularly low-growing habit, coriaceous texture and rather yellow-green colour. In more typical, sheltered situations, however, the fronds are larger and held more stiffly erect, and are mid-green in colour. In all habitats, the fronds have a rather dull surface, and the lower pinnae are seldom as inflexed as in other species.

Fronds are usually freely fertile, with sori present throughout the upper 1/4 to 3/4 of the frond, even in vegetatively luxuriant specimens. The sori are almost always of circular shape, even when at a young, green, stage, and turn *a conspicuous reddish-brown to bright-orange colour when mature*, with abundant yellow spores.

The annulus is of a *consistently darker, more reddish-brown colour* than that of other species, its colour contrasting sharply with that of the rest of the sporangium (see 'Technical confirmation'), and visible as a *shining thin brown line* even under a × 10 hand lens (Benoit, 1966). On the upper surface of the frond, slightly raised puckerings mark the position of sori beneath.

Fronds flush earlier in summer than do those of other species and hybrids (see 'Field notes') and this character can be particularly useful in the field when several species occur nearby.

Jan.	Feb.	Mar.	Apr.	May	Jun.	Jly	Aug.	Sep.	Oct.	Nov.	Dec.

Variation The size of fronds varies widely with exposure. Their form is subject mainly to minor inherent differences in outline shape.

Possible confusion The narrow, thick-textured, mid- to fairly dark-green fronds and calcifuge habitats, help to separate most specimens of *P. vulgare* from *P. cambricum* in the field. Plants are most likely to be confused with *P. interjectum*, but the smaller size, narrower, more parallel-sided frond, and more rounded red-brown sori all with distinct, dark annuli, help to identify most specimens of *P. vulgare*, but technical confirmation may be needed for doubtful material. The presence of good spores in quantity separate it from *P.* ×*font-queri* and *P.* ×*mantoniae*, of which *P. vulgare* is the common parent (and only the latter of which is at all frequent).

Technical confirmation Plants are tetraploids, with $n = 74$, $2n = 148$ chromosomes. The spores are all good, and their sori lack the large, branched paraphyses seen in *P. cambricum* (q.v.).

The combination of good spores and lack of paraphyses separate it from *P. cambricum*. The number of indurated cells of the annulus ranges from 7 to 17, with a mean in different plants varying from 10 to 14 (Shivas, 1961b; Roberts, 1970) or even higher in some populations (e.g. Peterken, 1962) (cf. range 4–19, mean varying from 5 to 10 in *P. cambricum*, range 4–13, mean varying from 7 to 9 in *P. interjectum*). Usually only a single, small, basal cell, little wider than the annulus itself, separates the end of the annulus from the sporangium stalk (cf. 2–3 basal cells in *P. interjectum*, 3–4 in *P. cambricum*). These features are strongly diagnostic of *P. vulgare*, clearly separating it even from *P. interjectum*. The annulus is also much narrower than that of the other two species, *c.* 60–80 μm (cf. 76–86 μm in *P. interjectum*, 81–100 μm in *P. cambricum*), although the length of each cell is similar to that of *P. cambricum* (Roberts, 1965, 1970).

In addition, the annulus is of a consistently dark reddish-brown colour compared with that of the other two species. The rhizome scales are also specifically distinctive: each is shorter and has a broader, more rounded base than in the other two species.

(All sporangial and numerical data updated by kind courtesy of R. H. Roberts.)

Field notes Polypodium vulgare is an extremely wide-ranging species, almost across the whole of the European continent from Iceland and Arctic Norway to the Mediterranean and eastwards through central Europe at least as far as the Black Sea. Plants of what may be the same taxon further occur discontinuously, mainly in more mountainous areas across northern Asia, such as the Urals. In Britain and Ireland, Common Polypody occurs very widely through Britain and Ireland, sometimes ascending to considerable altitude. It is a weak calcifuge, contrasting with the more calcicolous requirements of the other two species. It consequently occurs on a wide range of acidic rock types, such as grits and sandstones, but also on harder, metamorphic and volcanic ones such as gneiss and granites, especially where it can establish in small fissures. In inland acidic districts, it is often the only species of *Polypodium* present in natural situations. It seems also more exposure-tolerant and light-demanding than the other *Polypodium* species, as well as more winter-hardy. It is most frequently found on the tops and shoulders of rock-outcrops and rocky bluffs, and ascends in such habitats as stunted specimens to mountain cliffsides. In more lowland habitats, however, it can occur abundantly in moist drystone walls, its scaly rhizomes weaving amongst blocks, with fronds emerging. Plants become most luxuriant, however, where there is greater humidity, more shelter and light summer shade. Colonies may become particularly extensive on boulders and on overgrown walls in light-canopied woodland, such as that of Sessile Oak (*Quercus petraea*) over base-deficient soils. With increased humidity and greater moss growth, plants also become epiphytic on exposed roots and boles of trees, especially in valleys and ravines along stream courses. In darker, moist,

Polypodium vulgare: a–b, Harris, Outer Hebrides; *c*, Merionethshire; *d–e*, Mid Perthshire; *f*, West Argyll; *g*, Kirkcudbrightshire; *h*, Dumfries-shire, showing soral distribution.

Polypodium vulgare: *a*, Northumberland; *b–d*, West Cornwall.

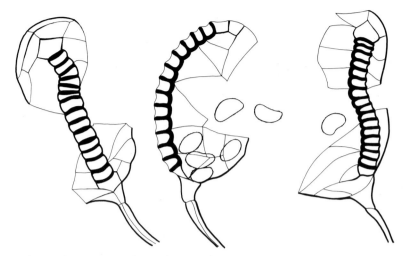

Polypodium vulgare, microscopic detail showing the high indurated cell number of the annulus (see the text).

acidic woodland, with more luxuriant growths of mosses and ferns, plants can become displaced to entirely epiphytic situations, becoming replaced on boulders by faster-growing specimens of Broad Buckler-fern (*Dryopteris dilatata*). In dark, coniferous plantations, *P. vulgare* usually becomes confined to places where outcropping rocks or old walls coincide with permanent breaks in the canopy.

Fronds of *P. vulgare* flush earlier in the season than those of other species of *Polypodium*, and are usually fully expanded by mid-July. They remain green until early in the following summer, and show minimal damage even after severe winter frosting.

Except for mosses, associated species are few in most habitats, but may include Honeysuckle (*Lonicera periclymenum*), Bramble (*Rubus fruticosus* agg.), Wood Sage (*Teucrium scorodonia*), White Stonecrop (*Sedum album*), Wall Pennywort (*Umbilicus rupestris*), Delicate Maidenhair Spleenwort (*Asplenium trichomanes* subsp. *trichomanes*), and sometimes Ling Heather (*Calluna vulgaris*) and Bracken (*Pteridium aquilinum*).

Polypodium ×font-queri Rothm.

Font-Quer's Polypody

(*P. cambricum* × *P. vulgare*)

Preliminary recognition Usually a rather small-fronded plant, with fronds broader than those of *P. vulgare*, more bell-shaped in outline than those of *P. cambricum* and of thinner texture than those of *P. interjectum*.

Occurrence Known in a very limited number of widely scattered sites in England, Wales and Scotland, chiefly in the west, usually growing on steep outcrops of base-rich rock, or on walls.

Identification P. ×*font-queri* is a smaller and usually less vigorous plant, forming smaller colonies than P. ×*shivasiae*, the other *P. cambricum* hybrid. Its fronds, mostly 20–30 cm long, are usually conspicuously wider than those of *P. vulgare*, and have longer, more linear pinnae. Small fronds are usually of a broadly deltoid outline, like a broad gothic arch, similar to small *P. cambricum* fronds. *Larger ones gradually adopt a more bell-shaped outline*, with the lower pinnae becoming long and exceeding those above in length, so that *the lower part of the frond has a flared-out silhouette*. These lower pinnae are usually held steeply forward of the rest of the frond, a character particularly inherited from *P. cambricum*. The fronds are slightly more rigid in texture than those of *P. cambricum*, but are notably thinner and more supple and flexible than those of *P. vulgare*. They usually inherit the *rather yellow-green colour and slightly glossy surface* of *P. cambricum*, or are intermediate in these features. They often show some of the slight twisting of the pinnae sometimes seen in *P. vulgare*, which, because of the thinness and length of pinnae of the hybrid, can give the fronds a rather crinkled appearance.

Plants are slow growing and colonies may have few fronds bearing sori. When present, sori are red-brown, and at first form only sparsely near the tip of the blade, but on more mature fronds, may eventually occupy much of the upper 2/3 of the blade, thus eventually becoming more extensive than is typical for *P. cambricum*, and more similar in this feature to *P. vulgare*.

Variation Plants inherit some of the variation of both parents, especially in details of frond and pinna shape. Occasional specimens with serrated lower pinnae occur.

Possible confusion The intermediacy of frond texture, frond outline and abortive spores, distinguish P. ×*font-queri* from both parents. These characters, plus the yellower-green, thinner-textured frond also help to separate it from *P. interjectum*. Technical characters will, however, usually be necessary to diagnose it with confidence from amongst smaller specimens of the other hybrids.

Technical confirmation Plants are triploid hybrids, with $2n = 111$ chromosomes, all of which remain unpaired at meiosis. Recognition of spore sterility confirms it as a hybrid: all the spores are highly mis-shapen, thus differing slightly from those of P. ×*mantoniae* (q.v.).

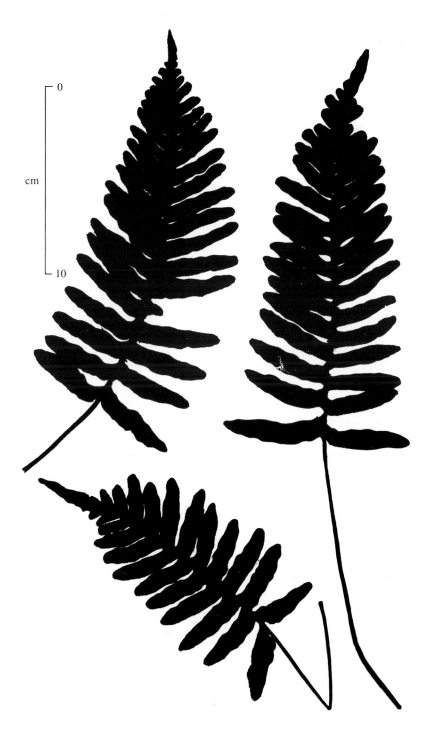

Polypodium ×*font-queri: all* Anglesey (courtesy R. H. Roberts).

Without the benefit of a chromosome count, diagnosis between the hybrids can be best undertaken by microscopic examination of the sporangia. In *P.* ×*font-queri*, the number of indurated cells in the annulus has a range of 8–20, with the mean in different plants ranging between 11 and 14 (cf. range 5–11, mean varying between 7 and 8 in *P.* ×*shivasiae* (the 4*n* hybrid) and range 2–14, mean varying between 9 and 10 in *P.* ×*mantoniae* (the 5*n* hybrid). The number of basal cells of the annulus ranges from 1 to 3, mean varying from 1.5 to 2.5 (cf. range 1–4, mean varying from 2.6 to 3.1 in *P.* ×*shivasiae*, range 1–3, mean varying from 1.5 to 2.0 *P.* ×*mantoniae*). These characters thus separate *P.* ×*font-queri* fairly clearly from *P.* ×*shivasiae*, but less clearly from *P.* ×*mantoniae* (which shares the same *P. vulgare* parent). With *P.* ×*mantoniae* there is more statistical overlap, and critical examination of sufficient sporangial material is needed. However, the length of the annulus cells averages only 18–24 μm in *P.* ×*font-queri*, and this is notably shorter than the 28–35 μm of *P.* ×*mantoniae*. In addition, the annulus, which is sometimes yellow or reddish-brown, but often of a fairly pale colour, approaches that of the *P. vulgare* parent in width (about 60–80 μm), and its basal cells are only a little broader than the annulus. Paraphyses have not been found in wild material of this hybrid.

(All sporangial and numerical data by kind courtesy of R. H. Roberts.)

Field notes Most stations for this hybrid occur at low altitude, in high-rainfall areas near western coasts, in areas of overlap of the two parents. They usually occur where there is outcropping base-rich rock in close proximity to moderately acidic habitats and where each parent grows nearby in some quantity. In such areas, the hybrid usually forms only relatively small colonies, probably mostly of independent origins.

At least some of its sites are on walls which retain earth banks, and are constructed of different local stones. In Anglesey, where a number of colonies have been found in such habitats, the walls include limestones and base-rich chlorite schists, plus other less basic rocks. Such conditions provide a mosaic of well-drained habitats of different base content, with abundant niches between the stones in which parental prothalli presumably establish in close proximity.

Polypodium ˣmantoniae Rothm.   Manton's Polypody

(P. interjectum × P. vulgare)

Preliminary recognition A vigorous hybrid, with fairly narrow fronds of appearance intermediate between those of the parents, arising in dense, leafy masses, which are usually more compact and closely spaced than in *P. interjectum.*

Occurrence The commonest of the three native *Polypodium* hybrids in Britain and Ireland, and present in at least 29 vice-counties, especially in the west. It sometimes forms extensive local colonies over boulders and on tree boles in low-canopied, fairly open, coastal woodlands and on roadside banks and walls, sometimes locally outnumbering its parents.

Identification Plants can be distinguished by their fronds of typically narrower outline than *P. interjectum*, and with more parallel-sided pinnae which broaden outwards less and have generally more rounded tips. Sometimes fronds approach those of *P. vulgare* fairly closely in form, but are usually of larger size (often 20–45 cm), and are *less parallel-sided in blade outline*, with the widest point occurring about 1/4 to 1/3 of the way from the base of the blade. They possess, however, the coarseness and thickness of texture of *P. interjectum*. In the field, their fronds usually begin to emerge about two weeks later than those of *P. vulgare*, but about two weeks ahead of those of *P. interjectum*, so that by mid-July, when just flushing, such colonies can be recognised relatively easily.

Sori usually occur over much of the lower surface of most of the mature fronds. The sori vary in shape from more or less oval near to the base of each pinna to a majority that are more nearly circular. Usually all are much smaller than those of *P. vulgare*, although they usually have a similar deep orange-brown colour, as most sporangia inherit the dark red-brown annulus of the *P. vulgare* parent (varying somewhat in this, see 'Technical confirmation').

The sporangia contain a very high proportion of abortive, shrivelled-looking spores (see 'Technical confirmation').

Variation Plants show some variation in overall frond outline between both parents, but other characters seem to remain more constant.

Possible confusion The abortive spores, intermediate seasonal behaviour and fronds combining the coarseness and thickness of texture of *P. interjectum*, together with the dark-coloured annulus more like that of *P. vulgare*, help to distinguish the hybrid from both parents, but technical confirmation is necessary to distinguish plants firmly from other hybrids.

Technical confirmation Plants are pentaploid hybrids, with $2n = 185$ chromosomes, forming approximately 74 pairs and 37 single chromosomes at meiosis. Recognition of spore sterility confirms it as a hybrid: the spores vary in appearance from some of more or less normal size and appearance and yellow colour, with a few that are exceptionally large, to the vast majority

Polypodium ˣ*mantoniae: all* West Cornwall.

that are small, shrivelled and whitish. In quantity, under a hand lens, these whitish-looking spores contrast with the uniformly bright-yellow spores of all the species (Benoit, 1966).

Without the benefit of a chromosome count, diagnosis between the hybrids can best be undertaken by microscopic examination of the sporangia. In *P.* ×*mantoniae*, the number of indurated cells of the annulus has a range of 2–14, with a mean in different plants varying between 9 and 10 (cf. range 5–11, mean varying between 7 and 18 in *P.* ×*shivasiae* (the 4*n* hybrid) and range 8–20, mean varying from 11 to 14 in *P.* ×*font-queri* (the 3*n* hybrid)). The number of basal cells of the annulus ranges from 1 to 3, mean varying from 1.5 to 2.0 (cf. range 1–4, mean varying from 2.6 to 3.1 in *P.* ×*shivasiae*, range 1–3, mean varying from 1.5 to 2.5 in *P.* ×*font-queri*). These characters thus separate *P.* ×*mantoniae* fairly clearly from *P.* ×*shivasiae*, but less clearly from *P.* ×*font-queri* (which shares the same *P. vulgare* parent). With the latter there is more statistical overlap, and critical examination of sufficient sporangial material is needed. However, the length of the annulus cells is similar to that of *P. interjectum* (28–35 μm) or only slightly shorter, and this is notably longer than the 18–24 μm of *P.* ×*font-queri*.

In addition, the characteristically dark-coloured annulus varies from reddish-brown, as in *P. vulgare*, to a deep orange-yellow or is occasionally variously coloured, and the annulus width (*c.* 74–80 μm) is intermediate between that of the parents. The walls of the indurated cells look thick compared with those of other hybrids. Paraphyses are absent.

(All sporangial and numerical data by kind courtesy of R. H. Roberts.)

Field notes *Polypodium* ×*mantoniae* may be by no means infrequent in western districts, and its occurrence has been particularly widely recognised by Welsh botanists in west Denbighshire, west Merionethshire, Caernarvonshire, Anglesey, Pembrokeshire and west Carmarthenshire. It would not be surprising to find many more stations for it in other western areas of humid climate where the parents grow together in abundance, particularly in Devon, Cornwall, the lowland parts of the Lake District or the south-west Scottish coast.

In the field, the vigour but compactness of growth of *P.* ×*mantoniae* often enables it to form dense colonies on mossy boulders over a wide range of rock types, varying from limestones to such acidic ones as granites, sandstones and grits, extending to the more acidic rocks mainly where these are modified by blown sand or salt-laden winds. Typical associated species on such rocks in coastal sites include Ivy (*Hedera helix*), Wood Sage (*Teucrium scorodonia*), English Stonecrop (*Sedum anglicum*) and Bramble (*Rubus fruticosus* agg.).

Observations made by Benoit (1966) on North Wales material, show that fertility varies between plants. In some, a proportion of normal-looking spores occur, and second-generation plants could thus arise. Some, which approach *P. vulgare* more in appearance, also have a particularly high proportion of apparently normal spores, suggesting the possibility of backcross segregants. Subsequently, Crabbe & Shivas (1975) have, indeed, reported cytological confirmation of one such known backcross to the *P. vulgare* parent, in material of Teesdale origin. Such plants are, however, generally scarce, and most wild 5*n* hybrids are probably of F₁ (first generation) interparental crosses.

Polypodium ˣshivasiae Rothm.

Shivas' Polypody

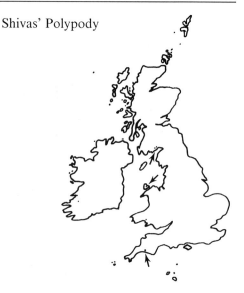

(*P. cambricum* × *P. interjectum*)

Preliminary recognition The most spectacular of the three hybrid polypodies, combining vigour with large size, looking like a larger, leafier, form of *P. interjectum*, with a wider frond outline, thinner, more flexible frond texture and yellower-green colour.

Occurrence An uncommon hybrid, occurring very locally in widely separated parts of England, Wales, Scotland and Ireland, mostly in western coastal districts and especially on sheltered, moist cliffs of basic rock, where both parents occur in quantity nearby.

Identification Plants may have fronds up to 40–65 cm or more in length, and these typically have an ovate outline that is rather wider than that of *P. interjectum.* They are also usually of a lighter, more yellow-green colour and more flexible texture, but are slightly more rigid than those of *P. cambricum* and are held away from the vertical rock surfaces on which they usually grow in a gracefully arching fashion. The fronds may have coarsely serrated pinna margins, especially on the lowermost pinnae.

   Sori are infrequent, especially in colonies that are vegetatively luxuriant, and are usually confined to a few of the upper pinnae. The sporangia contain spores of varied shapes, of which a few are large and spherical and many abortive; many sporangia fail to open.

Variation Plants vary in size according to luxuriance of the habitat, and can vary widely in degree of serration of the pinnae, even within a single colony.

Possible confusion The generally intermediate appearance of the frond outline, abortive spores and large size, distinguish *P.* ˣ*shivasiae* from either parent, as well as from *P. vulgare* (q.v.). The generally large size of the fronds helps to separate the plant from other hybrids, although technical characters may be needed to confirm its identity with confidence.

Technical confirmation Plants are tetraploid hybrids, with $2n = 148$ chromosomes, forming approximately 37 pairs and 74 single chromosomes at meiosis. Recognition of spore sterility confirms it as a hybrid.

   Without the benefit of a chromosome count, diagnosis between the hybrids can be best undertaken by microscopic examination of the sporangia. In *P.* ˣ*shivasiae*, the number of indurated cells in the annulus has a range of 5–11, with a mean in different plants varying between 7 and 8 (cf. range 2–14 (mean varying between 9 and 10) in *P.* ˣ*mantoniae* (the 5n hybrid) and range 8–20 (mean varying from 11 to 14) in *P.* ˣ*font-queri* (the 3n hybrid)). The number of basal cells of the annulus ranges from 3 to 4, mean varying from 2.6 to 3.1 (cf. range 1–3, mean varying from 1.5 to 3.5 in *P.* ˣ*font-queri*, range 1–3, mean varying from 1.5 to 2.0 in *P.* ˣ*mantoniae*). These characters thus separate *P.* ˣ*shivasiae* fairly clearly from both

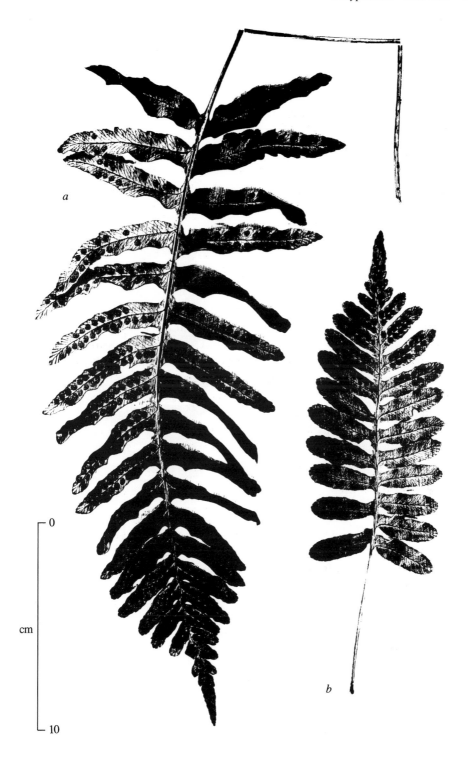

Polypodium [×]*shivasiae: a*, Anglesey; *b*, Kirkcudbrightshire (specimens courtesy R. H. Roberts and A. McG. Stirling), showing soral distribution.

Polypodium ×*shivasiae*: *a*, Kircudbrightshire; *b*, Anglesey.

0   10

cm

Polypodium [×]*shivasiae*: *all* east Cornwall.

P. [×]*font-queri* and *P.* [×]*mantoniae*. The length of the annulus cells averages from 24 to 26 μm, and this is intermediate between the 18–24 μm of *P.* [×]*font-queri* and the 28–35 μm of *P.* [×]*mantoniae*.

In addition, the annulus varies in colour from bright yellow to nearly colourless, through a range of shades similar to that of the *P. interjectum* parent. Paraphyses nearly as long as those of *P. cambricum*, but usually less freely branched and less abundant, are found in some plants of this hybrid but not in others (even in nearby populations). They have been illustrated from wild (Anglesey) material by Roberts (1970).

(All sporangial and numerical data by kind courtesy of R. H. Roberts.)

Field notes *Polypodium* ×*shivasiae* is a rare but vigorous hybrid, and under suitably mild, moist, sheltered conditions, may form extensive, leafy colonies. It occurs locally on limestone walls, but undoubtedly becomes most extensive where there are steep cliff-face exposures of basic rock, in humid districts at very low altitude near to the sea.

Under such conditions, colonies may form large, pure stands of arching fronds covering many square metres of moist rock faces, usually with both parents nearby, and with *P. interjectum* often in quantity. *Polypodium* ×*shivasiae* thrives mainly where there is some shade and shelter from light tree overgrowth, such as that of ash-dominated woodland. In such habitats, the hybrid probably competes more or less directly with both parents, and like the parents, its growth probably remains active during the winter months, when illumination levels are higher through loss of deciduous tree canopies.

Polystichum aculeatum (L.) Roth   Hard Shield-fern

(*P. lobatum* (Huds.) C. Presl)

Preliminary recognition A medium-sized fern, with leathery textured, coarsely bipinnately divided, lanceolate-outlined fronds, arising in a spreading fashion in rather sparse clusters. Their pinnules are very sharply pointed with distinct spinose margins.

Occurrence A mainly upland species, ranging throughout Britain and Ireland, always rather local, but becoming most frequent in northern England and Scotland. It occurs mainly in steep, damp rocky places, where lime-rich rock is exposed, particularly on rocky outcrops in river-valley woodland, and along narrow, shaded stream gorges, ascending to limestone pavements and rocky screes and fissures, on mountain slopes.

Identification Plants have medium-sized (to about 60 cm, occasionally more), bipinnately or less divided, *rather hard, rigid, leathery textured, shiny fronds*, which are of pale grass-green colour when flushing, darkening rapidly to a glossy deep green once fully expanded. They are of narrowly lanceolate outline, and arise in sparse, often asymmetric, shuttlecock-like clusters, from short, stocky, ascending rhizomes, which occasionally branch. Fronds in each crown are usually much less numerous than is typical for *P. setiferum* (q.v.), and mostly much more horizontally spreading in habit.

Jan. Feb. Mar.	Apr.	May	Jun.	Jly	Aug.	Sep.	Oct. Nov. Dec.

Polystichum aculeatum, habit of plant in life.

The stipes are about 1/4 to 1/5 or less the length of the frond (hence much shorter than those of *P. setiferum*), and *fairly thickly clothed with mixed large and small, chaffy, reddish-brown scales.* Mature fronds mostly have about 20–40, rather closely spaced pinnae on either side, the longest about 5–7 cm in length, situated at about the middle of the frond. Below this, the pinnae become progressively shorter, with the lowermost pair only about half the length of the middle ones and only pinnatisect (cf. *P. setiferum*, where the basal pinnae are not much shorter than those above and are fully pinnate). Each pinna usually curves slightly upwards towards the tip of the frond in a sickle-like manner, whilst tapering gradually into a very acute apex (much sharper than that of *P. setiferum*). In young plants, fronds have pinnae that are merely serrate or pinnatifid, but with increasingly mature specimens, the pinnae become progressively more distinctly cut into segments which are *short, acute and rigid,* and usually fewer than those of *P. setiferum*. All but the innermost few of these are joined by the whole of their bases to the pinna midrib (adnate). The innermost segment on the upper side of each pinna is usually conspicuously larger than the rest, and more deeply divided. All the segments have an outward-pointing lobe at their base, and the *tip of this lobe, as well as the outer tip of each segment, comes to an acutely angled point, which is drawn out into a fairly long, rather stiff (aristate) spine. Further spines also arise along the margins of each segment, giving the whole frond a spinose-margined appearance.*

Fertile fronds are similar to the vegetative ones, with each segment in the upper part of the

Polystichum aculeatum: a/a, Renfrewshire; *b–e*, Mid Perthshire.

frond bearing two rows of fairly large, rounded sori, each covered by a peltate indusium. At maturity, the indusia shrivel upwards, like inverted umbrellas, revealing the ripe sporangia below, like clusters of minute black grapes.

Fronds are winter-persistent, expanding in spring and remaining fully green until well into the following season, *usually long after the following year's fronds have fully flushed.* In early summer, specimens with two-year's green fronds thus often occur, the older fronds then dying but remaining for some time attached and hanging shrivelled below the current year's growth.

Variation There is wide variation in degree of cutting of fronds between juvenile and more adult specimens, and some variation between adults. Occasional specimens are so little cut they approach *P.* ×*illyricum* in appearance, whilst some more extensively divided ones can approach the appearance of *P. setiferum.*

Possible confusion Adult plants are likely to be confused only with *P. setiferum* (q.v.), but the harsher, more leathery fronds, with more closely spaced pinnae bearing more acute segments, pinnae reducing towards the frond base and usually shorter stipe length, clearly distinguish most specimens of *P. aculeatum* (q.v.). Juvenile specimens are often taken for mature plants of *P. lonchitis*, but their lack of sori, as well as their broader frond outline, more deeply cut pinnae and shorter spine length, distinguish *P. aculeatum*. The two hybrids *P.* ×*bicknellii* (with *P. setiferum*) and *P.* ×*illyricum* (with *P. lonchitis*), are also superficially rather similar, but the presence of good spores distinguishes *P. aculeatum* from both.

Technical confirmation Plants are tetraploids, with $n = 82$, $2n = 164$ chromosomes. Spores are larger (about 36–45 μm) and darker in colour than those of *P. setiferum*, and have a minutely papillate surface. *Polystichum aculeatum* is the allotetraploid derivative of the cross: *P. lonchitis* × *P. setiferum*.

Field notes *Polystichum aculeatum* is a predominantly west European species in range, its overall distribution stretching from extreme south-west Norway through Britain and Ireland to the Azores, the Iberian Peninsula, and the extreme North African coast, thence westwards discontinuously through mainly mid-European latitudes to at least the Caspian Sea. What are probably further related taxa occur in the Himalayas and Japan.

In Britain and Ireland, Hard Shield-fern is a calcicolous species, of widespread but often somewhat local occurrence, becoming commonest in the northern part of its range, especially in the valleys and glens of mountain districts of northern Ireland, northern England and most of Scotland. It occurs especially in areas of high rainfall, and becomes much more scarce in the more continental, increasingly summer-dry, climates of eastern England. Plants grow mainly in rocky places in river valley woodland, on steep rocky slopes, and on rock-face exposures along stream gorges and narrow ravines, often where permanently shaded,

and normally where lime-rich rocks are exposed. Plants also ascend in mountain habitats to around 760 m (*c.* 2500 ft) in the Scottish Highlands, where they grow on damp, shaded, slightly flushed ledges and in sheltered crevices in lime-rich pockets. In northern England, scattered plants occur also in deeper (probably damper) grykes of limestone pavements, and as scattered individuals around the mouths of limestone caves, sink-holes and mine-shafts. It is almost always a rather solitary species, occurring individually or in small numbers only, and is seldom as gregarious as *P. setiferum*. Associated plants are also often relatively few. In lowland ravines, these usually include plants typical of moist, shady places such as Wood Bitter-cress (*Cardamine flexuosa*), Opposite-leaved Golden Saxifrage (*Chrysosplenium oppositifolium*), Brittle Bladder-fern (*Cystopteris fragilis*) and Hart's Tongue (*Phyllitis scolopendrium*), whilst additional species in limestone grykes and limestone woods usually include Herb Robert (*Geranium robertianum*), Wood-sorrel (*Oxalis acetosella*), Common

Dog-violet (*Viola riviniana*), Mountain Melick (*Melica nutans*), Common Maidenhair-spleenwort (*Asplenium trichomanes* subsp. *quadrivalens*), and Wall Rue (*A. ruta-muraria*). In mountain habitats, it frequently grows with a richer assemblage of characteristic alpine calcicoles and woodland relicts. Some of those noted in Scottish Highland stations include Alpine Lady's-mantle (*Alchemilla alpina*), Moschatel (*Adoxa moschatellina*), Wood-sorrel (*Oxalis acetosella*), Wood Anemone (*Anemone nemorosa*), Bloody Cranesbill (*Geranium sanguineum*), Northern Buckler-fern (*Dryopteris expansa*), Green Spleenwort (*Asplenium viride*), Common Maidenhair-spleenwort, Brittle Bladder-fern, and Holly Fern (*Polystichum lonchitis*). Where *P. aculeatum* and *P. lonchitis* occur together (usual only in the lower margin of the range of *P. lonchitis*), the hybrid *P.* ×*illyricum* (q.v.) has been recorded as a rarity in both Scotland and Ireland. By contrast, in the lower part of the altitudinal range of *P. aculeatum*, there is an extensive overlap with *P. setiferum* (although the two are not often together), but the hybrid *P.* ×*bicknellii* (q.v.) is of not infrequent occurrence.

In most of its habitats, *P. aculeatum* grows where there is a small accumulation of calcium-rich mineral soil between boulders or in fissures. It thrives particularly where there is some seepage of moisture around the roots combined with very free drainage, with its roots in particularly close contact with the rock (much more so than is characteristic of *P. setiferum*). In lowland ravines, it often grows near trickling water, with its fronds in a hanging position, and in such situations, its smaller size and fewer-fronded crown than *P. setiferum* are perhaps of ecological advantage in enabling it to maintain a firmly rooted hold in such precarious habitats.

Through its greater winter-hardiness, *P. aculeatum* is much less confined to the Atlantic fringe of Britain and Ireland than is *P. setiferum*. But its habitats are mostly ones that are very sheltered and have a year-round high humidity, minimising the risk of winter desiccation.

Polystichum lonchitis (L.) Roth

Holly Fern

Preliminary recognition A usually small, narrow-fronded fern, of tough leathery texture and glossy appearance, once pinnately divided into simple, sharply angular pinnae set with distinctly spiny margins.

Occurrence An arctic-alpine species which is rare and local in occurrence in a few parts of the extreme west of Ireland, North Wales, northern England and southern Scotland, becoming more widespread in the mountains of central and north-west Scotland, in moist pockets of lime-rich mountain screes, occasionally descending to lower altitudes near extreme west coasts.

Identification Plants have dark, glossy green, stiff, leathery fronds, usually about 15–30 cm long, but sometimes much longer than this in favourable situations. The fronds, which arise in

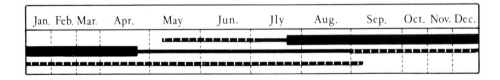

Jan.	Feb.	Mar.	Apr.	May	Jun.	Jly	Aug.	Sep.	Oct.	Nov.	Dec.

irregular shuttlecocks from the crowns of fairly stout, scaly, ascending rhizomes, are of spreading-ascending habit, and of *narrowly linear-lanceolate or nearly parallel-sided, linear outline*, often becoming twisted amongst the rock of the screes in which they grow. The fronds are once pinnately divided into about 20–40 *sharply angular, entire pinnae*, which are closely spaced along either side. The longest pinnae occur about the middle of the frond, tapering gradually above and below this point, but in many specimens, *the pinnae are of about equal length through the greater part of the frond*, only slightly shortening near the frond base, to leave a very short, chaffy-scaled stipe (becoming longer only in fronds growing in very deep clefts).

The pinnae are of markedly asymmetric, tapering outline, *each acutely pointed and curved falcately towards the tip of the frond*, like the blade of a scythe. The base of the upper edge of each is drawn upwards into an *acutely pointed angular lobe, usually overlapping and tucked beneath the next pinna above.*

The closely spaced or overlapping pinnae are of a rigidly leathery texture, and become frequently twisted. Their margins all round, are set with numerous, sharp teeth, drawn out into long, stiff, straight distinct spines, giving the fronds a prickly texture to touch.

Fertile fronds are similar to the vegetative ones, and bear orbicular sori as a single row down each side of the pinnae, with the sori usually restricted to the very upper part of the frond, and often to the outermost flanks of the pinnae. Each sorus has a peltate indusium.

Fronds are of bright grass-green colour whilst flushing, but darken rapidly to a glossy very deep green once fully expanded. In alpine situations, they flush rather late in the season, with young fronds not emerging until late spring, and becoming fully expanded by late July. The spores mature in late summer, and many appear to overwinter on the fronds, from which they are still being shed by July of the following year. The fronds are winter-green, persisting usually well into the following season in sheltered spots, and in the most favourable sites, still remaining green in their third year.

Variation There is much variation in size of individuals between different localities, both in terms of frond abundance and frond size (with fertile fronds from under 9 cm to often over 30 cm, and in at least one location to over 75 cm long). A large part of this is almost certainly environmentally induced and related to shelter, though there may also be some genetic basis. Probably also largely genetically determined are other features. These include considerable variation in pinna spacing along the frond, frond width and blade outline shape, with both narrow fronded and broader fronded forms often occurring together, each shape usually typical of particular plants. Further, unusually long-spinose margined populations occur in some localities, and these may have a regional trend.

Possible confusion Likely only with juvenile plants of *P. aculeatum*, and with the hybrid *P.* ˣ*illyricum*, of which *P. lonchitis* is one parent. The narrow frond outline, shorter stipe, tougher frond texture, and fertility even at small size, should help to separate most plants of *P. lonchitis* from *P. aculeatum*. The presence of uniformly good spores in *P. lonchitis* separates it from *P.* ˣ*illyricum.*

Technical confirmation Plants are diploids, with $n = 41$, $2n = 82$ chromosomes. Spores are smaller (*c.* 24–32 μm) than those of *P. aculeatum* (which are *c.* 36–45 μm, q.v.).

Field notes Plants of *P. lonchitis* occur widely through the cooler (especially montane and high latitude) parts of the north temperate zone, including somewhat disjunctly through the Canadian Arctic. In Europe the species has a distinctly arctic-alpine range, with its main centres of distribution separately

Polystichum lonchitis, fronds from typical populations: *a–c*, Argyll; *d*, *e*, Mid Perthshire; *f–i*, Forfarshire; *j*, Mid Perthshire.

0   10

cm

Polystichum lonchitis, fronds from unusually setose population: Central Perthshire.

throughout western Scandinavia and the European Alps, but with many smaller outlying stations in most of the more mountainous areas of the continent southward to the Iberian Peninsula, Italy, Greece and Turkey, of which those of Britain and Ireland are, with the exception of Iceland, the most western.

In Britain and Ireland, Holly Fern is a small, slow-growing, but probably long-lived calcicolous species, occurring mainly in upland habitats which are cool in summer and where there are available exposures of lime-rich rock (such as limestones and mica-schists). It seems most successful on steep landscapes, where habitats are well drained, shaded, and kept permanently cool and moist, by slight seepage of base-rich water. It grows most luxuriantly on the sides of shaded upland ravines, along shady lower parts of moist cliffs, on moist ledges, and especially between angular boulders of large block scree slopes, with cool easterly or northerly aspects. On cliffs and ledges, the plant is seldom gregarious, more often occurring as widely scattered individuals. In screes, however, it may occur locally in some numbers, with plants usually deeply recessed into sheltered hollows between large boulders, from which their fronds seldom emerge very fully into entirely exposed conditions.

In its headquarters in the central Scottish Highlands, plants occur chiefly at about 610–915 m (*c.* 2000–300 ft) elevation, descending very locally in a few steep, cool upland river valleys down to about 150 m (*c.* 500 ft). Its range seems to correspond roughly to a maximum summer temperature of about 27°C, and like many other arctic-alpine plants, in both Scotland and Ireland, descends much nearer to sea-level in extreme west coast habitats with low summer temperature maxima, where it avoids warm, dry, summer conditions. In limestone areas of northern England, plants occasionally also occur in moist pockets in deep grykes and around moist sinkhole mouths.

Associated with Holly Fern in fairly stable block-boulder screen slopes are chiefly other ferns. In the Scottish Highlands, these include mainly Brittle Bladder-fern (*Cystopteris fragilis*), Green Spleenwort (*Asplenium viride*), Common Maidenhair Spleenwort (*A. trichomanes* subsp. *quadrivalens*), Northern Buckler-fern (*Dryopteris expansa*), and more locally, Hard Shield-fern (*Polystichum aculeatum*), whilst on ledges, Moonwort (*Botrichium lunaria*) and Lesser Clubmoss (*Selaginella selagino-*

ides), as well as large numbers of rare alpine flowering plants, may grow beside it. In progressively more mature screes, it persists amongst considerable numbers of low-growing flowering plants, including alpines and lower-altitude woodland species, which find suitably cool, moist, outlying sites amongst moist, higher-altitude rocks. Associates noted include Roseroot (*Sedum rosea*), Moschatel (*Adoxa moschatellina*), Alpine Lady's-mantle (*Alchemilla alpina*), Mountain Avens (*Dryas octopetala*), Northern Bedstraw (*Galium boreale*), Purging Flax (*Linum catharticum*), Wood-sorrel (*Oxalis acetosella*), Wood Anemone (*Anemone nemorosa*), Herb Robert (*Geranium robertianum*), and often the moss *Ctenidium molluscum*.

In looser, more unstable scree, Holly Fern becomes replaced by Mountain Male-fern (*Dryopteris oreades*). In areas of altitudinal overlap with Hard Shield-fern (*Polystichum aculeatum*), the hybrid *P.* *×illyricum* (q.v.) has been recorded in both Scotland and Ireland. In much rarer areas of contact with Soft Shield-fern (*P. setiferum*), the hybrid, *P.* *×lonchitiforme* (q.v.) can occur and is known from a single locality in Ireland.

In many of these habitats, and especially in scree-crevice sites, Holly Fern grows in conditions of greatly reduced competition. The plant appears to develop very slowly, but succeeds in places in which few others can, because of its apparent better tolerance of low light levels between and beneath the boulders. The very deep blue-green colour of its fronds is perhaps ecologically significant in this connection, whilst their harsh, leathery texture and glossy surface may be significant for such a large-fronded and evergreen alpine in minimising desiccation during conditions of prolonged winter frost.

The rather slow rate of establishment and subsequent growth probably prevents Holly Fern from readily succeeding in the slopes of recent screes and rockfalls, or in old ones which remain too unstable and mobile. It occurs chiefly amongst older, more stable, more serially mature scree slopes, in which it too is probably eventually replaced by blanketing angiosperm vegetation. Nevertheless, once established, it can persist for appreciable periods amongst such vegetation, with large, old plants probably effectively marking the end of the pioneer phase of pteridophyte scree dominance.

Polystichum setiferum (Forsk.) Woynar

Soft Shield-fern

(*P. angulare* (Wild.) C. Presl)

Preliminary recognition A large, graceful fern, with soft-textured, arching, finely divided, lanceolate-outlined fronds, arising in fairly dense, shuttlecock-like clusters. Their pinnules are angular, and have finely spinose margins.

Occurrence A mainly lowland species widely spread throughout Ireland and western and southern Britain, thinning out northwards, becoming scarce in northern England and Scotland, except near the extreme west coast. It occurs mainly in river-valley woodland and along streambanks, but can also be common locally on sheltered lanebanks and in shady hedgerows in the south-west.

Identification Plants have large (to 1.5 m or more), *gracefully arching and drooping-tipped, fully bipinnately divided, rather flaccid fronds*, which are a bright pale yellow-green colour when freshly emerged, gradually maturing to a grass-green colour as the season progresses, and are *particularly soft-textured*. They arise in dense, shuttlecock-like clusters from short, stocky, ascending or erect rhizomes, which seldom branch. Fronds in each crown are much more numerous than is typical for *P. aculeatum*.

The fronds are of lanceolate outline, widest at or below the middle, *with the basal pinnae only slightly shorter than those above* and ceasing suddenly, typically leaving *a long, distinctive length of stipe*. The stipes are about 1/4 or more the length of each blade (usually more in juvenile specimens), and fairly thickly clothed with mixed large and small, *very chaffy, pale golden-brown scales*, the larger of which are often of quite cornflake-like appearance. Mature fronds have about 30–40 pinnae on either side, the longest about 6–8 cm in length, usually straight, and tapering gradually to a pointed apex (but blunter than that of *P. aculeatum*). The pinnae bear numerous, flat, short, rather blunt pinnules, most of which are distinctly stalked (especially the inner ones). The innermost pinnule on the upper side of each pinna is usually only slightly larger than the rest, but is often deeply cut. All the pinnules have a blunt lobe at their base, pointing outwards on the frond, and the tip of this lobe, as well as the upper tip of each pinnule, comes to an obtuse or approximately right-angled point, before ending rather suddenly, like each of the small toothed serrations along the pinnule margins, in a short, soft spine.

Fertile fronds are similar to the vegetative ones, with the pinnules, especially throughout the upper part of the frond, bearing a row of rounded sori, each covered by a peltate indusium, arranged down either side of the pinnule midribs.

Jan.	Feb.	Mar.	Apr.	May	Jun.	Jly	Aug.	Sep.	Oct.	Nov.	Dec.

Fronds are subpersistent, expanding in spring and remaining mostly green (especially in milder areas) until the following year's fronds are flushing. The young, expanding fronds, as in other species of *Polystichum*, pass through a stage where the nearly uncoiled croziers tip laxly backwards, which in this species gives each newly expanded frond a particularly flaccid-looking appearance at this stage, as if it is wilting.

Variation *Polystichum setiferum* is a genetically somewhat variable species in detail of pinnule size, shape, spacing and acuteness. Some of the grosser extremes have been selected as cultivated varieties, including ones bearing proliferous vegetative buds dorsally. Juvenile specimens of wild plants usually have proportionately considerably longer stipes than do adults.

Possible confusion Plants are most likely to be confused with *Polystichum aculeatum* or the hybrid between it and *P. aculeatum, P.* *×bicknellii* (q.v.). The soft texture, brighter-green colour, more blunt-tipped, narrower, more widely spaced pinnae with a long, fully pinnate basal pair, shorter, less acute-angled pinnules with a blunter appearance, and longer stipe, more numerous fronds and smaller spore size (see 'Technical confirmation'), distinguish *P. setiferum* (Sleep, 1971a). Small plants retain a considerable degree of pinna-cutting, and this, plus their longer stipes, distinguishes them from those of *P. aculeatum*. Only the presence of good spores separates *P. setiferum* from *P.* *×bicknellii*, its hybrid with *P. aculeatum*. The spiny margins to the pinnules and orbicular sori separate *Polystichum* from all species of *Dryopteris, Athyrium* or *Oreopteris* with similar large, lanceolate-outlined fronds.

Technical confirmation Plants are diploids, with $n = 41$, $2n = 82$ chromosomes. The spores are smaller (*c.* 30–38 μm) and paler in colour under a microscope than those of *P. aculeatum*, are non-papillate and have a perispore with a broad, wavy wing.

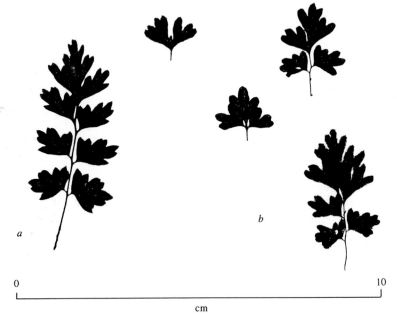

0   10

cm

Polystichum setiferum: a, South Devon; *b*, West Gloucestershire.

Polystichum setiferum: *all* West Cornwall.

Polystichum setiferum, young frond stages (from Orth, 1938).

Field notes *Polystichum setiferum* is a western and southern European species, with a somewhat discontinuous range eastwards to the Caspian sea and southwards to Madeira and the Canary Islands. It reaches its greatest abundance along the Atlantic seaboard from the western Iberian Peninsula to Britain and Ireland.

Within these islands, *setiferum* is an Atlantic species, growing mainly at low altitude. It ascends to about 245 m (*c.* 800 ft) in western Ireland, but elsewhere occurs mostly lower than this, and becomes confined to within a few feet of sea-level in the northern part of its range. It occurs mainly in sheltered, shaded river valleys of mixed deciduous woodland, but also grows in more open situations on sheltered banks near the sea. In south-west England especially, it occurs often in abundance, as a member of sheltered lanebank and shady hedgerow communities. Its range is chiefly through areas with warm, wet winter climates and year-round high humidity. Further inland, plants occur mostly within stream ravines, where the humidity is enhanced by local conditions. Plants also demand high shelter and moderate light, and seem intolerant of summer drought. In some areas of south-west England and south-west Wales, it is one of the most abundant ferns, becoming locally dominant

in suitable woodlands with a frequency of individuals seldom matched by the much more solitary *Polystichum aculeatum*. In such areas it re-establishes rapidly from spores in suitable situations, often pioneering damp banks where these are exposed by erosion.

Its habitats are almost always on sloping, permanently slightly moist ground, where there is moderate water movement. It is a moderate calcicole, on soils with a pH range of about 6.5–8.0, but which seems to become more closely limited to base-rich ground in northern England and Scotland, and more widely ranging in the south. It occurs sometimes on rich, light, loamy or sandy soils, but is perhaps most frequent on base-rich clays, such as the red clays of Devonshire and other parts of the West Country associated with the Old Red Sandstone system, as well as marl clays further north.

Frequently associated species in typical river-valley habitats include often dominant Wild Garlic (*Allium ursinum*), Bluebell (*Hyacinthoides non-scriptus*), Ivy (*Hedera helix*), Wood sorrel (*Oxalis acetosella*), Primrose (*Primula vulgaris*), Bramble (*Rubus fruticosus* agg.), Common Hemp-nettle (*Galeopsis tetrahit*), Wood Avens (*Geum urbanum*), Water Avens (*G. rivale*), Ground Ivy (*Glechoma hederacea*), Hart's Tongue Fern (*Phyllitis scolopen*

Polystichum setiferum, expanding crozier: May.

drium) and the mosses *Fissidens taxifolios* and *F. bryoides*. The Hart's Tongue usually occurs in nearby slightly wetter soil pockets, whilst the Soft Shield-fern grows where drainage is better. *Polystichum setiferum* and *Phyllitis scolopendrium*, as a pair, seem to replace *Dryopteris dilatata* and *Blechnum spicant*, respectively, in mild climates on less acidic ground.

Polystichum [×]bicknellii (Christ) Hahne

Lowland Hybrid Shield-fern

(*P. aculeatum* × *P. setiferum*)

Preliminary recognition Plants generally look like a more robust, darker green, somewhat more leathery form of *P. setiferum*, but have larger segments, a more tapering frond-base and a shorter stipe.

Occurrence Plants occur in widely scattered localities throughout England, Wales, Scotland and Ireland, usually as single sporadic individuals, but sometimes as small groups, in shaded situations where both parents grow together.

Identification Plants are often large and vigorous, with fronds commonly to a metre or more in length. They differ from *P. setiferum* chiefly in their stiffer more leathery texture, glossier surface, darker green colour, narrower frond outline which tapers more towards the base, shorter stipe and *more coarsely cut, often larger and more obliquely inserted pinna segments.* They differ from *P. aculeatum* mainly in their longer stipes, more erect frond habit, more numerous fronds per plant, longer first pinna pair, and *more fully and deeply cut pinnae*, of which the innermost pinnules are sometimes stalked. Plants shed mostly abortive spores.

Variation As with both parents, the hybrids are somewhat variable and to a certain extent, link between the parental morphologies.

Possible confusion Plants are likely to be superficially confused with either parent, but their large size, vigour and individuality of form when seen growing amongst a population of one or other parent helps to distinguish them in the field. The presence of mostly abortive spores separates it from either parent.

Technical confirmation Plants are triploid hybrids , with $2n = 124$ chromosomes, showing *c.* 41 pairs and *c.* 41 univalents at meiosis (Vida & Reichstein, 1975). Hybrids have been synthesised experimentally with ease using either parent as female; and the meiotic pairing observed strongly suggests that *P. setiferum* is part-parental to *P. aculeatum* (Manton, 1950; Sleep, 1971a, 1975).

Field notes This hybrid is known in a wide range of shaded, lowland, habitats, scattered throughout Britain and Ireland, mainly in rocky woodland, steep ravines, gorges, wooded river valleys, streambanks, shady lanesides, roadside banks and disused quarries. Most of its habitats are on basic, steeply sloping, ground and at least some of its stations occur in disturbed habitats, either where partially artificial, or where streamside erosion has kept habitats moderately open, and there has been active parental recolonisation.

The parent species have a considerable range of overlap in lowland, moderately basic habitats throughout Britain and Ireland and the hybrid can occur anywhere where the parents occur in quantity together, and may well be underrecorded.

0   10

cm

Polystichum [×]*bicknellii*, one frond: South Devon.

0   10
                          cm

Polystichum [×]*bicknellii*, one frond: Kirkcudbrightshire.

Polystichum ×illyricum (Burbas) Hahne

Alpine Hybrid Shield-fern

(P. aculeatum × P. lonchitis)

Preliminary recognition A medium-sized Shield-fern, with only partially sub-divided, angular, scythe-shaped pinnae, looking like a less divided, narrower-fronded form of *P. aculeatum*.

Occurrence Reported from a single station in north-west Ireland (Co. Leitrim) and one in north-west Scotland (West Sutherland). It could yet be found elsewhere where the two parent species grow together in quantity, especially perhaps in Scotland.

Identification Plants are of size and frond outline intermediate between those of *P. aculeatum* and *P. lonchitis*, with fronds to about 50 cm in length, but scarcely more than about 6 cm in width. Their *much narrower frond outline* provides their chief contrast with *P. aculeatum* in the field. The widest point is about the middle of the frond, but above and below this, the pinnae shorten gradually, to end above in a narrowly acute apex, and below reducing to about half the length or less of those in the mid-part of the frond, leaving a very short length of stipe.

The narrow pinnae can be numerous, with up to 40 or more on either side of large fronds, but usually less than this in smaller specimens. Each is slender and angular, and, as in *P. lonchitis* curved slightly towards the top of the frond in a scythe-like manner. In the upper part of the frond, the pinnae are usually more or less simple and entire, although deeply toothed, whilst throughout about the middle third of the frond, they become themselves usually subdivided near their inner ends, into a few, small, acutely angular, obliquely inserted pinnules. All the pinnae have an angular, acutely pointed lobe at their base, pointing upwards on the frond, and the margins of the pinnae throughout are set with long, spinose teeth. In texture, the fronds have a rather stiffer, more leathery and more prickly feel than those of *P. aculeatum*.

Fronds can produce abundant sori beneath their upper ends, especially on the outer parts of the pinnae, but their spores remain, for the most part, shrivelled and abortive.

Variation Even on a single plant, the degree of pinna-cutting varies much with the size of the frond, with smaller ones usually being much less divided than larger ones. The variability of the parents suggests that further variation between plants of the hybrid might well be expected.

Possible confusion Plants are very likely to be easily confused with either parent. The narrow frond outline should separate it from *P. aculeatum* in the field, whilst the deeply cut, even pinnatisect, pinnae, help to distinguish it from *P. lonchitis*. In doubtful cases, the abortive spores should separate it from either parent. It is also likely to be confused with the hybrid *P.* ×*lonchitiforme* (q.v.), to which it can look closely similar and which shares the common

Polystichum [×]*illyricum: a–c*, East Perthshire; *d*, West Sutherland.

P. lonchitis parent; confusion could arise if all three parents are present in the immediate vicinity. Under such conditions, *P.* ˣ*illyricum* can be separated with certainty only with a chromosome count (see 'Technical confirmation') (A. Sleep, personal communication).

Technical confirmation Plants are triploid hybrids, with $2n = 123$ chromosomes, showing *c.* 41 pairs and *c.* 41 single chromosomes at meiosis. Experimentally it has been shown that hybrids can be synthesised with ease, using either parent as female. The chromosome pairing observed strongly suggests that *P. lonchitis* is part-parental to the allotetraploid *P. aculeatum* (Manton, 1950; Sleep & Reichstein, 1967; Sleep, 1975). There is a high degree of spore abortion in hybrid plants, but, in European material, a limited number of good spores have been reported to occur which can give rise to triploid or hexaploid F_2 progeny, at least under experimental conditions (Vida & Reichstein, 1975).

Field notes The two parents of this hybrid occur widely throughout Scotland and the northern part of Ireland, with *P. lonchitis* mostly at higher altitudes and *P. aculeatum* relatively lower. On base-rich crags at moderate altitude, there is, however, occasional abundant overlap of the two parent species, in Scotland mostly on schists in the central Highlands and especially at low altitudes on limestones in the north-west Highlands and in more mountainous regions of western and north-western Ireland. The hybrid *Polystichum* ˣ*illyricum* was, however, unknown in Britain or Ireland until the early 1970s, when it was first reported from Ireland (Sleep & Synnott, 1972) and then from Scotland (Stirling, 1974). It is known from a limited number of sites within the generally sympatric portions of the range of its two parents in these islands, especially in western, relatively low-elevation sites of Ireland and Scotland, though is probably also present at modestly higher altitudes in more upland regions of Scotland.

In its main Irish locality in Glendale, Co. Leitrim, *P.* ˣ*illyricum* grows amongst loose limestone scree at moderate altitude between about 300 and 600 m (*c.* 1000 and 2000 ft) beneath tall and abrupt cool north-east-facing inland cliffs, a few kilometres from the sea. Here it grows in small local fans of mainly broken limestone rock, and in these sites and on the surrounding limestone rock faces it associates with a wide range of other base-loving ferns including Hart's-tongue (*Phyllitis scolopendrium*), Brittle Bladder-fern (*Cystopteris fragilis*) Common Maidenhair Spleenwort (*Asplenium trichomanes* subsp. *quadrivalens*), Green Spleenwort

(*A. viride*), and, in more exposed spots, Rusty-back Fern (*Ceterach officinarum*). These and many other ferns grow with particular profusion in the vicinity as well as do not only both parents of the hybrid but also Soft Shield-fern (*Polystichum setiferum*). The hybrid *P.* ˣ*lonchitiforme* (q.v.) also occurs nearby.

In its principally known Scottish localities (Stirling, 1974; Ferguson, 1986) *P.* ˣ*illyricum* grows mainly amongst sharp, loose limestone scree beneath tall, west-facing cliffs at altitudes of little more than 305 m (*c.* 100 ft). As well as the frequent presence of both parents in the area, it is also associated mainly with other fern species, each growing often deeply within the boulder scree, including Brittle Bladder-fern (*Cystopteris fragilis*), Common Maidenhair Spleenwort (*Asplenium trichomanes* subsp. *quadrivalens*) and Green Spleenwort (*Asplenium viride*).

In both its Irish and Scottish localities, a number of other rare alpines also occur nearby, including various alpine saxifrages (e.g. *Saxifraga aizoides, S. hypnoides, S. oppositifolia*), *Cochlearia alpina, Silene acaulis* and *Draba incana*, and, especially in Scotland, often locally abundant Mountain Avens (*Dryas octopetala*).

In both Ireland and Britain *P.* ˣ*illyricum* remains, however, very rare, although it can be particularly easily overlooked, and especially confused with a more juvenile form of *P. aculeatum* (q.v.). It is a markedly calcicolous hybrid, occurring usually as only very scattered or solitary individuals in markedly lime-yielding, cool, moist sites, usually where large numbers of both of its parents grow most profusely together.

Polystichum ˣlonchitiforme
(Halacsy) Becherer

Atlantic Hybrid
Shield-fern

(*P. lonchitis* × *P. setiferum*)

Preliminary recognition Plants look much like *P.* ˣ*illyricum*, and have very narrow-outlined, soft-textured fronds, with the outer halves of the pinnae much less cut than in *P. setiferum.*

Occurrence An extremely rare hybrid, recently discovered in a single locality in western Ireland, growing in limestone scree at low altitude.

Identification Plants differ from *P. setiferum* in having a very much narrower, more nearly strap-shaped outline, which tapers more below, and pinnae which become entire in their outer halves. They differ from *P. lonchitis* mainly in having the pinnae each cut in their inner halves in a manner approaching that of *P. setiferum*, and in having the more finely spinose margins of *P. setiferum* and soft frond texture. Sori occur, but the spores produced are entirely abortive.

Variation Insufficient material is known to assess this, but the variability of the parents and that shown by *P. aculeatum* (the allotetraploid derivative of this cross) suggests that considerable variation might well be expected.

Possible confusion The narrow frond outline, but soft-textured fronds, with abortive spores separate *P.* ˣ*lonchitiforme* from both parents, as well as from *P. aculeatum*, whilst the frond outline and pinnae entire in their outer halves also separate it clearly from *P.* ˣ*bicknellii*. It is much less easy to separate from *P.* ˣ*illyricum* (q.v.), which shares the *P. lonchitis* parent, and because of the likely variability in general appearance of each, technical confirmation is essential to distinguish them with certainty.

Technical confirmation Plants are diploid hybrids, with $2n = 82$ chromosomes (cf. $2n = 123$ in *P.* ˣ*illyricum*, q.v.) (Anne Sleep, personal communication).

Field notes This hybrid was unknown in the native flora until very recently (Sleep, 1976; Scannell, 1977).

At its single known site on the west coast of Ireland (Co. Leitrim), *P.* ˣ*lonchitiforme* occurs in broken limestone scree in an open, low-altitude habitat. It is an extremely rare hybrid anywhere, probably largely because the two parent species are normally widely altitudinally separated through most of their ranges, with *P. setiferum* a mainly southern and lowland woodland plant, *P. lonchitis* mainly a northern alpine. Only along the extreme west Atlantic coasts can alpines normally descend to near sea-level (because of cool summers) and southern plants spread north (because of mild winters), producing conditions under which these two species can naturally meet.

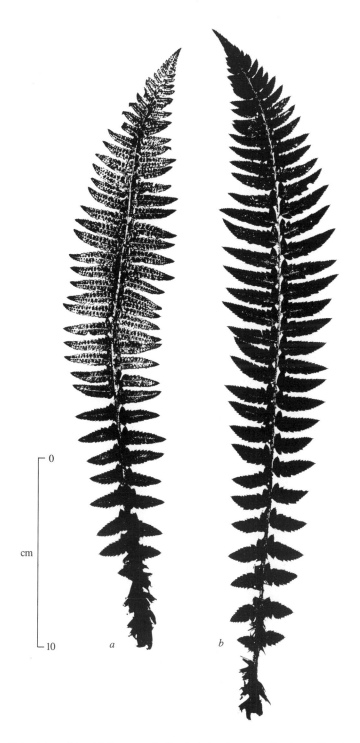

Polystichum ×*lonchitiforme: a–b*, Co. Leitrim (courtesy Dr Anne Sleep).

Pteridium aquilinum (L.) Kuhn   Bracken

Preliminary recognition Difficult to miss – the very large, triangular, tough, compound fronds, often forming dense stands in very large, sometimes impenetrable, masses, are quite distinctive. In mountain districts, they often give a characteristic colour to the lower hillsides – pale green in spring, dark green in summer, and a yellow-green eventually becoming a rufous 'bracken' brown in autumn.

Occurrence Widespread throughout every British and Irish county, from sea-level to the lower slopes of mountains and fells. Bracken occurs as a natural, but usually not obtrusive, component of dry, acidic woodlands of all types, especially in glades and along forest margins, but with removal of the forest cover, forms dense, extensive colonies, often to the near-exclusion of most other species.

Identification Established plants have large, compound fronds to (30)–80–200–(180) cm or more, borne on erect stipes that are of highly variable length and thickness, often semi-succulent at first, becoming toughly fibrous with age. The blades are triangular, triangular ovate to (mostly) ovate, typically with the lowermost increments of the rachis nearly erect, and the pinnae mostly horizontally rotated. Fronds are 2–3 × pinnate-pinnatifidly divided, with their ultimate segments of highly variable shape and dissection. On mature fronds, the margins of the segments turn downwards and under to form a narrow, membranous false-indusium.

Fronds arise individually from deeply buried, subterranean, often centimetre-thick, blackish creeping rhizomes. Young, emerging fronds in spring may push up through a considerable depth of soil, before emerging at the surface with a characteristic tomentum. The expanding croziers are at first tightly coiled but mostly rapidly loosen. They pass through stages of variously looking like inrolled, clenched fists, and, as the first pinna pair expand laterally, temporarily adopt something of the appearance of an eagle about to take to the air – presumably the origin of the name *aquilinum*. Thereafter, the whole blade expands in a distinctive acropetal progression of successive pinna pairs, which is unique amongst native ferns to this species.

Fronds persist for a single season, becoming increasingly leathery in texture with frond age and site exposure. They turn yellow-cream to reddish-buff and eventually largely collapse following the first severe frosts of autumn. Thereafter the overlapping layers of fallen fronds gradually compact to form a densely swamping frond blanket, darkening as they decay over winter, with the broken stipe bases usually persisting well into the following spring.

Jan.	Feb.	Mar.	Apr.	May	Jun.	Jly	Aug.	Sep.	Oct.	Nov.	Dec.

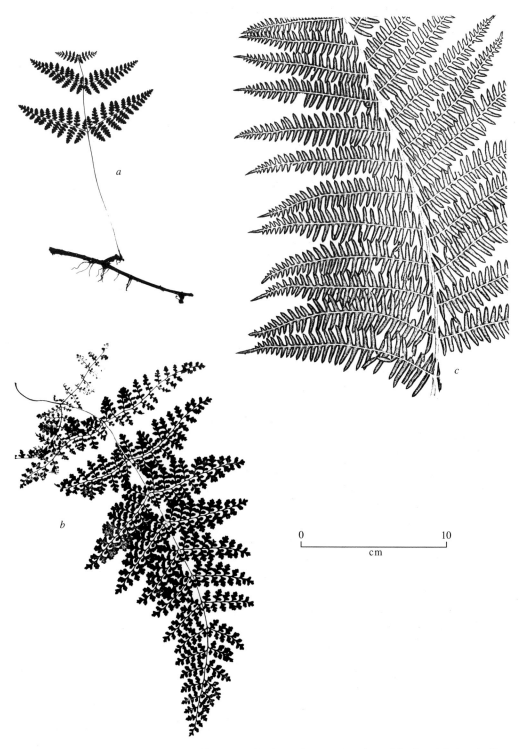

Pteridium aquilinum subsp. *aquilinum*: *a–b*, juvenile frond forms; *c*, portion of fertile adult frond. City of Edinburgh.

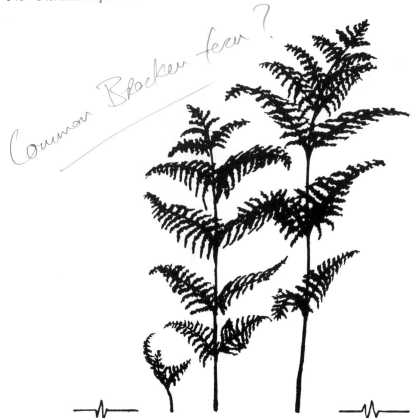

Common Bracken fern ?

Pteridium aquilinum subsp. *aquilinum*, habit of plant in life.

Pteridium aquilinum, young frond stages (after Orth, 1938).

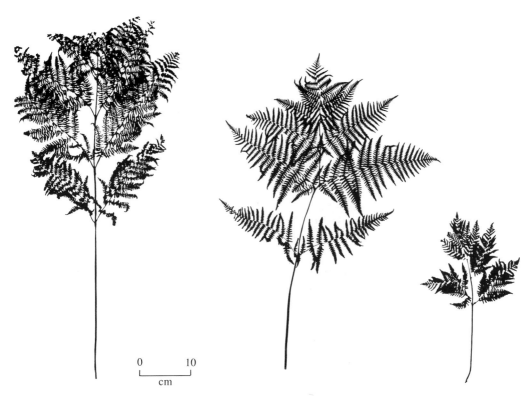

Comparisons of fronds of *Pteridium* taxa from nearby habitats collected at the same time: *left, P. aquilinum* subsp. *atlanticum*; centre, *P. aquilinum* subsp. *aquilinum; right, P. pinetorum* subsp. *osmundaceum* (all in comparative flat-pressed form). Perthshire, late July.

More confusing, however, is often the identification of juvenile plants, which, at first sight, are totally dissimilar to the adults. They are pale green, only very narrowly triangular or ovate in outline, of spreading habit, and of soft and lax texture. They have widely-spaced, rounded segments, giving the blades a delicate and lace-like appearance. They also have a distinctive pilose covering of soft, short white hairs over the surface of the blade and stipe. The almost complete unrolling of the first pinna pairs before the rest of the blade gives the best indication that these are bracken. The tips of their fronds thereafter uncoil extremely slowly over a very long period, the frond thereby appearing to have an almost indefinite habit of growth. Such sporelings are of common occurrence in places such as old damp mortar-courses of basement brickwork in many British cities and in mortar around leaking rainwater pipes at edges of roofs. Their usual absence from areas characteristically associated with bracken, their calcicolous habits, and their particularly soft-textured form, further leads to their usual total lack of recognition, and usually to sheer disbelief that these can be sporeling bracken plants at all!

A curious feature shown especially clearly by all stages of growth of *P. aquilinum* is the possession of nectaries on the outer flanks of the pinna–rachis junctions, and, in smaller form, at each pinnule–pinna midrib junction, first described by Darwin (1877). These can be seen as small, glossy, 'bald' patches when the frond is at a crozier stage, and the presence of a lateral line along either side of the stipe. The function of the nectaries seems to be to attract ants, who

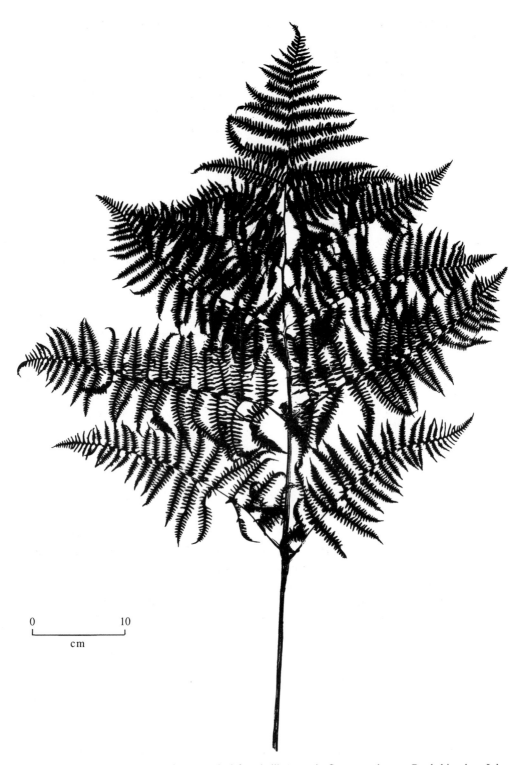

0   10
cm

Pteridium aquilinum subsp. *aquilinum*, typical frond silhouette in flat-pressed state: Perthshire, late July.

perform a protective function, and the lateral lines may act as guide-paths to the nectaries for ants (Page, 1982*b*).

The *P. aquilinum* complex is represented in Britain and Ireland by *P. aquilinum* (L.) Kuhn, with at least three subspecies, subsp. *aquilinum*, subsp. *atlanticum* and subsp. *fulvum* (Page, 1989*a*; Page & Mill, 1994, 1995). *Pteridium aquilinum* is the Bracken of the middle and southern latitudes of Europe, including most of the British Isles, but only subsp. *aquilinum* is an aggressive weed which is a widespread problem to human beings.

Pteridium aquilinum (L.) Kuhn subsp. *aquilinum* (Common Bracken)

Identification Plants typically have fronds that are of leathery texture when fully expanded, and which occur close together, forming tall, dense, vigorous, and frequently nearly impenetrable colonies of dark-green colouration.

When initially expanding, most surfaces are covered with a tomentum of short white hairs overlain by longer cinnamon-coloured hairs, giving the crozier a distinctive and characteristically brindled texture, and the croziers an overall brownish hue when viewed from a distance. These hairs persist for a while along most pinna and pinnule midribs, though are eventually shed as the whole blade eventually adopts a leathery and glabrous texture. The *acropetal succession of pinna pairs is typically completed by early summer*, with maturing pinnae moderately widely spaced and the pinna tips often remaining somewhat downward-arching. At full expansion, blades are triangular to (usually) triangular-ovate, with frond dissection characterised by the extreme variability of the shapes of ultimate segments between sites, between fronds, and even on different parts of the same frond. As fronds reach summer maturity, so the blade tips and often a larger part of the blade gradually curve in profile view to become nearly horizontal in orientation at its distal end, to the extent of eventually haphazardly leaning on one another within a colony. Such colonies of fronds *thus typically eventually form a billowing mass of untidy, variously orientated, arching and tangled fronds with lower pinnae eventually becoming senescent in the dense shade of the upper ones.* By autumn, fronds turn yellow-cream to buff-brown, and typically collapse and decay to a rotting frond blanket following the first severe frosts.

The tall, arching fronds eventually forming tangled masses and near-impenetrable colonies make this bracken a particularly characteristic feature of much marginal land in these islands. It is the typical, widespread and well-known form of Bracken in Britain and much of Europe, and when large, its fronds typically form vigorous colonies over extensive areas of exposed and semi-exposed sites, from coasts to open woodland to upland moorlands (see under 'Field notes'). Until recently, it was the only form of bracken known in these islands.

Pteridium aquilinum (L.) Kuhn subsp. *atlanticum* C. N. Page (Atlantic Bracken)

Identification Plants typically have fronds that remain of relatively soft texture with drooping pinnae and pinnules, even when fully expanded, collectively forming rather regimentedly upright colonies, of paler green colour than most bracken.

Fronds have stipes that are long, often notably thick (*c.* 6–12 mm or more at base) and especially succulent when young. Expanding croziers *have all surfaces covered with a dense tomentum of mainly or exclusively white hairs*, giving the expanding crozier tip a typically silver-white capping overall, conspicuous even from a distance. As the growing season

Pteridium aquilinum subsp. *aquilinum*, the only Bracken subspecies to become an agricultural problem. Typically extensive colonies, here of fairly stunted fronds: Dartmoor, Devon, early August.

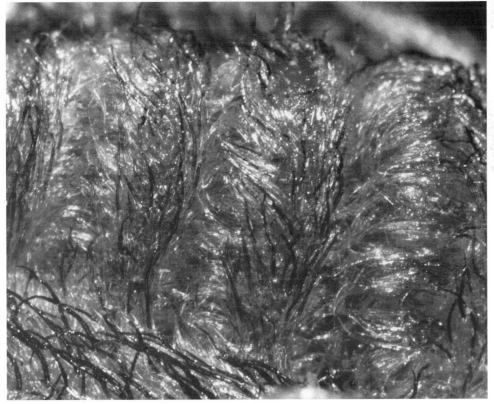

Pteridium aquilinum subsp. *aquilinum*: *above*, field view detail of expanding crozier showing sequenced vernation and brindled hair colouration; *below*, detail of crozier hair colouration showing large numbers of white hairs overlain by a smaller number of red hairs. Perthshire, May.

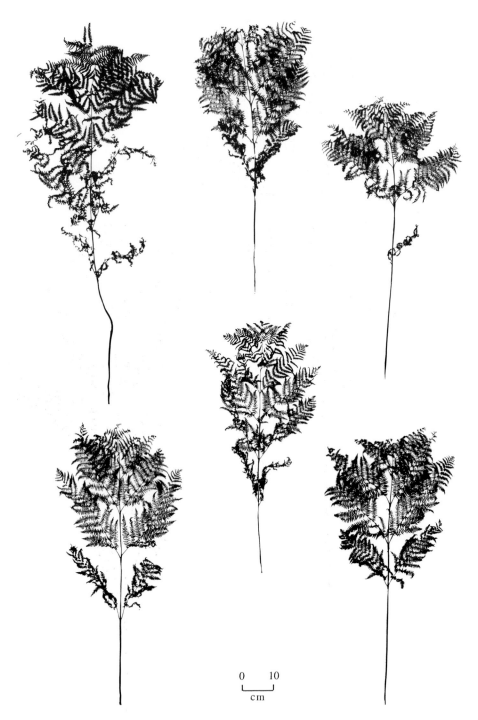

0   10
cm

Pteridium aquilinum subsp. *atlanticum*, typical frond silhouettes in flat-pressed state: note blunt-topped fronds with continuing apical vernation, yet lower pinnae already withering. Perthshire, late July.

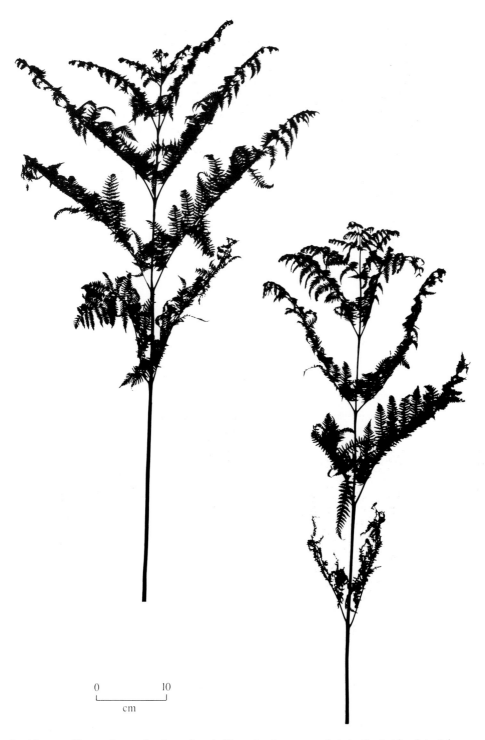

0   10

cm

Pteridium aquilinum subsp. *atlanticum*, frond silhouettes in unpressed state: Perthshire, late July.

Pteridium aquilinum subsp. *atlanticum*, field view of typical fronds during early summer expansion phase: note strongly sequenced (hypolepoid) vernation of expanding fronds, with white croziers and drooping-tipped laterals. Isle of Arran, July (type locality).

Pteridium aquilinum subsp. *atlanticum*, field view detail of expanding crozier showing abundance of white hairs and absence of red ones (tick, *Ixodes ricinus*, approximately 2 mm in length and common on all bracken fronds, shown for scale); Isle of Arran, June (type).

progresses, much of the white tomentum *usually persists long along the stipe, rachis and midribs*, visible as a pilose covering against the light. The pinna pairs *expand very slowly in a very strongly sequenced acropetal succession*, which is often incomplete by autumn, and which can appear almost indefinite. At frond maturity, the pinna pairs are typically very widely spaced, strongly horizontally rotated and long remain flaccid in texture, flexible and laxly drooping towards their tips, as do the tips of their pinnules. The fully expanded blades are very tall, ovate, never ternate, bipinnate–pinnatifid, and their ultimate segments mostly *short and blunt and perpendicularly inserted*. Several lowermost pinna pairs usually senesce particularly early and often while the apical growth of the blade is still expanding, the rachis typically remaining suberect throughout much or all of the blade length. Fronds turn pale tan in autumn, the tips often killed and blackened first by even the first cool nights of autumn, the remainder collapsing and decaying rapidly thereafter, following the first, even moderate, frosts.

The tall-growing, strongly upright-stemmed fronds, with laxly drooping pinnae and pinnules and white-capped croziers in spring, give colonies of this plant a conspicuous appearance even from a distance. Many apparent introgressants between it and the widespread *P. aquilinum* subsp. *aquilinum* are classed with that subspecies.

Pteridium aquilinum (L.) Kuhn subsp. *fulvum* C. N. Page (Perthshire Bracken)

Identification Plants typically have large but low-growing fronds that are of highly leathery, rigid and almost asperous texture when fully expanded, are full mid-green with orange-tinged stipes, and occur characteristically widely spaced in rather open colonies.

Fronds have stipes that are short, *fairly slender and rather wiry throughout*. On the flushing crozier, there are a large number of red hairs amongst those of paler colour, but as the fronds reach full expansion and harden, most hairs are rapidly shed so that the mature stipe and rachis typically become highly polished and glabrous or nearly so throughout. Fully expanded blades are *long (nearly twice as long as broad, or longer), and are of fairly constant triangular-ovate to triangular-elongate outline*, are sharply angled from the stipes and held rigidly in a straight *shallowly ascending* plane. Their usually 6–8 or more pinnae are slightly rotated and *somewhat obliquely inserted*, and bear straight and usually somewhat assurgently ascending pinnules with *rather uniform ultimate segments which are short and blunt with generally obtusely rounded tips*. The whole frond is of a dull mid-green colour through the growing season *with pale straw-yellow to (mostly) bright orange-brown* colouration to the stipes often extending to most of the rachides and pinna midribs, the whole frond typically adopting much of this orange-brown colour in autumn.

The obliquely ascending, rather low-growing blades with slightly forward swept and widely spreading unusually stiff and rigid pinnae (for *P. aquilinum*) borne on amber-yellow to orange-brown stipes and rachides, with fairly short and blunt ultimate divisions that are fairly uniform through the frond, make colonies of this plant fairly conspicuous in the field, even at first sight. Their large blade size and occurrence only, so far as known, in relatively sheltered lowland sites further make such patches distinctive from exposed upland forms of subsp. *aquilinum*.

Variation (all subspecies) Fronds vary much in size and degree of fertility with different habitats, much of which is probably environmentally induced. Once established, even stunted colonies in poor conditions bear fronds which are virtually scaled-down versions of larger ones.

Pteridium aquilinum subsp. *fulvum*, typical frond silhouettes in flat-pressed state: Perthshire, July (type).

Pteridium aquilinum subsp. *fulvum*, field views of typical fronds: note hand lens for scale (*below*). Perthshire, July (type locality and clone).

0   10
cm

Pteridium aquilinum subsp. *fulvum*, typical frond silhouette in unpressed state: Perthshire, July (type).

Possible confusion Established plants are so distinctive that confusion with any other plant is unlikely. Only sporelings look totally different (see above), but their nearly indefinitely growing frond tip and finely hairy covering help to distinguish them from other ferns.

Technical confirmation Plants of *P. aquilinum* subsp. *aquilinum* are normally tetraploids, with $n = 52$, $2n = 104$ chromosomes, though they are reported by Wolf *et al.* (1987) to behave as genetic diploids. Estimates of gene flow, genetic substructure and population heterogeneity in *P. aquilinum* subsp. *aquilinum* have been studied by Wolf *et al.* (1991). Further, the existence (in Wales) of an unusual (and presumably local) triploid clone has been additionally reported by Sheffield *et al.* (1993). The other taxa of Bracken described here are assumed also normally to be tetraploids. No diploid Bracken is currently known. Should a diploid Bracken exist, it might be perhaps most likely to present a morphology that would group it within *P. aquilinum* subsp. *atlanticum* (q.v.) as recognised here, the clones of which might be usefully checked for such a possibility.

Specific cautions *Stands of Bracken in the landscape are known to harbour especially large populations of arachnoid ticks (mainly* Ixoides ricinus *in Britain and Ireland), the bite of which can transmit the arthritic affliction Lyme disease to humans* (see Brown (1993, 1994) for recent summaries of this problem).

Fresh foliage yields the poison Prussic acid when crushed.

The whole plant of Bracken, including its airborne spores, is known to be significantly carcinogenic if ingested by either animals or humans, and two or possibly three separate chemicals with diverse carcinogenic effects are contained (see Evans & Mason (1965), Evans (1986, 1987), Taylor (1986, 1990, 1994), Ojika *et al.* (1987), Saito *et al.* (1989, 1990), Villalobos-Salazar *et al.* (1990, 1994), Evans & Galpin (1990), Galpin *et al.* (1990), Zu (1992), Potter & Pitman (1994) and Wells & McNally (1994) for diverse aspects and advances in knowledge of this extensive problem).

Bracken appears to be unique among native ferns in each of these respects.

Field notes Each of the different subspecies of *P. aquilinum* in these islands have specific ecological tendencies. Only subsp. *aquilinum* is an aggressive weed that is a widespread problem to human beings. *Pteridium aquilinum* subsp. *atlanticum* and subsp. *fulvum* are not.

Subsp. **aquilinum** is the only subspecies which, almost everywhere, is vigorous, invasive, and usually very abundant, forming the great bulk of aggressive, variable and weedy bracken in these islands. Overall, subsp. *aquilinum* occupies habitats from sea-level to about 610 m (*c*. 2000 ft) elevation, and is extensive over areas of deep, free-draining soils. In the uplands, its colonies may dominate whole hillsides, glens and fells, ascending to the limits of severe frost. Plants are, however, tolerant of a wide range of climatic conditions, including exposure to full daylight, although they become most luxuriant where direct sun is screened by frequent light cloud, and there is a high humidity and frequent light rain. It is a calcifuge subspecies, at least as a mature plant, occurring usually in ground with a pH from 4.5 to 6.5. Large stands are consequently rare in entirely limestone districts, except where deeper, moister, more acidic soils accumulate in pockets. Colonies become particularly extensive over sandstone-derived soils and, in parts of the northern Pennines in particular, dense stands of bracken mark well the geological outcrops of fell sandstones, in contrast to the heavier soils of clays and shales. Subsp. *aquilinum* also succeeds well in nutrient-poor soils, and itself forms a highly nutrient-deficient, acidic, peaty humus from its enormous annual accumulation of dead frond litter, which decomposes only slowly.

In heathy situations and upland woodland margins, subsp. *aquilinum* is commonly found with Ling Heather (*Calluna vulgaris*), Bell Heather (*Erica cinerea*), Bilberry (*Vaccinium myrtillus*) and grasses such as Wavy Hair-grass (*Deschampsia flexuosa*), Common Bent-grass (*Agrostis tenuis*) and Sheep's Fescue (*Festuca ovina*). When in dense, solid stands, however, Bracken excludes most other species from beneath its own frond canopy, with such associates reduced to isolated patches or much-weakened individuals. Other species which do sometimes survive in Bracken stands that are not too dense, however, frequently seem to include Foxglove (*Digitalis purpurea*). Heath Bedstraw (*Galium saxatile*) and Sheep's Sorrel (*Rumex acetosella*).

Subsp. **atlanticum** is a plant of limited occurrence which occurs in relatively 'pure' form only in certain low-altitude habitats widely scattered through mainly western and southern regions of these islands, and especially in more coastal sites and in soils of higher base status. Although imperfectly known, its habitats appear to be chiefly in slightly moist flushes in coastal or valley-bottom sites at low altitude chiefly on limestone and other basic soils, principally in mild, oceanic climates. In such sites, many of its stands abut directly on to adjacent more arching-fronded stands of typical subsp. *aquilinum*. Its currently known distribution includes scattered localities in lowland parts of Argyll, Isle of Arran, Kintyre and the Dumbarton-shire and Ayrshire coasts. On a morphological basis, subsp. *atlanticum* apparently shows extremely wide introgression into the common subsp. *aquilinum*, into which it seems likely that there has been especially wide gene incorporation in these islands. Outside the British Isles, evidence suggests that stands of relatively 'pure' subsp. *atlanticum* are present also in at least the Iberian Peninsula and southward to the west African coast. In these islands, its known oceanic distribution, its apparent frost-sensitivity (growing in many sites near to *Hymenophyllum* (q.v.) on nearby more acidic rocks, for example), and apparent long growing-season requirement, suggest it to be among the Atlantic-southern elements in our flora.

Subsp. **fulvum** has been recognised only recently. Its colonies are known as yet only from a relatively few lowland inland woodland and woodland margin sites scattered over shallow slopes in mostly Birch,

Birch–Oak (*Betula–Quercus*) and mixed broadleaf vegetation in central Scotland, especially from the vicinity of Loch Tummel and Rannoch in Perthshire. Colonies are characterised edaphically by their occurrence with a relatively rich and diverse herbaceous understorey of grasses and flowering plants. No colonies nor herbarium specimens have as yet been detected elsewhere in Britain or mainland Europe, and it may hence be a Scottish endemic.

As a species, members of the *Pteridium aquilinum* alliance range widely, from generally the southernmost limits of *P. pinetorum* (with only small areas of overlap, as the two species are largely allopatric), across much of southern Europe, to the Mediterranean and eastward to the Caucasus, and probably just south of the Mediterranean along the North African coast, spreading further south in extreme Atlantic Africa as far south as at least the Ivory Coast (as subsp. *atlanticum*). Indeed, *P. aquilinum* is, through much of its range, the Bracken species that is most characteristically associated with generally mild winter conditions (seldom severely cold ones), and vegetationally originally associated with mainly broadleaf woodland vegetation.

The origins and inter-relationships of the taxa within *P. aquilinum* remain as yet little understood and somewhat enigmatic. All of the subspecies outlined here produce spores in appropriate seasons in their native habitats. Experimental investigations show that the spores are normally viable, and that it is possible to hybridise between them. Where the subspecies of *P. aquilinum* meet, field evidence suggests the presence of widespread hybridisation along contact fronts, to the extent that, were it not for edaphic differences helping to maintain the pure identity of subsp. *atlanticum* and subsp *fulvum* patches, each might have become entirely introgressed into already much more extensive and always vegetationally more aggressive subsp. *aquilinum*.

Identification and location of these various different bracken subspecies is today particularly significant in terms of research relating to the active evolution of the members of this unusual fern genus, and their exact location in relation to more widespread and indiscriminate attempts at general bracken control, for it would be a scientific loss to eliminate these unusual and local taxa. Indeed, these more local brackens may also have significant field indicator value, for those of *P. aquilinum* (and especially subsp. *atlanticum*) may mark particularly relictual sites in which

diploid brackens especially could yet be found.

Pteridium aquilinum (mainly subsp. *aquilinum*) has undoubtedly increased enormously in abundance in Britain through prehistoric and historic times. It owes its present-day success partly to its own innate abilities of exploitation of edaphically poor situations, coupled with considerable sporophyte longevity and effective vegetative spread, and partly to interference with forest, woodland and moorland habitats by human beings (Page, 1976, 1986). In the latter category, especially important have been the long history in Britain of forest clearance, woodland and moorland burning, and grazing by domesticated animals (particularly sheep). Together, these factors have reduced or removed much natural woodland and also removed Bracken's natural competitors. Indeed, such clearances of the natural forest cover go back to Mesolithic and Neolithic occupation, where palynological (fossil spore) records show great increases locally in the frequency of Bracken along with other weeds around areas of human habitation. In such areas, Bracken showed a very quick response to forest opening, probably resulting from burning as well as felling, and to the subsequent use of areas as pasture (Turner, 1964, 1965, 1970; Smith, 1970; Godwin, 1975). Clearance of forest has, indeed, continued through the whole history of human occupation in Britain, but it was probably not until the great forest fellings of the last few centuries and the extensive replacement of former forest by grazing sheep in the uplands, that the problem of Bracken, as we see it today, became particularly rife (Page, 1976, 1986; Rymer, 1976). Sometimes, individual, extensive plants of considerable age may be present, although it seems very likely that most such stands are the result of multiple, co-extensive, intergrown, individuals from separate original invasions.

Spore production in Bracken varies very widely, but is almost always greater in open situations than in woodland habitats, where plants are rarely fertile. Sporeling plants are known mostly in artificial habitats, and in particular these include old brickwork (see e.g. Ridley, 1936; Thompson, 1939). Young plants have been observed to establish vigorously in, for example, the damp mortar rubble of bomb-damaged sites in London during and after the air-raids of the Second World War (Lousley, 1944). By contrast, establishment from spores in the wild is probably of infrequent occurrence compared with the frequency of occurrence of sporelings in old brickwork, but here burning of vegetation probably acts as one of the principal means by which sufficiently virgin habitats for prothallial establishment are normally created (Page, 1976). The oddly calcicole tendencies of sporeling plants contrast sharply with the markedly calcifuge habits of the mature plants. The young plants certainly seem base-tolerant, and may, indeed, be base-demanding, and it seems possible that this may be an adaptation on the part of the plant to enable young plants to pioneer burned-over sites, where minerals, including many bases such as potash, must initially be in particularly rich supply, but which rapidly become lost from the habitat as the young plants begin to establish.

Bracken is thus especially ecologically adaptable as well as polymorphic. In modern agricultural practice, it is certainly a serious weed, especially of upland districts and in some coastal areas. Not only does it greatly depress the productivity of infested land, but it is also carcinogenic to stock, a great resource of ticks and thus tick-borne disease (including Lyme disease in human beings). and a source of carcinogenic airborne spore clouds in season, whose human effect is under current investigation. It is also spreading at an estimated rate of up to 1–3% per year. Extensive applied studies have consequently been directed towards control of this plant, both in Britain and overseas (see Thomson & Smith, 1990; Smith & Taylor, 1994). A herbicide claimed to be effective against Bracken, containing the active ingredient Asulam, is, unfortunately, also completely toxic to many other ferns on which it has been experimentally tested, as well as several other plants. Valuable as such a herbicide may be for control of Bracken, its potentially lethal effect on other species of ferns

which may become subject to spray-drift, if not specifically avoided, could raise considerable conservation concern, and other biological controls are being sought.

Evolutionarily, however, Bracken, especially in the form of subsp. *aquilinum*, has become a very successful ecological opportunist. Accumulating scientific evidence suggests that it is a highly complex, strategically diverse, rapidly responding, geographically mobile and genetically polymorphic organism, with a number of pteridologically unusual innate mechanisms of considerable environmental tolerance, insect mutualism and fire adaptation, which enable it to score strategic advantages at otherwise vulnerable stages of its life-cycle in ways that are so subtle we have scarcely yet awoken to them (Page, 1986).

One effective and long-term biological control for subsp. *aquilinum* does, however, exist. This control need not affect the persistence of the other non-aggressive Bracken taxa, nor harm other ferns or other woodland biota. It is to put back the original woodland cover. Long-term enclosure experiments have already shown that by excluding grazing and by creating conditions favourable to renewed tree establishment, growth and natural tree regeneration, a new woodland canopy can be directly created, much like that of the original woodland cover. Beneath this canopy, the vigour of Bracken becomes again naturally suppressed, while renewed opportunity simultaneously develops for other native woodland species, including other non-aggressive woodland ferns, to return again in the Bracken's stead.

Note added in proof (*Pteridium aquilinum* subsp. *aquilinum*).
Unusually among native ferns, a large percentage of the spores of *Pteridium aquilinum* subsp. *aquilinum* have been shown to be capable of germination in darkness (Lindsay *et al.*, 1995), and it has been presumed that this, coupled with an ability to germinate and grow faster than other ferns, gives the plant an 'edge' over other ferns whose spores normally require the stimulus of light to germinate and subsequently mostly grow more slowly. As Sheffield (1997) has pointed out, the price that must be paid for this rapid dark germination must be the absence, in this taxon, of a long-lived soil spore bank.

The sites in the field in which such an 'edge' might arise have not previously been made clear. These adaptations might be collectively interpreted as advantaging the plant to respond especially rapidly to rain downwash of spores into the dark vesicles and interstices of porous cooled surficial ash deposits arising directly from dry late-summer heathland fires. This gives the spores, the gametophytes and resulting sporelings the necessary ability to exploit fully the sudden availability of such temporarily ephemeral situations and the low vegetative competition combined with high mineral nutrition that results.

Pteridium pinetorum C. N. Page & R. R. Mill

Pinewood Bracken

Preliminary recognition *A structurally distinctive, very winter-hardy, subspecies of Bracken, with a low-growing, wiry-textured, broadly triangular and sometimes nearly ternate-tripartite blade, which has large and widely spaced elongate ultimate segments, all the pinnae of which expand simultaneously in spring (i.e. it is not of highly sequenced vernation).* Northern Bracken lacks the vigour and invasive tendencies of Common Bracken.

Occurrence Known only from a small number of widely scattered sites in Perthshire and Inverness-shire, in the neighbourhood of Scottish native pinewoods, from which it is almost certainly relictual.

Identification *Pteridium pinetorum* ('bracken of the pinewoods') is the native member of the *P. latiusculum* complex, of high northern boreal latitudes (Page & Mill, 1995). All forms of this Bracken are recognisable in the field by a combination of characters which in total (and usually in isolation) distinguish them from members of the *P. aquilinum* group. Even in sheltered sites, most fronds have characteristically much more slender stipes, rachides and pinna midribs than have those of *P. aquilinum* (stipes usually less than 7 mm at the base and 6 mm at the first pinna junction, rachides under 5 mm at their widest, and pinna midribs less than 3–4 mm at the base). Fronds may reach 80 cm but are often much less; their stipes, which are usually under 30 cm in length, are only weakly mucilaginous when broken, have very few white hairs when flushing, and become particularly rapidly glabrous as the frond matures. When fully expanded, their fronds are usually subternate and of regularly triangular blade outline (often subternate), and are simpler in overall structure, more evenly and boldly pinnate, *with fewer pinnae than those of* P. aquilinum *and with large and widely spaced simple, long, usually* straight or very slightly falcate *oblong-elongate ultimate segments with broad obtusely rounded tips.* Whole blades thus have a characteristically more boldly divided pattern of division than do those of *P. aquilinum*, are bright grass green on all upper surfaces when freshly expanded, and are ultimately glossy above.

In spring, all fronds complete full expansion from flushing *at a time similar to or earlier than the fronds of nearby stands of* P. aquilinum *in the same localities, with the frond tip and all the pinnae expanding very rapidly and almost simultaneously throughout the frond. Their expanding croziers are regularly and tightly coiled* (cf. the markedly sequenced successive pinna expansion of the *P. aquilinum* group, in which the expanding crozier is typically loose and fist-like). All

Jan.	Feb.	Mar.	Apr.	May	Jun.	Jly	Aug.	Sep.	Oct.	Nov.	Dec.

have a low proportion of white hairs to the expanding crozier, but a relatively large number of ephemeral red hairs, which, when dense, give a characteristic deep red-brown colour to the expanding crozier in spring. While expanding, the pinnae are assurgent (stiffly inclined) and rigid (never drooping at the tips as is common in *P. aquilinum*), and when fully expanded, the pinnae in plan-view are usually swept steeply forward (at *c.* 45°) along the rachis, but, throughout the blade, the pinnae are maintained *largely in the plane of the blade* (i.e. not rotated to a subhorizontal position as is common in most *P. aquilinum*). The frond texture is tough, coriaceous and leathery, and the pinnae wiry in texture when fully expanded. Fronds (especially of subsp. *pinetorum*) turn a *rich, particularly deep cinnamon colour in autumn*, and thereafter characteristically *remain largely standing throughout much of the winter months*, with the lamina gradually eroding from the still aerial blades.

The especially distinctive features of this bracken are thus: its extremely early-season and rapid frond expansion rate; harsh and wiry texture of the stipe and frond from the outset; nearly simultaneous pinna pair expansion throughout each blade; inclined blade orientation from the first pinna pair; obliquely ascending rigid pinnae to the expanding fronds; and the very abundant presence of numerous, long, cinnamon-coloured hairs over the frond and pinna croziers, giving them a conspicuously red-brown colouration during their brief expansion phase. Indeed, the simultaneous flushing of the pinnae, the bright green colour of the blade, the angled blade and pinna orientation and the cinnamon-coloured croziers combine to make this plant distinctive and easily recognisable in the field even from a distance.

At least two subspecies of *P. pinetorum* are so-far recorded in these islands: subsp. *pinetorum* and subsp. *osmundaceum* (Page & Mill, 1995). Neither of these either is vegetationally aggressive nor a problem to human beings.

Pteridium pinetorum C. N. Page & R. R. Mill subsp. *osmundaceum* (Christ) C. N. Page. (*Osmunda*-like Bracken)

This subspecies is characterised in the field by its small *apparently narrowly triangular blades (caused by the lower pinnae being held inward and upward*, but wide when flattened out in herbarium material), sparse covering of ephemeral red-brown hairs over the expanding croziers in spring, and by its *especially bold pattern of blade division into few, conspicuous oblong segments, which are proportionately extremely large for the size of the expanded frond* (hence giving the frond in life its distinctive *Osmunda*-like appearance, even at first sight).

Variation Few sites for each of these Brackens are yet known, but the main variation in the field is in the size of the (always rather small) blades.

Possible confusion The small blade size, bold dissection patterns, slender wiry stipes and wiry textured fronds and large segments distinguished *P. pinetorum* from even small fronds of *P. aquilinum* in exposed positions.

Technical confirmation Little, as yet, studied.

Pteridium pinetorum C. N. Page & R. R. Mill subsp. *pinetorum* (Cairngorm Bracken)

This subspecies is characterised by its often *particularly dense covering of red-brown hairs in the expanding croziers in spring*, by its particularly *broadly triangular (typically equilateral)*

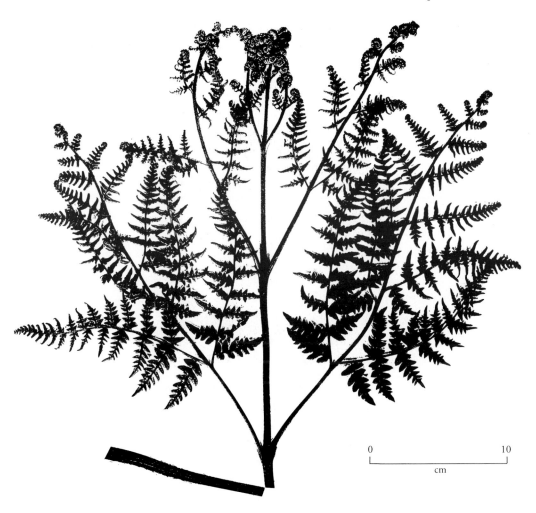

Pteridium pinetorum subsp. *pinetorum*, silhouette of expanding crozier in unpressed state showing near-simultaneous vernation of all frond segments: Inverness-shire, late June (type).

expanded blade held rigidly at an angle of c. 30° to the horizontal from a single point of inflection at the first pinna pair origin, and by its ultimate segments which are large, but still smaller than in subsp. *osmundaceum* (q.v.).

It is known with certainty from a single general area near to Aviemore, Inverness-shire, although the occurrence of other similar colonies in the same region has been reported. Type: Scotland: East Inverness-shire, Rothiemurchus Forest near Loch an Eilean, 4 June 1983, *C. N. Page* 17049 (holotype E) cited in *Watsonia* **17**: 431 (1989) as voucher of *P. aquilinum* subsp. *latiusculum*.

Field notes The finding of these several taxa of Bracken in Scotland contrasts starkly with the traditional view that all Bracken in these islands, and especially in Scotland, is merely a single, simple and uniform taxon.

The totality of each of the Bracken colonies seen in Scotland attributable to this alliance are colonies which are of relatively small size and have a large number of morphological features, seasonal behaviour patterns and ecological tendencies in

Pteridium pinetorum subsp. *pinetorum*, typical frond silhouette in flat-pressed state: Perthshire, August (type).

Pteridium pinetorum subsp. *pinetorum*, field view of typical fronds: Inverness-shire, late July (type, locality and clone).

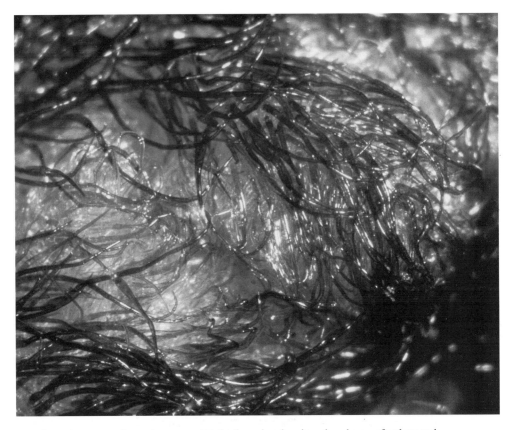

Pteridium pinetorum subsp. *pinetorum*, detail of crozier showing abundance of ephemeral cinnamon-red hairs: Inverness-shire, June.

common. All, furthermore, are notable in lacking the general ecological vigour associated with members of the *P. aquilinum* group. All have a low (typically under 50 cm) frond height, with a low frond density within their stands, giving colonies a relatively sparse and open aspect almost throughout (whether under trees or in the open).

The simultaneous unrolling (vernation) of all the units (tip, pinna and pinnules) of the frond and the early frond full expansion date thereby achieved (usually up to 2 weeks earlier than those of nearby stands of *P. aquilinum*) are distinctive features that make colonies of this species particularly possible to locate initially during late May and early June.

All are plants of cool, generally moist, soil sites, of highly acidic character, mainly growing in shallow sandy peats overlying fluvio-glacial outwash sands, gravels and rock detritus, including over granitic areas on the flanks of the Cairngorm massif. In several areas, colonies are associated with other nearby typically northern pinewood understorey relicts. At the type site these include *Juniperus communis* L., *Trientalis europaea* L., as well as *Erica cinerea* L., *Anemone nemorosa* L., *Vaccinium myrtillus* L., *Viola riviniana* Reichenb. and *Veronica chamaedrys* L. Subsp. *osmundaceum* seems able to spread into wetter, peatier ground with *Myrica gale* than is usual for any other Bracken, and, indeed, in this respect behaves much more like *Osmunda* too.

All colonies seem to be very frost-hardy, with their early expanding fronds remaining generally undamaged by the occurrence of late spring frosts (contrasting, in this respect, with members of the *P. aquilinum* agg., and especially strongly to *P. equilinum* subsp. *atlanticum*). Their low frond density and long-standing frond habit into autumn and winter, with slow loss of the blades, means that they thus fail to form the dense and swamping dead frond blanket of *P. aquilinum*, allowing a relatively rich associated flora to exist amongst the colonies.

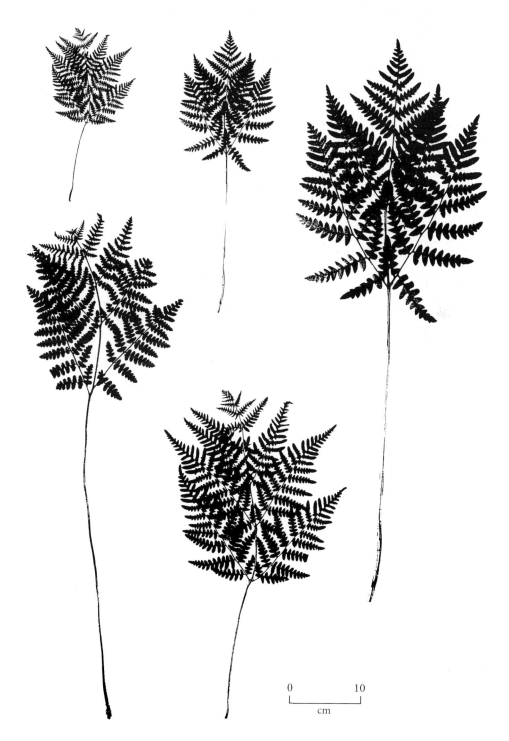

Pteridium pinetorum subsp. *osmundaceum*, typical frond silhouettes in flat-pressed state: Perthshire, July.

Pteridium pinetorum subsp. *osmundaceum*, field view of large lowermost pinna of typical frond: Perthshire, July.

All the plants of this species are parts of the Bracken alliance, which is widespread outside these islands through the far northern boreal coniferous forests of northern Europe and far northern Asia, especially throughout the range of Scots Pine (*Pinus sylvestris* L.). That the scattered native sites of this Bracken are all in native pinewood or former native pinewood sites within the original Scots Pine range of the Caledonian forest of Scotland seems entirely in keeping with this general ecological picture. For such present or past sites, it appears to be a distinctive indicator species.

Thelypteris palustris Schott   Marsh Fern

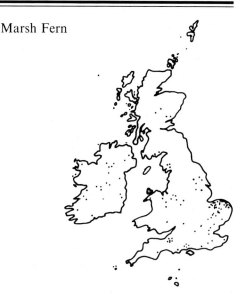

Preliminary recognition A medium- to large-sized fern, with erect, pale green, rather delicate-textured fronds, with widely separated pinnae, arising singly or in small groups and growing in marshy places.

Occurrence Of scattered distribution, chiefly through England, Wales and Ireland, nowhere widespread, with its main concentration of stations in the low fenland areas of East Anglia (especially Norfolk). Absent from all mountainous areas, and hence local in Wales and Scotland.

Identification Plants have *strongly erect fronds which arise at shortly spaced intervals* of usually a few centimetres, from a *thin, creeping rhizome* just beneath the surface of the ground. The fronds have lanceolate outlines, widest near the middle, and are divided into narrow, often widely spaced, pinnae. The apex of the blade often narrows rather abruptly, but below *the pinnae cease suddenly without greatly shortening*. The blunt-tipped pinnae are themselves cut nearly to the midribs, have sinuous non-toothed margins, and occur in *nearly opposite pairs*. Fronds of two types are usually present. The sterile vegetative ones are the shorter (to about 80 cm) and broader, with shorter stipes. The fertile ones are taller (to about 150 cm) and narrower with relatively much longer stipes (up to twice the length of the blade), and their pinnae can be particularly narrow.

In all fronds, the pinnae are borne in a more-or-less flat plane (i.e. neither markedly inclined nor reflexed forward). They are a *pale, very slightly bluish, green*, and the stipe and rachis are also pale, the stipe becoming darker only at the extreme base. The texture of the fronds is always rather *soft and papery*, especially where shaded and sheltered amongst tall reedswamp vegetation. The stipes are brittle once expanded, and are virtually without hairs or scales. The taller fertile fronds are less delicate than the vegetative ones, and stand the most stiffly erect. Their pinnae have down-rolled margins, which usually partially conceal the small sori on their undersides.

New fronds arise in spring, the sterile ones first about late May, the fertile ones about a month later. Spores mature and are shed during July and August, and the fronds thereafter die rapidly with the onset of frosts, usually earlier than those of *Dryopteris cristata* when in the same habitat.

Jan.	Feb.	Mar.	Apr.	May	Jun.	Jly	Aug.	Sep.	Oct.	Nov.	Dec.

Thelypteris palustris: a–c, East Norfolk; *d*, Forfarshire; *e*, Co. Kerry.

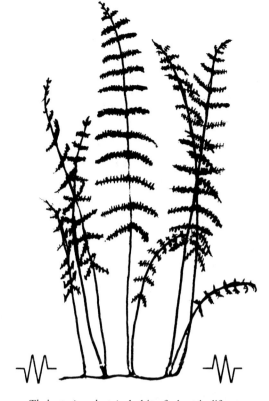

Thelypteris palustris, habit of plant in life.

Variation This is not a very variable species, and plants differ mainly only in size.

Possible confusion The fronds arising singly from a creeping rhizome, their pinnae ceasing suddenly below, the long, slender stipe and lack of distinct smell distinguish *Thelypteris palustris* from the similarly textured and divided *Oreopteris limbosperma* (q.v.). The rhizome habit also distinguishes it clearly from *Athyrium filix-femina* and all forms of the *Dryopteris filix-mas* group.

Technical confirmation Plants are diploids, with $n = 35$, $2n = 70$ chromosomes. The surface of the lamina is without glands, and the indusia, which are delicate and of irregular or slightly reniform shape, with gland-fringe margins, are shed at an early stage.

Field notes The genus *Thelypteris* is a taxonomically small genus of ferns with two related species in the northern hemisphere and one (*T. confluens* (Thunb.) Morton) in the southern. The latter has a distinct distribution through Africa south of the Sahara, southern India, parts of South-East Asia and New Zealand. Of the northern hemisphere species, only *T. palustris* Schott occurs in the Old World, with what is usually regarded as two varieties of a separate single species (*T. thelypteroides* (Holub) in China, Japan and North America). All species grow in usually permanently wet ground in rather exposed places.

Our single native species, *Thelypteris palustris*, ranges widely from Britain and Ireland to western Siberia, through predominantly cool temperate latitudes, but especially through areas of hot summers and distinctly continental climate. In Europe, plants are frequent in distribution through mostly mid-latitude climates, but stretching more sparsely as far north as Finland and as far south as Greece.

Within Britain and Ireland, plants grow mainly in exposed fen and reedswamp conditions which are permanently wet and organic-rich, but not highly acidic, and where deep silty mud becomes trapped. Plants are colonists, particularly at early seral stages of vegetation closure and where floating vegetation mats form over dykes and ditches, ultimately surviving to a stage where they have almost scrambling fronds amongst a vigorous annual growth of high reeds. Under such conditions, the long, dark, slender rhizome creeps for considerable distances, adding 30 cm or so of growth per season, and branching occasionally. These rhizomes extend near the surface of soft mud or 2–3 cm below the surface of lightly compacted recent decomposing litter accumulation, with the actively growing tips above wholly waterlogged conditions. The roots may, however, descend into more anaerobic and stagnant, sulphuretted and methanous submerged layers.

Heath (1876: 250) picturesquely described the plants then growing in a wet wooded site near Newton Abbott in Devon as 'growing in great abundance' and having 'black creeping rhizomes immersed in black bog water, above which the delicate light green fronds' which were 'exceedingly delicate and fragile' were 'beautifully waving' throughout the wooded moss and sedge-rich areas where 'the substance of the bog was more than usually liquid'.

In the fenland communities of East Anglia, commonly associated species may include abundant reedswamp-forming dominants such as Common Red (*Phragmites communis*), the Reed-grasses *Glyceria maxima* and *Phalaris arundinacea*, Saw Sedge (*Cladium mariscus*), Blunt-flowered Rush (*Juncus subnodulosus*); with more scattered Fen Bedstraw (*Galium uliginosum*), Yellow Loosestrife (*Lysimachia vulgaris*), Ragged Robin (*Lychnis flos-cuculi*), Brooklime (*Veronica beccabunga*), Marsh Helleborine (*Epipactis palustris*), Bogbean (*Menyanthes trifoliata*), Marsh Willow-herb (*Epilobium palustre*), Buckthorn (*Rhamnus catharticus*), Marsh Marigold (*Caltha palustris*), Pennywort (*Hydrocotyle vulgaris*), Water Violet (*Hottonia palustris*), and Yellow Flag (*Iris pseudacorus*), as well as several sedges (*Carex* spp.). Other ferns, mainly Fen Buckler-fern (*Dryopteris cristata*) and Royal Fern (*Osmunda regalis*), may also occur nearby.

Fronds of all plants die down in autumn, and begin to emerge usually through accumulated winter silt in about mid-May (but as early as late April in sunnier niches and milder springs). In experimental cultivation, those of juvenile plants typically remain green and presumably photosynthetically active to much later in the season than do the foliage of many nearby adult plants.

As Marsh Fern is a plant of essentially continental climatic conditions, it becomes more scarce in the mild winter, cool summer conditions of the western fringe of Britain and Ireland, where it is extensively replaced by *Osmunda regalis* in this habitat.

Marsh Fern appears to have become much scarcer over the last century through steady drainage of marsh and fenland areas, especially in central England and in Ireland, and its continued wellbeing in many of its habitats may well require particular conservation vigilance.

Trichomanes speciosum Willd.

Killarney Fern

Preliminary recognition A medium-sized fern, with deep-green translucent fronds, of membranous texture and very delicate appearance, mostly growing in a hanging position, in dark, humid, extremely sheltered, permanently moist situations, mainly on the sides of rocky gorges by moving water.

Occurrence A plant of extremely oceanic western range in Britain and Ireland, formerly known chiefly from south-west Ireland, with a smaller number of widely scattered stations, mainly in southern and north-west Ireland, and the oceanic fringe of north Cornwall, west North Wales, north-west England and western Scotland (on and near Arran). Now extinct or extremely rare in most of its former localities and specifically protected from collection by law (see p. 380).

Plants brought into cultivation from native sources by various Victorian collectors have, in many cases, been maintained in several botanic gardens, for nearly a century.

Identification Fronds arise singly from long, slender, wiry, surface-creeping rhizomes (up to 3 mm diameter), which are conspicuously covered with dark, articulating, hair-like scales. The rhizomes branch frequently, and in old, undisturbed colonies may form a considerable network, with numerous, fine roots spreading closely against the rock surface. On vertical rock surfaces, the fronds are borne in a gracefully drooping manner. They are about 20–45 cm in length, of broadly ovate to ovate-lanceolate outline, and the stipe occupies about 1/3 of the length. The fronds are finely dissected throughout, with acutely tapering pinnae, and a very thin (one cell thick) lamina, which extends down either side of the thin, rigid, pale-green stipe as a distinct green wing. The diagram on p. 377 illustrates the habit of the plant.

Fronds have a bright (when young) to deep (when mature) translucent green colour, and a distinctive thin, membranous, but firm, pellucid texture. Throughout the frond, the venation is conspicuous, with the ultimate branchlets of the veins extending to the tips of the lobes of the pinnae. In life, the fronds have a distinctive, delicately crisped texture, which is completely lost in herbarium material.

When fertile, sporangia develop within small, entire, urn-shaped receptacles on the upper edge of each pinna lamina. Within the receptacles, the sporangia are attached to a central, hair-like bristle. As the sori mature in a gradated fashion, so the bristle extends out of the receptacle, exposing the sporangia which then discharge and are shed. On old fronds, these dark-coloured bristles can often be seen remaining, protruding several millimetres from the receptacles.

Possible confusion The comparatively large, membranous fronds are totally distinctive amongst British and Irish ferns. The winged stipe and thicker rhizomes distinguish even small fronds from those of *Hymenophyllum* (which have thin unwinged stipes and thread-like

Trichomanes speciosum: a, Co. Kerry; *b–c*, Caernarvonshire.

Trichomanes speciosum, habit of plant in life.

rhizomes, q.v.). Poorly annotated herbarium specimens are known to have been taken for marine algae!

Variation In more favourable growing conditions, fronds are proportionately larger and more finely cut. There are also geographic differences in detail of frond form. Plants from some Irish stations with narrower frond outlines have been separated as var. *andrewsii* Newm.

Technical confirmation Plants have $n = 72$, $2n = 144$ chromosomes. Unusually amongst pteridophytes, the gametophyte generation of *Trichomanes speciosum* is filamentous and bears gemmae. This gametophyte can become independent and perennial, eventually growing to form a network of much-branched filaments over rock faces (see under 'Field notes'). They have the appearance of a bushy dull-green minute turf, and their filaments can look similar to some species of green algae (e.g. *Cladophora* or *Vaucheria*) or to the protonemal stages of mosses. Rumsey *et al.* (1990) have shown them to differ from these in the possession of short brown unicellular rhizoids, and in having gemmae and 'gemmifers' (the specialised cells on which the gemmae are produced). The gemmae are multicellular vegetative units that are specialised for dehiscence from the parent thallus, with the capacity to establish new gametophyte colonies. All of these are visible with a hand lens. Under microscopic examination, *Trichomanes* filaments can be seen to have cells never more than 3 times longer than broad, with each cell containing very numerous discoid chloroplasts. This contrasts with algae, which have either much longer cells and similar chloroplasts or only 4 very large chloroplasts.

Field notes Trichomanes speciosum is a plant chiefly of extreme western European range, occurring from the Cape Verde Islands, the Canaries, Madeira and the Azores in the North Atlantic northward to the extreme fringes of the Iberian peninsula (widely scattered and very local), western Brittany, Ireland and Britain (Jalas & Suominen, 1972). It is currently thought to be extinct (at least as a sporophyte – see below) in its former stations in Portugal (Jalas & Suominen, 1972; Ratcliffe *et al.*, 1993) but is reported recently to have been found on the west coast of Italy (Ferrarini, 1977). Some of its most successful surviving stations perhaps remain those of the oakwoods of Kerry in south-western Ireland (see e.g. Doyle, 1987). What are probably further related forms occur highly discontinuously in several other more distant (and mostly more tropic) regions of the world, especially in the wetter coastal fringes of west tropical Africa, in southern Central and northern South America, and in Japan, Malaya and the Sino-Himalayan region (as *T. orientalis* C. Chr.). Close allies of *T. speciosum* are known as Tertiary fossils in Europe and the former western USSR (see e.g. Fataliyev, 1960), and this evidence plus the highly fragmentary nature of the present range indicates the members of this alliance to be considerably relictual.

Mature sporophyte colonies of *T. speciosum* occur in a widely discontinuous array of always local stations within Ireland (especially in southern and western counties), and western Britain, notably in Cornwall (very locally), north-west Wales, northern England and south-western Scotland. It is reported from altitudes of up to 500 m in Kerry, but most of its widely scattered stations are at much lower elevations than this, and mostly in steep, sheltered and well-shaded stream ravines near to western Atlantic coasts.

In such sites the plant occurs mostly on steep-sided rock faces, beneath overhangs or in the neighbourhood of the splash zone of permanent waterfalls, or from the roofs and sides of shallow caves, almost exclusively in natural conditions of almost unvarying high humidity, permanent seepage, usually running groundwater, and extreme shelter. Recently Ratcliffe *et al.* (1993) have summarised its habitats as being in deep shade, in large dripping caves, in crevices and cliffs and in small caves, and on dripping vertical cliffs and rocky banks in wooded ravines. It grows chiefly on hard, predominantly acidic rocks, including sandstones, grits and schists, but is occasionally epiphytic on tree boles within ravines. Fronds are apparently winter-hardy through freezing conditions, even though these are usually short-lived in most of its sites. The association of the plant with areas of greatest climatic oceanity in these islands perhaps reflects its requirement not only for conditions of high humidity, but also for cool, moist summers and the long length of growing season constantly available in such sites.

Killarney Fern is always a very slow-growing fern in Britain and Ireland, and each frond can take several months to unfurl fully. Once expanded, however, non-fertile fronds are evergreen and can last for several years. In extremely humid situations, their blades are known occasionally to bear bulbils, capable of growth into a new plant. Plants have the ability to survive in very low light, and are perhaps largely confined to such situations by a low competitive ability in modestly better-illuminated situations, resulting from its inherent slow rate of growth.

Despite such a growth rate, scattered colonies of this fern on steep, dark, rock faces are sometimes quite large, and such patches may easily be one to two centuries or much more in age. Patches of mature sporophytes may intermingle with or grow adjacent to very numerous bryophytes and locally with one or both native species of *Hymenophyllum* (q.v.). Nearby bryophytes specifically noted include *Mnium hornum, Ctenidium molluscum, Scapania undulata, Riccardia pinguis, R. multifida, Thamnobryum alopecurum, Rhizobium punctatum, Fiddidens taxifolios, Calliergon cuspidatum, Brachytheciane rivulare, Amphideum mongeotii, Dichodentium pellucidum* and *Concephalum conicum.* Numerous

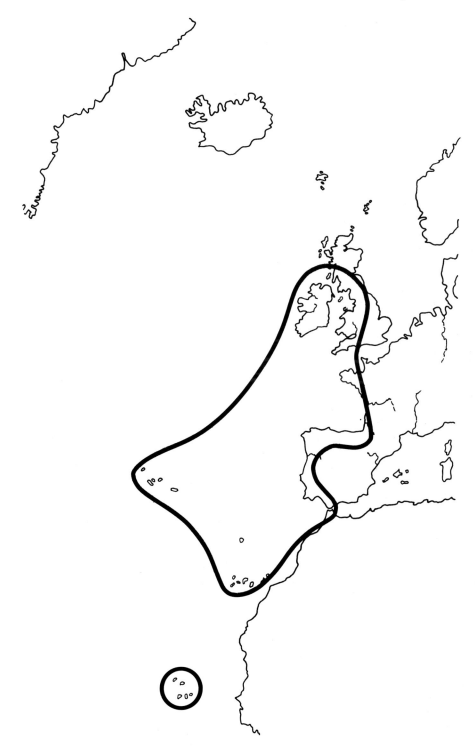

Trichomanes speciosum, north Atlantic range of species as known from wild sporophyte populations. An additional station in Italy has also been reported, with others further afield (see the text).

large fronds of *Trichomanes* may occur in such larger patches, arching from the surface of steep rock faces on short wiry stipes, then drooping gracefully down, forming overlapping curtains of similarly inclined blades. In such situations, the translucent texture and fine dissection of each blade allows filtered light to pass to other fronds and bryophytes beneath, while the pointed pinnule tips steadily drip with slowly percolating water running over the delicate and pellucid rigid frond surfaces, which nod spasmodically with each drip released.

In Britain and Ireland, plants of *T. speciosum* seem rarely to produce spores, and the few fertile fronds are said to wither after spore discharge, which happens perhaps only in occasional dry summers. Spores, when produced, mature between July and September. The gametophytes arising from them, which probably establish in small rock crevices or on the surfaces of dark rock faces, are said to take 12 months to produce a sporophyte. Recent explorations for its gametophyte confirm the theory originally suggested by Farrar (on the basis of similar observations in America – e.g. Farrar, 1967, 1985), of the presence of relictual populations of its gametophytes at locations well beyond the known occurrence of sporophyte colonies. In Britain, such perennial gametophyte colonies have indeed been found, and are now known to include sites in North Wales and in northern Britain (notably in the Lake District and in the Yorkshire Dales (e.g. Rumsey *et al.*, 1990). Some of the sites include ones in which sporophytes have been known, but at which they have been extinct for over a century. Such gametophytes are filamentous and produce wefts over rock surfaces in areas where the light intensity is probably too low for much bryophyte competition. They can reproduce indefinitely by gemmae, thereby maintaining themselves more or less permanently in such sites in the absence of sporophytes. These mats of much-branched filaments look particularly similar to patches of green algae or to the protonemal stages of some mosses (see 'Technical confirmation') and are hence easily overlooked in the field. In analogous habitats in the USA, the independent gametophyte generations, and particularly their gemmae, are much more tolerant of desiccating and freezing conditions than are their sporophytes (Farrar, 1985: 362; Farrar *et al.*, 1983). Throughout Britain and Ireland, the presence of such gametophytes of *T. speciosum* in sheltered dark ravines and cave mouths in moist, acidic sites may thus be geographically much more widespread than is yet known. This relationship between gametophyte and sporophyte generations has been termed 'facultitive independence' by Rumsey & Sheffield (1997).

This extremely attractive sporophyte plant, probably always rare, was hunted almost to extinction last century in both Britain and Ireland, largely through the collection of whole and probably ancient colonies, for which their slow rate of growth is totally unable to compensate. Ratcliffe *et al.* (1993), in a survey of the floristic composition of a total of 43 sites in Britain and Ireland, have estimated that the total number of surviving clones of the plant within these islands may number no more than between 50 and 100 individuals, monitoring data for which suggest that the frond number of many remains more or less constant. Plants should on no account be removed from the wild and are, indeed, specifically protected by law*. But, even in the absence of collecting, the restriction of most established sporophyte colonies of *T. speciosum* in Britain and Ireland to local and geographically highly isolated patches in a limited number of widely discontinuous stations, has probably long imposed considerable genetic isolation, resulting in local inbred populations persisting vegetatively for considerable periods. Further, the infrequency of spore production and the independence and isolation of the known perennial gametophytes, which presumably largely removes them from the actively breeding gene pool of the species, are collectively almost certainly factors contributing to a substantial loss of opportunity for natural outbreeding between individuals. Lack of such outbreeding may be one reason for the slow growth and the apparent absence of vigour seen in most surviving colonies. Some direct scientific intervention may eventually be required to assay and help to restore the degree of genetic heterozygosity and vigour of wild colonies of a species which has virtually become the emblem of pteridophyte conservation in these islands.

---

* Conservation of Wild Creatures and Wild Plants Act, 1975; Irish Wildlife Act, 1976; Wildlife and Countryside Act, 1981.

Woodsia alpina (Bolton) S. F. Gray

Alpine Woodsia

W. hyperborea (Lilj) R. Br.

Preliminary recognition A small or very small-sized fern, looking like a diminutive form of *Cystopteris fragilis*, but of more compact, rosette habit, with more robust, articulated stipes and a slightly scaly frond.

Occurrence A rare arctic-alpine species known in Britain only in a few mountain stations in North Wales and the central Scottish Highlands, in crevices and shallow ledges of steep rocky alpine cliffs, almost always in precipitous places.

Identification Plants have fronds of generally smaller size than those of *W. ilvensis* (q.v.), usually about 3–8 cm in length, but occasionally to 15 cm in exceptional individuals.

The fronds form rather spreading rosettes arising from short, sparsely scaly rhizomes, which branch occasionally on old specimens to produce *a few, closely spaced crowns.* The blades are borne on short, fairly thick, robust and rigid, brown-coloured stipes, usually little more than about 1/4 to 1/2 the length of the blade, and have a jointed articulation about 1/3 of the way up, unique in this position to this genus amongst British or Irish ferns.

The fronds have a fairly broad, oblong-linear, outline, and are divided into rather few (generally about 5–10) pinnae, *which are quite closely spaced and usually not oppositely arranged. Each pinna is fairly short and broad, and characteristically of triangular shape.* The pinnae are pinnatifidly subdivided into about 3–7 rather obtuse segments.

The whole frond is clothed with *rather sparse*, long, zigzag hairs, and on the frond underside there are usually also *a few* short scales, *but these are never dense and are confined to the rachis and do not occur on the pinna midribs.* The lamina is usually of a rather pale-green colour, and has a somewhat delicate appearance and texture.

Most fronds on mature plants are fertile, and in the upper part bear several sori near the margin of each ultimate segment. The sori are orbicular, and have a cup-shaped indusium surrounding the base of the sorus, with the upturning edges divided into numerous, long, filiform projections, which incurve over the sorus when young. When mature, these are usually darker coloured than those of *W. ilvensis.*

Jan.	Feb.	Mar.	Apr.	May	Jun.	Jly	Aug.	Sep.	Oct.	Nov.	Dec.

Woodsia alpina: *all* Mid Perthshire (from nineteenth century herbarium material).

Variation Plants vary considerably in size and, with this, in degree of division of the frond and pinnae. Occasional specimens with longer, narrower, more deeply divided pinnae have been reported in the past, particularly in Scotland, and could have been hybrids between the two native species (= *W. × gracilis* (Laws.) Butters.), but further investigation of such specimens is needed.

Possible confusion The usually smaller, more delicate, less scaly and less hairy frond, with less oppositely set, broader, shorter pinnae, lacking scales to the pinna midribs, and arising more numerously from fewer crowns, separate *W. alpina* from *W. ilvensis* (q.v.). Plants are also likely to be confused with young specimens of *Cystopteris fragilis* (q.v.), or other ferns at a very young stage, but the presence of hairs, the fimbriate-edged, cup-shaped indusia, and the thicker stipes with a distinct articulation confirm *Woodsia.*

Technical confirmation Plants are tetraploids, with $n = c.$ 82, $2n = c.$ 164 chromosomes. The spores are about 51–57 μm (cf. 42–49 μm in *W. ilvensis*).

Field notes *Woodsia alpina* has a widespread circumboreal range and in Europe show a markedly arctic-alpine distribution, with outlying western stations in Iceland, Britain and the Pyrenees.

Alpine Woodsia is absent from Ireland, and in Britain is predominantly northern and western in distribution, occurring discontinuously from north Wales to Highland Scotland, where it is a rare fern growing only in high mountain sites between 580 m (*c.* 1900 ft) and about 915 m (*c.* 3000 ft) or possibly higher. It occurs in small fissures on exposed cliff ledges, steep craggy cliffs and rock faces, with a range of aspects, but often southerly. Such habitats occur usually in precipitous sites, where few other vascular plant competitors ever gain a hold. It is generally restricted to areas where there are low summer maximum temperatures, usually not exceeding about 28 °C.

It is a moderately basiphilous species, growing on a range of hard volcanic or metamorphic rock types, including basalts and slates, and is reputed to be more exacting in habitats than is *W. ilvensis*. Plants occur in places where there is extremely free drainage and minimal competition, and are usually crowded tightly into very small rocky fissures in vertical rock, with minimal soil. Plants also seem to require very free air movement about their fronds.

Often they occur in niches where there is some overhead projection giving modest shelter from battering by the most heavy, driving rain. Plants are, however, tolerant of considerable winter freezing, and at some of their Scottish sites, in winter become heavily encrusted with ice.

Plants occur naturally as gregarious small colonies, spreading along small fissures by rhizome growth of old individuals, and re-establishing in neighbouring fissures from spores. Undisturbed, they have formerly formed numerous multi-crowned colonies over cliff faces, but survive today only in one or two places in such numbers. Like *W. ilvensis*, *W. alpina* is probably a relict species in the British flora from a more widespread population in post-glacial time. Like *W. ilvensis*, its few surviving stations seem very likely to have suffered from considerable genetic impoverishment even before widespread collection of specimens in the nineteenth century. *Woodsia alpina*, like *W. ilvensis*, is now protected and specifically cited by law in an attempt to protect it (see p. 380), but as with *W. ilvensis*, the number of individuals in some sites is now so few, and hence their genetic basis so limited, that doubt must now surround their ability to reform viable breeding populations, and more positive measures may be needed.

Woodsia ilvensis (L.) R. Br.

Oblong Woodsia

Preliminary recognition A generally small-sized fern, looking like a smaller, tougher form of *Cystopteris fragilis*, but with more robust stipes, and distinctly hairy fronds with a scaly undersurface.

Occurrence A rare arctic-alpine species, limited in Britain to a few mountain stations in Wales, northern England, southern Scotland and the eastern Scottish Highlands, in crevices and shallow ledges of steep rocky alpine cliffs, mostly in precipitous places.

Identification Plants mostly have small fronds about 5–10 cm in length, occasionally to 15 cm, *which look distinctly hairy and have a scaly undersurface.* They arise in ascending or spreading rosettes from short, sparsely scaly rhizomes. The rhizomes branch occasionally, and in old plants eventually build up a number of small, closely packed crowns, with the older parts covered in persistent bases of previous years' fronds. The fronds have short, fairly robust stipes, usually about half the length of the blade. About 1/3 of the way up each stipe is a distinct articulation marking an abscission zone at which the old frond eventually breaks cleanly away, like that at the base of a deciduous tree leaf. The stipes have chaffy pale-brown scales at the base and numerous narrow hair-like scales on their upper parts.

The blades have *an oblong-linear or oblong-triangular outline, which is generally narrower than that of W. alpina,* and their 7–15 rather widely spaced pinnae on each side, are usually more oppositely arranged. The lower pinnae are usually much longer than wide, and all of them are of oblong or triangular-oblong outline, but *much less obviously triangular and more distinctly oblong than those of W. alpina.* The pinnae are deeply pinnatifidly divided (or are completely pinnately divided near the base) into a fairly large number (about 7–13) of oblong-ovate, rather obtuse segments. *On the undersurface the rachis bears numerous, long, linear acute scales which extend on to the mid-veins of the pinnae. Both surfaces of the blades are also fairly densely covered in distinct, long, jointed, zigzag hairs.* The fronds are of a dark, dull green colour and rather firm texture, and their distinctive pilose covering is at first whitish in colour as the frond expands, but becomes a deep red brown when mature, giving the rather tough-textured fronds a velvety appearance.

Most fronds on mature plants are fertile, and in the upper part bear several sori near the

Jan.	Feb.	Mar.	Apr.	May	Jun.	Jly	Aug.	Sep.	Oct.	Nov.	Dec.

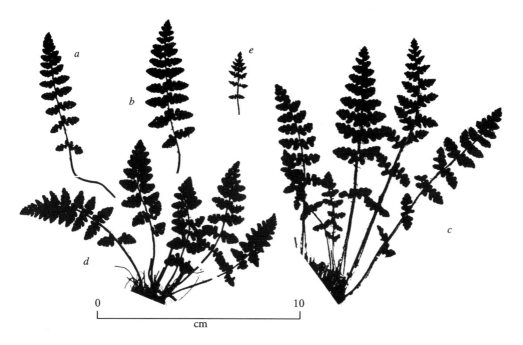

Woodsia ilvensis: *a–c*, Moffat area, Dumfries/Selkirkshire; *d–e*, Teesdale area, Co. Durham/North Yorkshire (all prepared from nineteenth-century herbarium material).

margin of each ultimate segment. As with *W. alpina*, the sori are orbicular, and have a cup-shaped indusium surrounding the base of the sorus, with the upturning edges divided into numerous, long filiform projections, which incurve over the sorus when young. These are usually paler coloured than those of *W. alpina*.

Variation Plants vary greatly in frond size, even from year to year and, with this, vary in degree of division of the frond and pinnae. The degree of separation of pinnae is also variable.

Possible confusion The usually more robust, more hairy and scaly frond, with more oblong pinnae that are more oppositely set and have scales along their midribs, distinguish *W. ilvensis* from *W. alpina* (q.v.). Plants are also likely to be confused with young specimens of *Cystopteris fragilis* (q.v.) or other very young ferns of all types, but the presence of hairs, the fimbriate-edged cup-shaped indusia, and the thicker stipes with a distinct articulation confirm *Woodsia*.

Technical confirmation Plants are diploids, with $n = 41$, $2n = 82$ chromosomes. The spores are about 42–49 μm (cf. 51–57 μm in *W. alpina*).

Field notes Woodsia ilvensis has a widespread circumboreal range with many discontinuous high-latitude stations across northern Asia, North America, Arctic Canada and southern Greenland. In Europe, it shows a markedly more northern trend than that of *W. alpina*, with the bulk of its stations in Iceland and throughout Fennoscandia, with only more outlying southern stations in central Europe, the Alps and in northern Britain.

Oblong Woodsia is absent from Ireland, and in Britain is northern and western in distribution, occurring discontinuously from north Wales and the Lake District to Highland Scotland, where it is a rare fern growing only in mountain sites between about 365 m (*c.* 1200 ft) and about 716 m (*c.* 2300 ft). It thus occurs at overall slightly lower

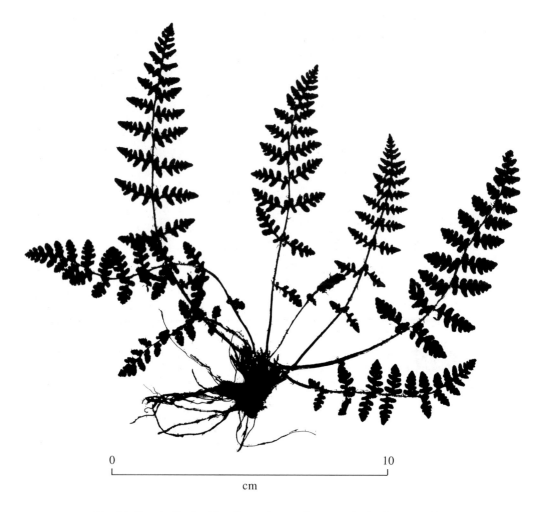

0   10
|_____|
                         cm

Woodsia ilvensis: Forfarshire (from nineteenth century herbarium material).

elevations than Alpine Woodsia, though above about 580 m (*c.* 1900 ft) there is a wide range of overlap, and the two species sometimes occur side by side. It occurs, like *W. alpina*, in small fissures on mountain cliff ledges, and on sheer, precipitous rocky faces, where there are steep, crumbling 'rotten' rocks on steep craggy mountains. Like Alpine Woodsia, Oblong Woodsia is restricted to habitats where there are relatively low summer maximum temperatures, but also ones of relatively high rainfall. Mitchell (1979) gives rainfall figures of 63 in. per annum (1540 mm) for its central Scottish Highland sites to more than 142 in. (3500 mm) for those in Snowdonia, whilst Ratcliffe (1960) has noted that plants usually fare best during years of wetter summers. Although it seems to have a greater tolerance of south-facing sites,

which are exposed to long periods of direct sunlight, than does *W. alpina*, it is clearly limited by severe drought spells (Mitchell, 1979), and plants can take several years to recover fully from very dry summers. Its niches are sometimes also ones where the local surrounding rock provides some degree of shelter from the most severe winds and driving rain. Plants probably, however, require fairly free air-movement about their fronds, helping the woolly surface to remain dry, although in winter, plants seem tolerant of periods of considerable ice encrustation.

Plants in different sites occur on a range of mostly volcanic and hard sedimentary rock types, which include basalts, pumice tuffs, Silurian grits and slaty shales. It seems to avoid both extremely acidic and, perhaps, strongly basic rocks, and is

Woodsia ilvensis: Grampian Mountains, Inverness-shire, June.

regarded by Mitchell (1979) as a plant of subacidic to neutral sites, with a tolerance, rather than a demand, for small amounts of lime. In its rocky habitats, associated species are normally few, but those recorded have included Ling Heather (*Calluna vulgaris*), Thyme (*Thymus drucei*), Roseroot (*Sedum rosea*), Wavy Hair-grass (*Deschampsia flexuosa*), Sheep's and Viviparous Fescues (*Festuca ovina* and *F. vivipara*), Parsley Fern (*Cryptogramma crispa*), Fir Clubmoss (*Huperzia selago*), *Rhacomitrium, Hypnum, Dicranum* and *Polytrichum* mosses and *Cladonia* lichens (Mitchell, 1979).

Plants of Oblong Woodsia, like Alpine Woodsia, seem potentially long lived, but are clearly, very slow growing, Yet they also seem to be in a particularly delicate balance with their environment, and this balance is easily upset. Undisturbed in the past, they have formed colonies of numerous clumps over nearby cliff faces. Plants seem to have occurred in such form in modest numbers in its Teesdale locality, as well as in several places in the Moffat Hills of southern Scotland, where they survived up to the mid-nineteenth century (see Rickard, 1972), and the history of their subsequent extermination in the Moffat area has been studied by Mitchell (1979).

Although it has been traditional to rightly blame Victorian botanical collectors for this sudden demise in historic time, biologically, such collection has probably been a final blow to a long period of natural decline. Like Alpine Woodsia, Oblong Woodsia is probably a relict in the British flora from more numerous and widely ranging British populations in post-glacial time. As with so many arctic-alpines, its remaining sites have become progressively smaller and more isolated from one another over many centuries. Even before botanical collection, in their increasingly local patches, numbers must have already fallen close to a size where they were genetically considerably impoverished as well as substantially inbred, probably contributing significantly to their lack of vigour. As with *W. alpina*, although now specifically secured from collection by law in an attempt to protect surviving plants (see p. 380), even if untouched, doubt must surround their natural ability at some sites to be able again to form viable breeding populations, and more positive measures may eventually be needed.

Clubmosses

(alphabetically)

pp. 391–422

Quillworts

pp. 422–432

Diphasiastrum alpinum (L.) Holub   Alpine Clubmoss

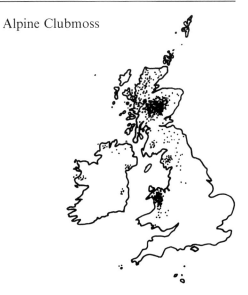

(*Lycopodium alpinum* L.)

(*Diphasiastrum complanatum* (L.) J. Holub subsp. *alpinum* (L.) Jermy)

Preliminary recognition A low-growing much-branched stiff moss-like plant of compact habit, with small, erect, scaly branches resembling twigs of a cypress tree, arising in even-topped, tight clusters.

Occurrence Fairly frequent at altitude as scattered plants amongst grass and heather over peaty soils, mainly throughout the mountainous parts of Wales, northern England and especially Scotland. More scarce in a few scattered localities on Irish Mountains.

Identification Plants consist of perennial creeping stems from which the conspicuous tufts of erect branches arise at intervals. The inconspicuous creeping stems are slender and tough, may be as much as 50 cm in length (though commonly much less), are whitish in colour, bear roots, and have small scale-like leaves at distant intervals. These stems, which occur partly buried in the ground, often hidden by mosses and surrounding vegetation, give rise (often from alternate sides) to frequent branches which bear the aerial stem clusters. These clusters arise as a series (commonly 3–5) of *repeated nearby dichotomies*, commonly in a single plane, forming *compact tufts of stiff, dark blue-green*, slightly glaucous, more or less erect, parallel shoots, *with a distinctive level-topped appearance in the field*, and looking like miniature shrubs.

The individual shoots of these clusters are each densely covered with minute (2–4 mm long and about 1 mm wide) scale-like leaves, *tightly appressed to the shoot and arranged in four distinct lengthwise ranks, giving each shoot a superficially squared appearance.* This shape can usually be seen clearly through the upper few centimetres of each shoot. On older parts of the shoots, however, and especially on the main running shoot and the bases of the primary branches from it, the shoots are somewhat more dorsiventrally flattened, *with the leaves on the underside small and like the shape of a bricklayer's trowel*, and these are usefully diagnostic in comparison with those of *D.* ×*issleri* (q.v.).

Some of the aerial shoots (often in groups) become longer than others and each terminates in one small (1.5–2.0 cm long) cylindrical stalkless cone, *only a little thicker in diameter than the shoot which bears it.* Early in the season, the cones are olive-green in colour and have tightly appressed cone scales, but later (by mid-summer) become a more conspicuous pale yellow-brown with spreading scales.

Jan.	Feb.	Mar.	Apr.	May	Jun.	Jly	Aug.	Sep.	Oct.	Nov.	Dec.
							███				

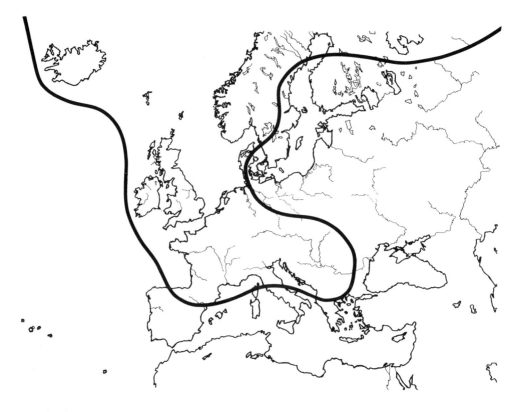

Southern boundary of arctic-alpine range of *Diphasiastrum alpinum* in Europe.

Variation This is not a very variable species.

Possible confusion Only likely with the rare hybrid *Diphasiastrum* ˣ*issleri* (q.v.), of which *D. alpinum* is one parent. Plants of *D. alpinum* differ mainly in their smaller size, more compact habit and much less dimorphic leaves.

Technical confirmation Plants have $n = c.$ 24–25, $2n = c.$ 48 chromosomes. The sporophylls of the cone are of a long, lanceolate and tapering shape, at least twice as long as the sporangium, with good spores which are about 45 μm in diameter (A. C. Jermy, personal communication).

Field notes Plants are frequent members of short grassy heath and dwarf shrub (mostly Heather) communities on mountains, ascending to nearly 1220 m (*c.* 4000 ft) and becoming infrequent below about 457 m (1500 ft). Plants are particularly characteristic of well-drained slopes where there is only a very thin covering of compacted, peaty ground over rock, but also succeed sometimes in patches where the plant competition of dense-turf vegetation is reduced by grazing pressure and by high exposure. They are considerably exposure-resistant, and in the most exposed sites typically adopt a particularly low-growing and compact habit.

Its habitats vary from species-poor to sometimes fairly species-rich ones, but its main associated species are ones typical of exposed, acidic, mountain-flank communities, and usually include Ling Heather (*Calluna vulgaris*), Bell-heather (*Erica cinerea*), Alpine Lady's-mantle (*Alchemilla alpina*), Heath Bedstraw (*Galium saxatile*), occasional Least Willow (*Salix herbacea*), and scattered low plants of Sheep's Fescue (*Festuca ovina*), Wavy Hair-gress (*Deschampsia flexuosa*), and Common Bent (*Agrostis tenuis*). Stagshorn Clubmoss (*Lycopodium clavatum*) and

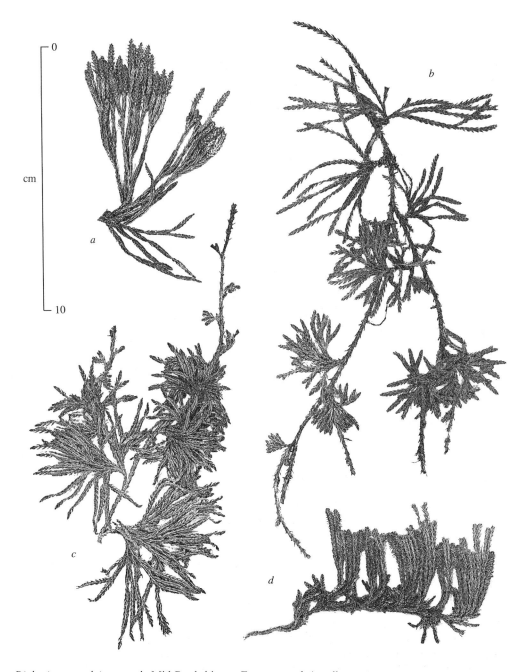

Diphasiastrum alpinum: *a*, *b*, Mid Perthshire; *c*, Easterness: *d*, Argyll.

Fir Clubmoss (*Huperzia selago*) frequently also grow nearby. Alpine Clubmoss also occurs in mountain patches where there is long snow-lie, along with Bilberry (*Vaccinium myrtillus*), Stiff Sedge (*Carex bigelowii*) and *Cladonia* lichens. Plants frequently occur too in herb-rich vegetation on cliff ledges, where the shallow soils limit the density of competition.

The spores of Alpine Clubmoss are said to have highly resistant walls and are presumably slow to germinate. The prothalli occur in the ground, at depths of up to 4 cm, where they form a mycorrhizal association with soil fungi, on which they are partly or entirely dependent. Prothalli found in Wales and attributed to this species, have been described as 'off-white, carrot-shaped structures about 3 mm long', with their outer layers containing healthy, intracellular, fungal hyphae (Thomas, 1975).

Diphasiastrum ˣissleri
(Rouy) Holub

Hybrid Alpine
Clubmoss

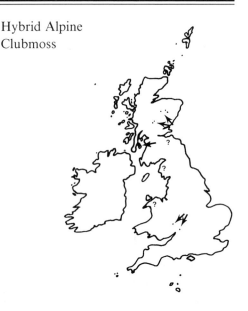

(*Lycopodium alpinum* L. var. *decipiens* Syme ex Druce)

(*D. complanatum* (L.) J. Holub subsp. *issleri* (Rouy) Jermy)

Preliminary recognition Plants generally look like a larger, less-compact form of *D. alpinum*, with broader, more flattened shoots, and a laxer habit of growth.

Occurrence A rare plant occurring sporadically in a few widely scattered stations in southern England and the Midlands, in some of which it has been known for over a century, but in many of which it may have become extinct. Specimens also occur at low elevations in more mountainous areas of Highland Scotland.

Identification Plants consist of perennial creeping stems, which extend just above soil level through moss and lichen, or just beneath it, and give rise at intervals to alternate side branches. The side branches each grow for 2–3 years, and form open, ascending, flattened fans of branchlets. The main shoot as well as the branchlets are *more dorsi-ventrally flattened* than those of *D. alpinum*, and *have an overall looser and laxer habit* and are less obviously flat-topped. Their leaves are also of *much less tightly appressed habit*, and are less greyish-bloomed than in *D. alpinum*.

Some of the side branchlets bear a single cone at their tips, each cone 2.0–2.5 cm long and raised on a sparsely bracteate pedicel. The branchlets raise the cones to a height of about 5–8 cm above the ground, and the long length and sometimes more irregular height of these pedunculate cones, often branching at their bases, can serve as a useful guide in the field to the presence of *D.* ˣ*issleri* in areas of mountainside where the dwarfer, more flat-topped cones of *D. alpinum* may be present as well.

Additional useful characters for distinguishing *D.* ˣ*issleri* from *D. alpinum* are particularly the characters of the leaves. The broad lateral leaves of the flattened shoots have *more of their length fused to the stem* (up to 2/3 of the length of the leaf fused in *D.* ˣ*issleri*, cf. about 1/2 in *D. alpinum*) and the *upper free portion of the leaf less strongly incurved*. The ventral leaves are less well developed than in *D. alpinum*, standing out only slightly, and are of *elliptic-lanceolate shape*, with the free portion of the blade about 0.8 mm long (cf. standing well away from the stem, trowel-shaped, with the free portion about 1.2 mm long in *D. alpinum*).

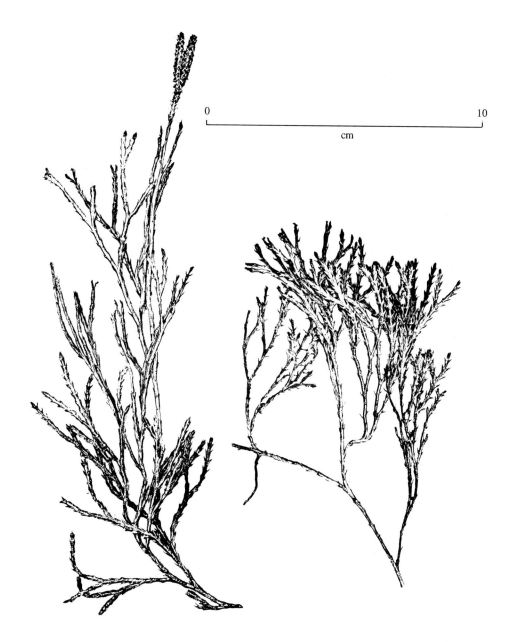

0   10

cm

Diphasiastrum ˟*issleri*: East Gloucestershire (courtesy A. C. Jermy).

Plants of *Diphasiastrum* ˟*issleri* are the presumed hybrids (or progeny of presumed hybrids) between *Diphasiastrum alpinum* (q.v.) (= *D. complanatum* subsp. *alpinum sensu* Jermy & Camus, 1991) and *D. complanatum* (presumed *D. complanatum* subsp. *complanatum*) (Wilce, 1965; Dostal, 1984; Jermy, 1989 – but see also below), the latter taxon of which is present throughout northern Europe, but not known in Britain. These intermediates may be partly fertile.

However, an alternative view that I tentatively record here is that some of the non-*D.*

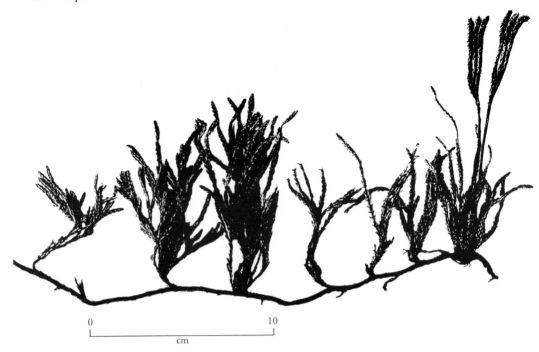

Diphasiastrum ˣ*issleri:*, profile view of shoot: Inverness-shire.

alpinum plants of *Diphasiastrum* from Scotland may indeed be hybrids with *D. alpinum*, but perhaps with *Diphasiastrum tristachyum* (not *D. complanatum*) as the other palaeo-parent. This view is based (see also 'Field notes') on both morphology of the Scottish plants (especially on uprightness of vegetative growth and habit of cone form) as well as comparison with the ecology of this other potential parent. In mainland Europe *D. tristachyum* is a species of more intermediate altitudes, overlapping with both *D. complanatum* at generally lower altitudes and *D. alpinum* at usually higher ones. If once present in Britain, it would seem to have been the more likely to have come into regular contact with *D. alpinum*, at least in Scottish upland sites.

By this argument too, *D. complanatum*, by contrast, may have been the palaeo-parent of mainly the lowland English material.

Variation Specimens of southern English origin seem to be more or less intermediate between *D. alpinum* and *D. complanatum* subsp. *complanatum*. Those of more northern origin are more variable, and possible differences in the parentage of these might be involved (see above).

Possible confusion Likely only with *D. alpinum*.

Technical confirmation The sporophylls have been described as of ovate-acuminate shape, 1.5–2 times as long as the sporangium (cf. lanceolate and gradually tapering, at least twice as long as the sporangium on *D. alpinum*), with spores which are only partly abortive and of 32–38 μm diameter (cf. all good and about 45 μm in diameter in *D. alpinum*) (A. C. Jermy, personal communication). The chromosome number of British material has not been determined.

Field notes *Diphasiastrum* ˣ*issleri* is known in both Europe (Jalas & Suominen, 1972) and North America (Lellinger, 1985), where potential parents occur. Within Europe, *D.* ˣ*issleri* is recorded from a wide range of localities throughout the Alps and northern central parts of the continent (Hegi,

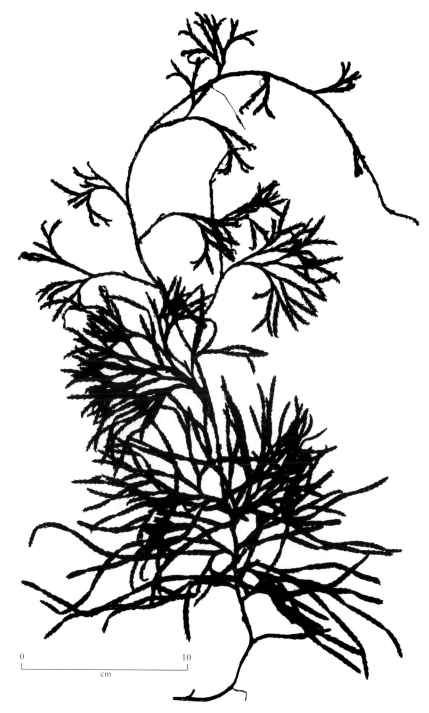

0   10
|_____|
                cm

Diphasiastrum ×*issleri*, vertical view of shoot in flat-pressed condition: Isle of Arran.

1984), and in Britain can be regarded as a northern Continental element. A particularly excellent field photograph of it is given in Europe by Rasbach *et al.* (1976: 187).

In Britain, plants of confirmed taxonomic status occur on warm, sunny, open ground, amongst dwarf shrub communities on low-altitude heathlands usually in deep Ling Heather (*Calluna vulgaris*) in a few, widely scattered stations through the southern half of Britain and in the Scottish Highlands. In its Malvern Hills (Worcestershire) locality, scattered plants occur locally in *Calluna–Deschampsia flexuosa* and *Calluna–Erica cinerea* communities, on cool, moist, sloping banks where *Vaccinium myrtillus* and *Potentilla erecta* are common, with occasional *Galium saxatile* and *Luzula multiflora*, several mosses including *Campylopus fragilis, C. introflexus, Dicranum scoparium* and *Pleurozium schreberi*, and *Cladonia* lichens.

Additionally, in several particularly interesting recently discovered localities in south Aberdeenshire, very distinctively non-*alpinum* plants of *Diphasiastrum* tentatively attributed *D.* ×*issleri* are locally extensive and form quite large but well scattered patches on north-east slopes at about 300–400 m (thus in much higher, cooler sites than the southern English ones, but nevertheless still well below the normal local altitude for *D. alpinum* in the area). These occur on shallow, peaty, stunted, exposed heathlands of *Calluna–Vaccinium*, sometimes associated with *Lycopodium annotinum* and *Huperzia selago*, and with nearby *Diphasiastrum alpinum*, or in partially open gaps where there is a little *Vaccinium vitis-idaea* and *Arctostaphylos uva-ursi*, *Antennaria dioica* and *Blechnum spicant*. Here they appear strikingly different in appearance from local *D. alpinum* in the field (Prof. C. H. Gimmingham, personal communications).

Usually treated as hybrids, plants attributed to *Diphasiastrum* ×*issleri* in Britain may also be at least partly self-reproducing within their own right, through their ability to produce a proportion of apparently good spores in their somewhat infrequently produced cones (e.g. Dostal, 1984; Jermy, 1989). This behaviour presumably opens opportunity for this taxon to produce at least limited numbers of local progeny and to found new colonies by such means, and, presumably too, to become involved in further hybridisation, from which a more complete spectrum of introgression might well eventually arise. This seems to accord fully with field observations on several lower mountain slopes in the central Scottish Highlands, where plants which are by varying degrees intermediate between non-*D.*

alpinum and nearby true *D. alpinum* occur locally.

The presence in Britain at all of non-*D. alpinum* but clear *Diphasiastrum* plants suggests the probability that another parent or parents in the same genus must have once also been present here. This missing, presumed palaeo-parent has been widely assumed to be *D. complanatum*. In both Europe and North America, however, *D. complanatum* is a species mainly of semi-open forest floor sites with much raw litter in (usually) coniferous or mixed conifer-broadleaf rolling forests, where it colonises relatively dry, but typically humus-rich, acidic soils. In such sites, in eastern North America, I have seen it build sometimes extensive colonies over lowland woodland floors, probably indicating relatively long periods of lack of terrestrial disturbance of such acidic woodland sites. In mainland Europe, *D. tristachyum* is also present, and this species is associated principally with original coniferous woodlands, growing in thin acidic and stony soils, often at generally higher altitude than does *D. complanatum*.

Based on this ecology, *D. complanatum* (and/or perhaps also mainland European *D. tristachyum*, see above) could well each have been once former associates of our native *Pinus sylvestris* pinewood or Pine–Birch (*Pinus–Betula*) wood in the British Isles (which itself ranges through considerable altitude), with either or perhaps both taxa perhaps long remaining so in the persisting pinewood fragments. It thus still remains notable how the distribution in Britain of the non-*alpinum* forms of *Diphasiastrum*, at least in Scotland, largely correspond in their occurrence with the distribution of shallow granitic soils, especially in cooler and climatically drier and more continental climates on and around northern granitic masses, and thus especially so around the granites of Dalradian origin in Scotland from central Perthshire to the northern Grampians, and around those of Moine origin stretching northward to east Sutherland and Caithness.

Such an overall history might, indeed, have shown close parallels to that of *Pteridium pinetorum* (q.v.) in Scoland, which today probably still shares a similar wider Eurasian boreal ecology and range those of *D. complanatum* and *D. triquetrum*, and which is certainly still present in fragments of the Scottish native pinewoods. Each such now-continental taxon probably declined for similar reasons under the influence of increasing mildness of the British and Irish climate, perhaps initially and largely through the sub-Atlantic or Atlantic periods of post-glacial time.

Huperzia selago (L.) Bernh.   Fir Clubmoss

(*Lycopodium selago* L.)

Preliminary recognition A small plant with rigid, stiff, strongly upright, slightly prickly leaved stems, arising in close groups, like small shoots of Juniper.

Occurrence Fairly frequent in moorland and heath vegetation and on bare mineral ledges in mountainous areas of Wales, northern England and Scotland, and in widely scattered mountain areas throughout Ireland. Also in lowland heathland habitats elsewhere in Britain and Ireland, but much less common than formerly.

Identification Plants have perennial leafy stems, with usually *a short, decumbent, basal portion, which gives rise to a dense tuft of upward-growing shoots.* Each shoot usually forks several times, to give rise to a cluster of thick, stiff stems, about 5–10 cm or more in height. Each stem puts on a similar annual growth increment of about 2–4 cm, thus maintaining a height similar to that of all others in the cluster, and giving each plant *a more or less constantly flat-topped appearance.* Numerous tough, wiry roots arise from amongst the old, pale-coloured decumbent parts of the plant, holding the elongating shoots stiffly erect. Each shoot has a rather Juniper-like appearance, and the whole plant one of a small, stiff, shrubby bush.

The erect shoots are densely clothed with small (2–8 mm long), *sharply pointed, evergreen leaves,* without evident midribs, which are *all similar and mostly arranged in a spreading manner, all around the shoot.* The cones are not structurally distinctive. Instead, the *upper parts of many of the shoots* (sometimes over the greater part of their length) have *conspicuous rounded sporangia in the axils of the leaves,* which are at first green but become yellow-brown when mature in late summer.

The upper parts of most stems also bear small, leafy, trident-like, flattened buds ('bulbils' or 'gemmae') which readily detach, disperse and take root as new small plants.

In the field, plants present a texture which is *stout, tough and slightly prickly to touch,* and of slightly glossy bright-green colour. Sometimes whole plants are sterile, or sterile and fertile branches occur intermixed.

Variation Leaves are shortest in plants in most exposed conditions, and may also be more appressed to the stem. Plants in most luxuriant situations are larger and laxer, with more-spreading leaves.

Jan.	Feb.	Mar.	Apr.	May	Jun.	Jly	Aug.	Sep.	Oct.	Nov.	Dec.

Huperzia selago var. *selago*: *a, b, d*, Clyde Isles; *c*, Inverness-shire.

Huperzia selago: *a*, *b*, from populations identified as var. *recurva*; *c*, from population identified as var. *appressa*. *All* South Aberdeenshire.

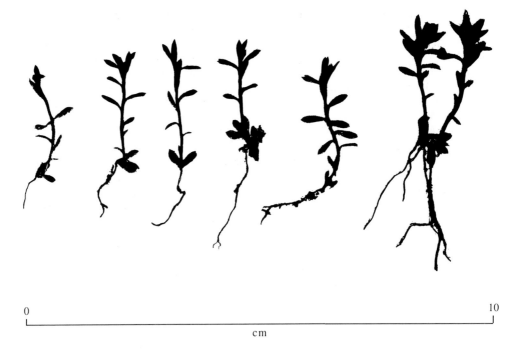

0   10
cm

Huperzia selago, germinating bulbils, horticulture, ex Mid Perthshire.

In addition to these environmental influences, amongst mountainous areas, at least in Eastern Scotland but possibly also elsewhere, are individuals and possibly small populations which have been taxonomically separated from the more widespread var. *selago*, and may well represent distinct genotypes. These include var. *recurva* Kit, which tends towards having an especially widely spreading to nearly recurved leaf habit and which seems to form particularly large specimens, and var. *appressa* Desv., a form with tightly packed forward-swept foliage throughout, which appears always to have rather slender shoots and to form only relatively small plants. Further research on the exact status of these plants, each of which appear to be geographically limited in Britain, seems justified.

Possible confusion A distinctive species, not likely to be easily confused. The stiffly erect habit with little creeping stage of growth, absence of a separate cone and the presence of bulbils distinguish *H. selago* from other native clubmosses.

Technical confirmation Plants have $2n = c$. 264 chromosomes.

Field notes *Huperzia selago*, in various locally differing forms, is a widespread northern circumboreal species. In Europe it has a widespread generally western and high to mid-latitude distribution, with concentrations in abundance especially in Iceland, throughout Fennoscandia, mountains of central and western Europe and especially the Alps, the Pyrenees and Britain and Ireland.

In these islands, Fir Clubmoss occurs mainly in more mountainous and northern districts, where it is plant of exposed situations without tree or shrub overgrowth, especially in grassland turf and dwarf-shrub communities on mountains and on heaths, and on rocky mountain ledges, ascending to 1065 m (3500 ft) or more. It occurs chiefly on peaty hummocks, around boulders, on exposed ridges and well-drained summit plateaux where there is thin peaty soil over rocks, in bare mineral soil patches and in areas of long snow-lie, chiefly over hard acidic rocks such as gneiss and granite, but also over schists. In such habitats, it is often associated with other lycopods, chiefly Alpine Clubmoss (*Diphasiastrum alpinum*) and Stagshorn Clubmoss (*Lycopodium clavatum*). Other common associates include Ling Heather (*Calluna vulgaris*), Bell Heather (*Erica cinerea*), Crowberry (*Empetrum nigrum*), Bilberry (*Vaccinium myrtillus*), Heath Bedstraw (*Galium saxatile*) and Alpine Lady's-mantle (*Alchemilla alpina*). Tormentil (*Potentilla erecta*), Wavy hair-grass (*Deschampsia flexuosa*), occasional *Salix myrsinites*, and mosses such as *Dicranum scoparium, Hylocomium splendens, Hypnum cupressiforme* and *Pleurozium schreberi*. Occasionally it also occurs on exposed lowland heaths in England, and on lightly grazed coastal turf in the west of Scotland.

Huperzia selago appears to be a relatively short-lived plant, which slowly builds up its tufts over several seasons and then probably reaches an abrupt and fairly rapid demise. Reproduction of plants probably relies heavily on the establishment of its wind-blown bulbils, which are produced in quantity on the mature plants every year, and which detach and take root readily once dispersed. Questions surround the frequency of its reproduction by spores. Plants seem tolerant of wide temperature extremes, wind exposure and probably long snow-cover, but are reduced by heavy grazing and are highly fire sensitive, being eliminated by burning.

Lycopodiella inundata (L.) Holub

Marsh Clubmoss

(*Lycopodium inundatum* L.)

Preliminary recognition A short, creeping, small, bright green, moss-like plant, clothed with stiff, upward-curving, cylindrical, tapering leaves, each stem mostly decaying after only a single year's growth.

Occurrence Widely scattered in Britain and Ireland, but scarce everywhere, often extremely local and diminishing, in acidic, boggy, periodically flooded, exposed sites, mainly on lowland heaths or in upland valleys, forming patches over bare, peaty or sandy ground.

Identification Plants have strongly prostrate, *short, creeping stems*, which cling very closely and firmly to the muddy surfaces on which they grow, often partly immersed in them. Even in summer, their sites are often periodically shallowly flooded, particularly after steady rain and the plants may occur partly submerged. Stems achieve *conspicuous bright green growth increments of only a few (2–4) centimetres a year*. The new growth forms a continuation of the previous season's stem which, however, *largely dies away in its first winter*, and persists into the second year only in a brownish, semi-decayed state, with a green growing tip. From this, the new year's growth resumes. Each new season's stem turns slightly upwards near its extending

| Jan. Feb. Mar. | Apr. | May | Jun. | Jly | Aug. | Sep. | Oct. Nov. Dec. |

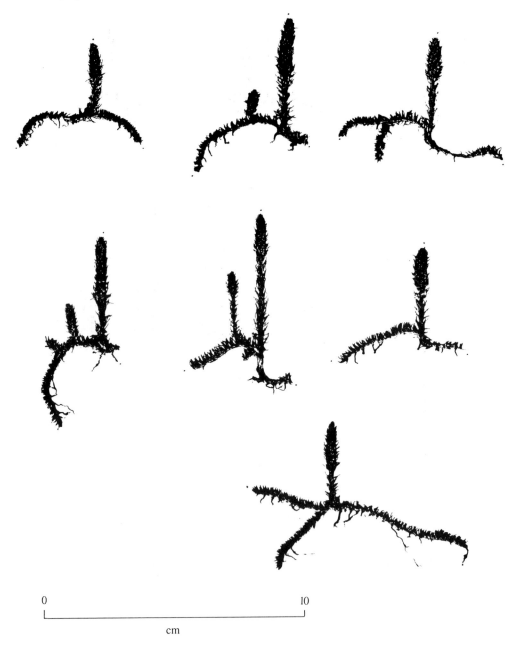

0   10
cm

Lycopodiella inundata, variation in a single population: *all* South Aberdeenshire.

tip, and is clothed throughout with 4–6 mm long, slender, *cylindrical-tapering, awl-shaped, stiffly assurgent leaves*, which curve upwards from near the base and have a fleshy texture and glistening cellular appearance under a hand lens. Each leaf ends in a soft, acute, flexible point. Weak shoots usually remain simple; stronger ones give off an occasional branch shoot, which becomes the weak shoot of a new independent plant in the following season as its junction with the old shoot rapidly decays away.

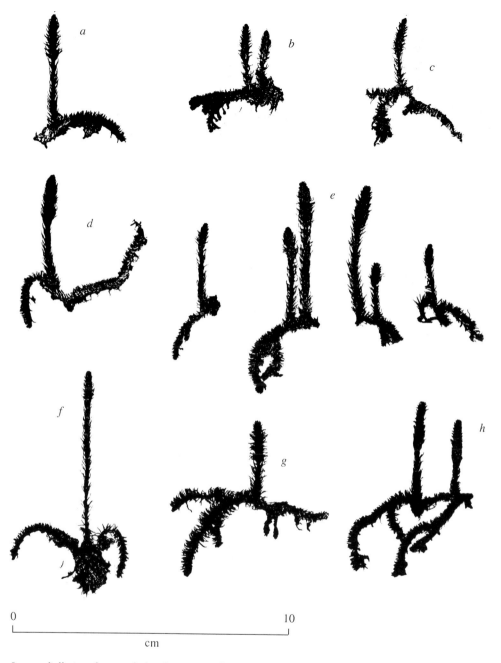

Lycopodiella inundata, variation between native populations: *a*, Sussex; *b*, *c*, West Inverness-shire; *d*, *e*, Surrey; *f*, Kent; *g*, Nairn; *h*, Dorset.

Occasional strong shoots also give rise from their dorsal surface to usually a single 2–3 cm high, radially symmetric, stiffly erect, cone shoot. Each cone shoot has a slender, upright basal section clothed with stiff, tapering, but more closely appressed leaves. Above this, the cone shoot widens into an *elliptical cone, which is only poorly differentiated* from that below. The cone bracts closely resemble leaves, but are longer and more spreading. The cone shoots appear during June and July. They are at first a similar green colour to the rest of the plant, but mature to an olive-yellow colour by October, after which they very rapidly decay.

Possible confusion The curious annual die-back of growth makes *L. inundata* unlike any other species of native clubmoss. Its cone is somewhat similar to that of *Selaginella selaginoides*, but larger, fatter and stiffer, and the habitats of the two plants are totally different (see 'Field notes').

Variation Within single populations, individuals show mostly environmentally induced differences in growth habit, including in the length of growth and branching achieved by the creeping vegetative parts and in the number of cones borne on each separate plantlet. Beyond this, plantlets in single populations usually show a distinctive sister-uniformity about them. Greater variation, however, occurs between widespread individual populations, especially in characters of shoot diameter, apparent plantlet vigour, average cone frequency and cone size achieved. How much of this variation is genetic and the result of a high degree of population isolation or how much is environmentally induced remains a subject for further research. We are, however, at the very periphery of the range of this mostly more subtropical group, and extremes of growth form adapting to extremes of habitat conditions might well be expected.

Technical confirmation Plants have $n = 78$ chromosomes.

Field notes *Lycopodiella inundata* is a widespread northern clubmoss of both Old and New Worlds. Other close relatives occur sometimes extensively in more southern, hotter and lower-altitude peaty, sandy and gravelly hollows such as those over the sandier soils of the south-eastern USA, often in open patches associated with nearby pinewood and other natural conifers, as far south as Florida. In Europe, it is a widely ranging but predominantly western species of many lowland areas at mid-latitudes, with outlying stations through central Europe, spreading southward mainly to the Alps and Pyrenees and with outlying stations in Brittany, the north-western Iberian Peninsula and the Azores.

Marsh Clubmoss is present sporadically and somewhat discontinuously throughout many parts of Britain and Ireland, where it is the only clubmoss able to grow in acidic wet ground. (*Selaginella selaginoides* grows in basic, damp ground.) It occurs mostly at low altitudes on damp heaths and commons or in valley bottom sites in mountain areas. It has a widely scattered distribution, and has always been a local and scarce plant, but is now quite rare.

It has a somewhat curious ecology, demanding moist, peaty, sandy or muddy sites, which are almost competition-free. It has been recorded from damp sandy and peaty depressions on wet heaths, marshy edges of lakes, silty patches in valley bogs, sandy flats, bare wet stony heaths, and wet dune slacks. It is known to have established anew also in certain artificial habitats, including depressions bared by removal of heathland turf and in old tank-track ruts! Its sites are mostly on sandy or peaty substrates, and are usually highly acidic ones, which are base and mineral deficient. It thrives in habitats which remain wet in summer and can be flooded to a depth of several centimetres in winter.

In such habitats, plants pioneer open, often silty, peaty or sandy surfaces, and in favourable sites, can form nearly pure local stands, with established colonies sometimes covering several square metres. Each such patch is probably most often developed by steady vegetative growth and fragmentation, from one or a few original individuals. In vegetatively luxuriant colonies, normally only a fairly small number of stems produce cone shoots each season. It seems possible that different stems bear cones in different years, and that it may take several years of steady growth before the same shoot cones again.

In habitats which remain relatively open, plants probably live for many years, and individual colonies have been known to survive for over 30 years. Patches may, however, change considerably in size and vigour in different years, presumably largely under the influence of the preceding summer and winter's weather, as well as that of the present. The plant also has a reputation for disappearing easily, although it can often be difficult to refind one small colony on a large, open heath at all! Nevertheless, colonies do seem to become eliminated naturally wherever habitats close over with more vigorous flowering plant or bogmoss growth. Its preference for sites that are periodically flooded probably reflects the lack of strong competition by other plants for these, especially where flooding also results in a winter deposition of peaty silt. The annual surface regrowth of this species, and the relatively expendable nature of that of the previous year, seem to adapt this species to such conditions.

Just as it may disappear from former sites, so it can appear in new ones or reappear in old ones in which it has not been known for many years, even when there are no other known populations nearby. It seems very probable that such new colonies establish mainly from prothallial growth, and that these arise from spores that have lain submerged and dormant in muddy layers for considerable periods, although this needs experimental verification.

Associated plant species in most habitats are usually few, although several may occur around the general periphery of the clubmoss patches. Observations on Scottish habitats suggest that the most usually associated are the Sundews *Drosera anglica* and *D. rotundifolia*, which often occur intermixed with the clubmoss, whilst nearby may be Bog Asphodel (*Narthecium ossifragum*), Common Cotton-grass (*Eriophorum angustifolium*), Deer-grass (*Trichophorum caespitosum*), White Beaksedge (*Rhynchospora alba*), Purple Moorgrass (*Molinia caerulea*), *Carex* spp., including Few-flowered Sedge (*C. pauciflora*), and various bogmosses (*Sphagnum* spp.). In nearby more permanently wet situations, Lesser Spearwort (*Ranunculus flammula*), Shore-weed (*Littorella uniflora*), Bladderworts (*Utricularia* spp.), Common Spike-rush (*Eleocharis palustris*) and Jointed Rush (*Juncus articulatus*) have been recorded. Some of its sites occur in or around *Carex rostrata* swamps.

As with several of the other clubmosses, the acid-loving preferences of *L. inundata* and its general affinities with moist, peaty habitats, give it a distribution that, in northern Britain at least, follows many of the peaty and gravelly lower valley habitats originally probably forming a mosaic of semi-open areas within woodland either largely of Scots Pine (*Pinus sylvestris*) or of a mosaic of pine, birch and other tree species. Further south, it may have been formerly associated also with earlier open pinewoods and more latterly probably largely with open Birch and Birch–Oak (*Betula–Quercus*) Woodland vegetation, the clubmoss growing in damper, less wooded and more open, semi-permanent hollows, disturbance patches where rain has accumulated, and in more natural heathy patches in amongst the more general former woodland canopy.

Today, *L. inundata* has disappeared from many of its former sites on heaths in the Midlands, as well as from some sites in southern England, Wales and Scotland. Reasons for this general decline are not fully known. Drainage of wetland sites has probably also contributed to general habitat fragmentation (Jermy *et al.*, 1978). But the strong coincidence of the areas of greatest loss with regions of high atmospheric pollution suggests that it may also be particularly pollution sensitive, since it depends heavily on incoming rainwater and immediate surface run-off for its modest mineral supplies.

Lycopodium annotinum L.   Interrupted Clubmoss

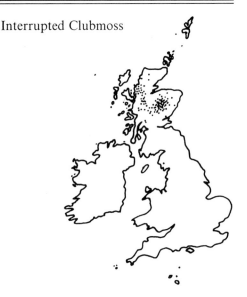

Preliminary recognition A relatively large and leafy moss-like plant, with stiffly spreading leaves which are rather harsh to the touch, distinct but stalkless cones and the vegetative portions showing constrictions of growth at regular intervals.

Occurrence A strictly mountain plant, local and rare in open subalpine vegetation in the Scottish Highlands, extremely rare or extinct in a few scattered localities in northern England and North Wales, and unknown in Ireland.

Identification Plants have long (50 cm or more) tough and flexuous creeping, rooted perennial stems which give rise to ascending side branches at intervals of a few centimetres. The creeping stems have small (3–5 mm long) lanceolate leaves, arranged somewhat distantly all around the stem, most turning upwards but with some of those on the lowermost side appressed to the stem. The side-branch stems, which are also perennial, are themselves initially mostly unbranched (or with an occasional dichotomous fork). These side-branch stems are at first ascending or erect (often to around 10 cm high, sometimes more), but as they gradually lengthen, they recline and become prostrate, themselves often producing roots and giving rise to fresh side branches. Leaves on the branch stems are slightly larger (5–7 mm long), than those on the creeping stems, and are more crowded, *spreading to ascending, acutely pointed and somewhat prickly to touch. The end of each growing season is marked by a ring of leaves that are smaller and more appressed to the stem, hence marking the vegetative growth into a series of annual increments separated by constrictions or 'interruptions' to growth.* The leaves on all stems persist for several years.

Cones, which are *normally without any stalks* (sessile), are borne one per shoot (solitary) from the end of some of the longer erect side-branch stems. Each cone is up to 3 cm in length, about as broad as the broadest part of the shoot which bears it, and is at first greenish becoming yellow-brown when mature (mid-summer).

Variation This is not a very variable species.

Possible confusion Likely only with Stag's-horn Clubmoss (*Lycopodium clavatum*, q.v.), from which *L. annotinum* differs in its solitary cones borne singly on each shoot (cf. stalked cones

Jan.	Feb.	Mar.	Apr.	May	Jun.	Jly	Aug.	Sep.	Oct.	Nov.	Dec.
							███				

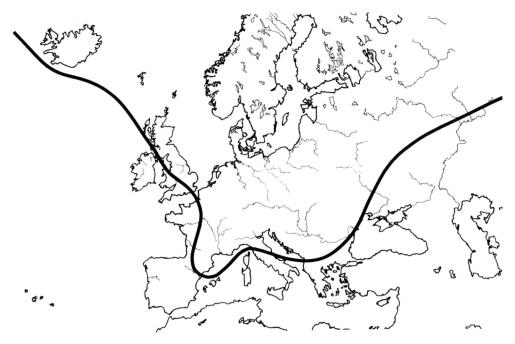

Southern boundary of northern-montane range of *Lycopodium annotinum* in Europe.

often borne in pairs), more-spreading leaves gives the shoots an overall more leafy appearance, interrupted growth, more prickly feel and absence of long hair-like silvery tips to the leaves.

Technical confirmation Plants have $n = c$. 34, $2n = c$. 68 chromosomes.

Field notes *Lycopodium annotinum* is a widespread northern circumboreal species, its distribution largely following that of the boreal pine forests of the Earth, with local variants in different parts of its range. In Europe, it is an essential northern species, most abundant in the coniferous forests of Fennoscandia but stretching as far south as central Europe, the Alps and Pyrenees.

In these islands, Interrupted Clubmoss is a northern and mainly mountain species, absent from Ireland and rare other than in Highland Scotland. It occurs chiefly in moist woods, especially those of pine, and now remains as a relic in occasional moist hollows in cool, damp areas of more upland moorlands. In such sites, bright-green patches of the plant can occasionally be found occurring usually above 455 m (1500 ft), often much higher, and scarcely ever descending to the lower valleys as do the other mountain clubmosses (*D. alpinum*, *L. clavatum* and *H. selago*). It is a strongly calcifuge plant, seldom occurring in any abundance, with its chief centre in the Cairngorms area of Highland Scotland. It grows chiefly amongst other acid-loving species such as Ling Heather (*Calluna vulgaris*), Bilberry (*Vaccinum myrtillus*) and Crowberry (*V. vitis-idaea*), in exposed moist, but well-drained situations, especially where there is only a thin covering of peat over rock debris, amongst rocks or on wet ravine edges. Other associates noted in the Cairngorms area include Linnaea (*Linnaea borealis*), One-flowered Wintergreen (*Moneses uniflora*), Hard Fern (*Blechnum spicant*) and numerous mosses, including *Hylocomium splendens*, *Hypnum cupressiforme*, *Pleurozium schreberi*, *Rhytitiadelphus loreus*, *Thuidium tamariscinum* and *Dicranum scoparium*. It also occurs in deep Heather associations in relatively dry situations, in areas of high rainfall.

Lycopodium annotinum is normally probably a very long-lived plant, which slowly builds up colonies, perhaps over the course of a century or more, with what are probably usually mainly single clones. The more aged of these may eventually spread to cover patches as large as 10–20 square metres, achieving such size through continuous steady growth over long periods of time (perhaps a

Lycopodium annotinum: All Cairngorm Mountains, East Inverness-shire.

century or more). Within such patches, long and ancient horizontally running stems spread widely amongst or below accumulated surface leaf and moss debris, occasionally branching and rooting at intervals, and giving rise to sparsely branched leafy ascending green stems, each taking several years to mature before terminating in loose clusters of single cones of varied and uneven height. In the field, the annual constrictions to the length of the leaves along the upright portions of shoot can be particularly conspicuous. These mark the point of resumption of each spring's growth from the semi-dormant overwintering positions, the very much shorter leaves of which contrast sharply with the long, spreading, lance-shaped and well-separated leaves of the summer's growth of each shoot.

The acid-loving and largely northern preferences of *L. annotinum* and its affinities for moist, generally peaty substrates, give it a distribution which closely parallels that of the original range of Scots Pine (*Pinus sylvestris*) in northern Britain, the two species very probably being formerly closely associated in many mid-mountain areas, and for the original distribution of which type of forest, this clubmoss is probably a particularly valuable indicator.

Callaghan *et al.* (1986) 'regard' *L. annotinum* as a species that is successful in both colonising and surviving in a spatially and temporally heterogeneous habitat. Despite this potential, colonies of such size are rare and usually exist only in sites that appear to have remained largely undisturbed. Overall the species has disappeared from a number of former sites, partly through collection but probably largely through changes in land use, especially in modern times. Amongst the latter, widespread deer-grazing preventing the return of the pinewood forest cover and virtually equally widespread burning of upland areas for heather management for grouse are probably the two single most important factors in the widespread loss of this particularly beautiful and potentially valuable pinewood indicator species.

Lycopodium clavatum L.

Stag's-horn Clubmoss

Preliminary recognition A low-growing but long-sprawling plant, giving the appearance of a much overgrown and burry moss, with bright-green leafy shoots and distinctive, creamy-yellow upright cones borne mostly in forked pairs, on tall, slender stalks.

Occurrence Local, usually as small patches in moist open situations mainly on mountains, moorlands and sandy heaths, especially in Scotland, northern England and north Wales, but with widely scattered locations also throughout the rest of England, Wales and Ireland. It is the most widely distributed and best-known clubmoss in Britain, but has a much more local range in Ireland.

Identification Plants have long (to 1 metre, sometimes much more) *sprawling, perennial main stems* which give rise to *abundant very conspicuous bright-green partly sprawling side-branches*, which are angled forwards. The prostrate main stems are tough and wiry, pale yellowish-green in colour, surrounded by abundant, 5–7 mm long, lance-shaped leaves, and grow vigorously through or over low-growing (often mossy) surrounding vegetation, rooting rather shallowly at intervals. The branch stems arise at frequent (usually 2–8 cm) intervals, and are at first ascending but as they elongate become procumbently sprawling with only their tips ascending,

Jan.	Feb.	Mar.	Apr.	May	Jun.	Jly	Aug.	Sep.	Oct.	Nov.	Dec.

Lycopodium clavatum: a–e, Mid Perthshire; *f*, young plants, Ayrshire.

and are themselves sometimes irregularly pinnately or dichotomously branched. All branch stems bear *abundant and densely crowded, perennial, overlapping, narrowly lance-shaped, incurved and rather appressed leaves all round the stem*, each leaf being 3–5 mm long, a vivid bright slightly glossy green colour, with a distinct midrib and *ending in a conspicuous long, slender colourless or silvery (hyaline) hair-like (filiform) tip, which incurves towards the stem.* These tips give the whole shoot a characteristically silvery-green and hoary appearance, even at a distance.

Linearly cylindrical cones (strobili) are *usually borne in pairs* (but sometimes singly or in threes) from the tips of taller (8–15 cm high) side-branch stems, each cone on a *long (3–5 cm), slender, pale-yellow stalk (peduncle)* bearing only minute remote leaves. The *peduncle typically forks* one or more times, with each branch of the fork ending in a separate cone. The 2–4 cm long, *shuttle-shaped, catkin-like cones* stand erect and are at first pale green, becoming bright creamy-yellow by mid-summer, eventually yielding large quantities of bright-yellow spores.

Variation This is not a very variable species.

Possible confusion The brightly coloured cones standing high, in pairs, on long peduncles above bright green leafy shoots, give *L. clavatum* a characteristic and distinctive appearance in the field. These features, plus the silver hair-like tips to the leaves and the uninterrupted annual growth, readily separate *L. clavatum* from *L. annotinum* (q.v.), the only other species with which it is likely to be confused. *Lycopodium clavatum* is the only species of native clubmoss in which the cones are always borne on such long, slender stalks.

Technical confirmation Plants have $n = 34$, $2n = 68$ chromosomes.

Field notes *Lycopodium clavatum* is a widespread northern circumboreal species, with various regional forms recognised within its overall range. In Europe, it is a widespread, mainly western European, species with a relatively continuous range from northern Fennoscandia to the Alps and Pyrenees.

In these islands, *L. clavatum* is a mainly northern species, which occurs locally on mountain sides, moorlands and heaths, ascending to about 820 m (*c.* 2700 ft), always in open situations, mostly above the tree-line, but descending more rarely to lower mountain valleys and other open sites at low level, such as lowland heaths and occasionally on old, well-stabilised sand-dune slacks, and disused mine-workings.

In mountain habitats, it occurs chiefly on northerly aspects and in areas of high rainfall and sometimes long snow-lie. It generally grows in the peaty humus of mossy hollows where it may sometimes occur in small dense patches forming a compact matted turf. Most usually plants sprawl and spread far over surrounding vegetation, over thin peaty soils, on hidden boulders and across tracks in heather moorland. It occurs mostly in dwarf shrub heath vegetation with Ling Heather (*Calluna vulgaris*), Cross-leaved Heath (*Erica tetralix*), Bilberry (*Vaccinium myrtillus*), Crowberry (*Empetrum nigrum*), Tormentil (*Potentilla erecta*) and many mosses. It is also frequently associated with the other mountain clubmosses *Diphasiastrum alpinum* and *Huperzia selago* (q.v.) in such communities, but does not ascend to such high altitudes. *Lycopodium clavatum* is present chiefly in calcifuge vegetation over old hard rocks such as granite and gneiss, but sometimes also on naturally open slopes over basalt or schist rock in more species-rich grassland communities. It invades these especially where they occur on thin soils over rocky substrates, and these are very freely drained, such as on the sides of railway embankments in upland areas. Sometimes plants locally re-invade patches on these which have been subject to light burning, probably growing in from surrounding unburned areas, temporarily forming dense stands which have been observed to then cone vigorously (A. C. Jermy, personal communication).

Lycopodium clavatum is normally a fairly long-lived plant, which slowly builds up sparse but widely spreading colonies in which, as with *L. annotinum* (q.v.), long horizontally running stems spread widely amongst, or arch vigorously over, accumulated surface leaf and moss debris, occasionally branching and rooting at intervals, and giving rise to sparsely branched leafy ascending green stems terminating in slender perfectly vertical portions of shoot bearing its typically forked (occasionally trifid) cones. The cones appear to have a particularly long spore-shedding season, in most years continuing to liberate clouds of yellowish spores from the large cones until at least the end of July (cf. at least a month earlier in *Diphasiastrum alpinum* (q.v.), with which it most frequently grows). The whole of such plants appears, however, to be highly sensitive to elimination by fires. Where there is extensive burning of heather in upland moorland areas, such as 'muirburn' in Scotland, *L. clavatum* seems to be eliminated extensively from these, surviving, if at all, only in occasional patches of unburned vegetation where these have proved too damp to burn. *Lycopodium clavatum* is thus today found extensively only in old, unburned, heather moorland, typically with old, shrubby heather plants, with their old parts densely encrusted with lichens.

Elsewhere, especially in lowland England and Ireland, the plant has decreased considerably and disappeared from many of its former situations through intensification of agriculture and utilisation of rough marginal ground. The plant was also extensively collected in former times. The bright-yellow spores ('Lycopodium Powder') were formerly in demand as an item of pharmaceutical and theatrical use, whilst whole plants have also long been used in rural folk-tradition to make garlands for rustic personal adornment.

Selaginella selaginoides (L.) Link   Lesser Clubmoss

Preliminary recognition A small and delicate, low-growing plant, of moss-like habit and form, but differing in bearing one or more erect, leafy, club-shaped cone shoots.

Occurrence Locally frequent throughout the mountains of north Wales, northern England and Scotland, descending to sea-level in the Western Isles, and scattered throughout much of Ireland, except the south.

Identification Plants are moss-like perennials, producing annual cone shoots. They have a low-growing vegetative phase, with thin, weak-looking, trailing stems, a few centimetres in length, which usually turn upwards near their growing tips. These stems *branch irregularly and wander decumbently over mosses and amongst the bases of surrounding herbage*, and bear sparse, small (1–2 mm long), acutely pointed triangular leaves arranged all around, *but usually pointing irregularly and untidily in all directions.*

Jan.	Feb.	Mar.	Apr.	May	Jun.	Jly	Aug.	Sep.	Oct.	Nov.	Dec.

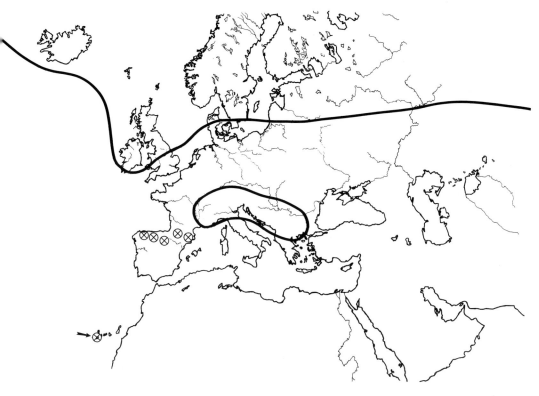

Main southern boundary and outlying southern stations of the arctic-alpine range of *Selaginella selaginoides* in Europe.

From these rather weak-looking vegetative beginnings arise surprisingly strong-looking *foxtail-shaped fertile shoots, borne usually singly* (occasionally two or more) per plant, and *which stand rather starkly erect from a short decumbent base*. Most are about 3–6 cm high, and about 4–6 mm broad, although exceptionally tall specimens in moist, well-sheltered situations, may reach 10 cm. Much of the length of each is composed of an *ill-defined, leafy cone*.

The cone is composed of a large number of weakly differentiated, leaf-like sporophylls arranged radially around the erect cone axis, bearing, in the axils of many, an assortment of microsporangia and megasporangia. Sometimes sporangia of each type occur more or less randomly within a cone. More often microsporangia and megasporangia are zoned within each cone, with the megasporangia always nearer the base and the microsporangia always above them, and sometimes there are multiple zones of micro- and megasporangia, especially in larger cones. At maturity, the simple, globose, creamish-yellow-coloured microsporangia each liberate a large number of small, orange, microspores; the 4-lobed white megasporangia each liberate only 4, large, shining-white megaspores, the latter each sufficiently large to be clearly seen with the naked eye (see 'Technical confirmation').

All parts of the plant are of very soft texture. The vegeative parts and the cone early in the season are a bright green colour. The cone usually becomes a pale yellowish-green colour by maturity, and often a conspicuous pinkish-cream before dying down in winter. All leaves and sporophylls are acute tipped, and have a distinct midrib, as well as *finely denticulate, delicately*

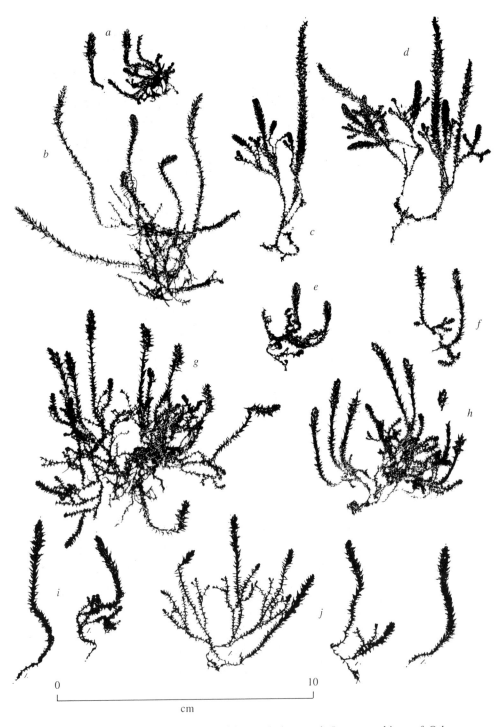

Selaginella selaginoides, relatively fragile-habit populations: *a*, *b*, Inverness-shire; *c–f*, Orkney; *g*, Deeside; *h*, Sutherland; *i*, *j*, Easterness.

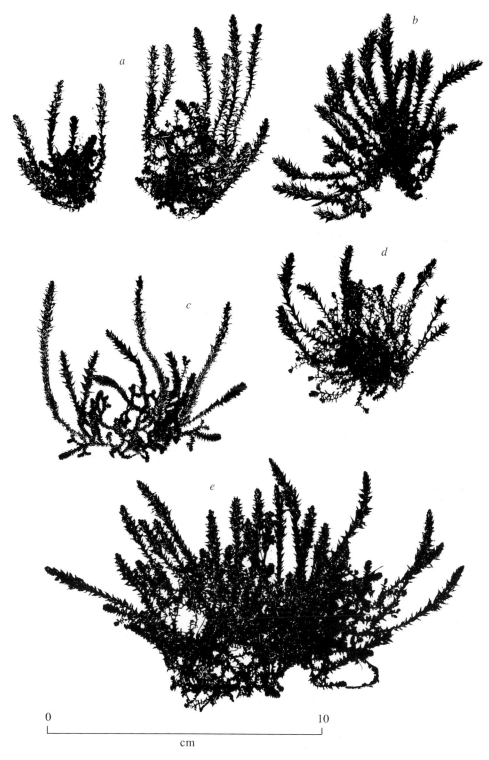

Selaginella selaginoides, relatively robust-habit populations: *a*, Forfarshire (Clova); *b*, Isle of Arran; *c*, *d*, Deeside; *e*, Glen Isla.

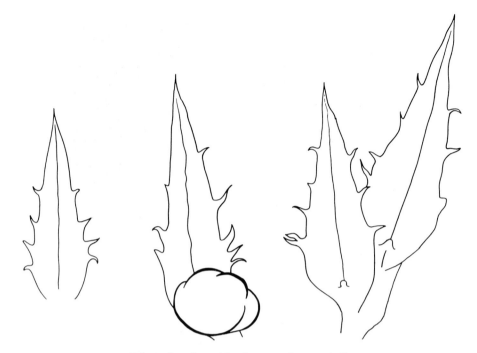

Selaginella selaginoides, leaves and sporophyll.

spinulose margins, easily visible with a hand lens and usually with the naked eye. Each leaf also has a small, membranous, flap of tissue (the 'ligule'), close to its base and the angle between leaf and stem.

Variation Especially between populations, plants vary very considerably in size, in overall fragility versus robustness of the vegetative shoots, in individual vigour, in the overall abundance of branching pattern achieved and hence eventual plantlet size, and in the size and robustness of the cones typically formed. The latter also vary widely between individuals in some populations in terms of the megaspore: microspore ratio, and the position of these in the cones (with repeat sequences occasionally occurring in some populations). How much of this variation might be genetic (especially possible considering the generally bimodal distribution of the species and the remote geographic positions of some of its sites) and how much is the result of local environmental influence remains to be determined.

Possible confusion The small size of the plant, soft texture, sparsely leafy vegetative shoots and heterosporous cones distinguish *Selaginella selaginoides* from other clubmosses. The presence of erect cone shoots and a ligule to each leaf distinguish plants of *Selaginella* from true mosses (which, by contrast, always have only small capsules borne on slender stalks, and always lack ligules to the leaves). Ligules are indicated in the right-hand diagram above.

Technical confirmation Plants have $n = 9$, $2n = 18$ chromosomes. Microspores are *c.* 30 μm diameter, megaspores *c.* 690 μm diameter, the megaspores thus being approximately 23 times the diameter of the microspores, and over 12 000 times their volume. *Selaginella selaginoides* is the only native clubmoss that is ligulate, and also the only one that is heterosporous.

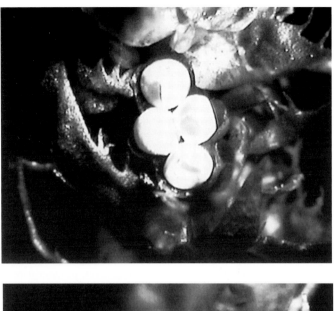

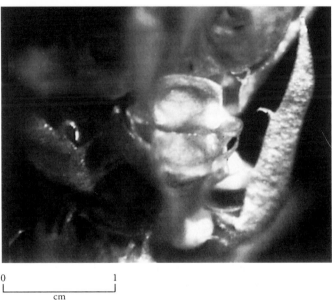

Going, going, gone! *Above*, megasporangium walls fully withdrawn from around a quartet of megaspores, to the point of being fully primed, with the megaspores about to be suddenly and violently released. *Below*, the moment immediately following discharge, with the now empty megaspore cradles formed from the original megasporangium wall, showing the large upper and lower lobes and the much smaller lateral ones (modified from Page, 1989*b*).

Field notes *Selaginella selaginoides* is a plant of circumboreal range (Polunin, 1959), restricted in Europe mainly to Iceland, northern Russia, Finland, Scandinavia, northern Britain and Ireland, but with smaller outposts mainly in the Pyrenees and Alps (Jalas & Suominen, 1972) and with a single extremely southerly outlying station 2100 km (1300 miles) south of its nearest European station in a mossy, mountain fissure at *c*. 1200 m on the Pico del Malpaso on the Canary Island of Hierro (Page, 1971, 1977).

In Britain and Ireland, *S. selaginoides* is a

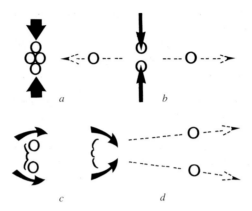

Diagrammatic representation of the megaspore discharge sequence of *Selaginella selaginoides* described here. *a–b*, Frontal view: *a*, quartet of megaspores with central pair becoming squeezed by compression action between the upper and lower megaspores in the axil of a deflexed megasporophyll (thick solid arrows indicate main compression force directions); *b*, central megaspore pair suddenly ejected laterally as a result of this compression action. *c–d*, Side view: *c*, larger arms of megaspore cradle carrying the upper and lower megaspores snap forward as an immediate result of loss of the central megaspore pair; *d*, the upper and lower megaspores are hurled violently from the cone by the resulting slingshot action (slender solid arrows indicate direction of initial megaspore and cradle movement; thin dotted arrows indicate the ultimate direction of megaspore trajectory) (after Page 1989*b*).

frequent component of base-rich, usually open and exposed habitats, where it is mainly a montane species occurring above the tree-line (*c*. 610 m, 2000 ft) from where its principal range extends to at least 914 m (3000 ft). More occasionally it descends also to lower altitudes, but only in sites that are fully exposed and hence lack tree overgrowth. Plants usually grow as small, scattered individuals, in a wide range of habitats, all of which are damp and usually show some signs of being calcareous. These include rocky turf by the sea (sometimes in shellsand in 'machair' turf in western Scotland), sand-dune slacks and pastures, base-flushed patches in upland heath communities, along rocky and gravelly margins of upland streams in basic areas, ravine-sides, calcareous wet flushes, crevices of damp schistose rocks, damp shallow-soil ledges of high-altitude cliffs, and basic patches in alpine meadows and upland valleys, often in areas of considerable snow-lie.

The plant's ecological preferences are sufficiently strong for it to be a reliable base indicator, but its size is sufficiently small to confine it to damp places that are not densely overgrown by tall, rank herbage. It is thus rare in the lowlands, except in short, coastal turf near the sea in the extreme west, but it becomes more frequent again at higher altitude.

In lowland habitats, it associates with a great many other generally calcicolous species, especially in habitats such as machair turf. In flushes and on base-rich ledges in mountains, it grows mainly in sites kept fairly open by steady erosion or on shallow soils where other vegetation is mostly sparse and low-growing. Associated species in such habitats commonly include Yellow Saxifrage (*Saxifraga aizoides*), Starry Saxifrage (*S. stellaris*), Mossy Saxifrage (*S. hypnoides*), Hoary Whitlow-grass (*Draba incana*), Alpine Lady's-mantle (*Alchemilla alpina*), Sibbaldia (*Sibbaldia procumbens*), Alpine Meadow Rue (*Thalictrum alpinum*), Moss Campion (*Silene acaulis*), Common and Marsh Violets (*Viola riviniana* and *V. palustris*), Alpine Scurvy-grass (*Cochlearia alpina*), Three-flowered Rush (*Juncus triglumis*), various small sedges (especially *Carex demissa* but also *C. pulicaris*, *C. dioica* and *C. nigra*), Variegated Horsetail (*Equisetum variegatum*) and mosses such as *Ctenidium molluscum*, *Bryum pseudotriquetrum* and *Campylium stellatum*. Common Butterwort (*Pinguicula vulgaris*) often occurs on damp rock ledges nearby.

Its montane habitats, as well as its coastal dune ones, are sites that are tree free, and probably were never covered by extensive tree growth in post-glacial times. It is very probable that *S. selaginoides* was a once-extensive species of the early post-glacial, before the widespread development of tree vegetation at mid-altitudinal levels (Page, 1988*b*).

For such a diminutive plant, production of the relatively large cone shoot presumably requires considerable effort and, to produce one at all, plants seem to need to make a considerable start to growth during the winter months. In its Scottish mountain habitats, the plant seems to achieve this by making considerable headway with the annual cone's growth beneath regular coverings of winter snow. Although, under such conditions, the light must be weak, temperatures are not too severe (see p. 45). Sporne (1962: 87) notes that a mycorrhizal association has been demonstrated with this species, and this association may be significant in assisting the clubmoss to grow under such conditions. It is much rarer, even at low altitudes, in the extreme east, presumably thriving less well in climates where there is least reliable winter snow and where winter temperatures and desiccation would certainly be more severe. Southward it is presumably limited in range by high summer temperature maxima. In the Scottish mountains, crops of cone shoots seem particularly abundant in summers following severe winters, perhaps because in these, the snow has lasted longest. Each cone matures (in Scottish Highland sites) by mid to late August or early September. At cone maturity, the microspores are released from the cone in a totally passive fashion, to be dispersed by the agency of wind currents alone.

By contrast, megaspore dispersal takes place by an active discharge mechanism, termed 'compression and slingshot ejection' (Page, 1989*b*). This process involves the once-only (for each ripe megasporangium), sudden and virtually explosive release of the slowly accumulated tension acquired by successive megasporophylls – each megasporangium, with its adjacent megasporophyll, acting as an independent discharge system.

Tensioning of the system appears to be initiated by the shrinking and eventual splitting of the ripe megasporangium walls. As these progressively retract from around the megaspore quartet, so the whole megasporangium becomes drawn progressively deeper into the axil of its subtending megasporophyll blade, compressing the four megaspores within tightly together. This axial retreat of each mature megasporangium causes a steady backward flexing from its axil of the whole subtending megasporophyll blade, which consequently becomes forced into a divergent angle from the axis of the cone and thereby increasingly tensioned.

Indeed, even casual observation of almost any fully mature cone of this species in the field typically shows one or more deflexed megasporophyll blade standing at an abruptly divergent angle from those of the rest of the cone. At such a stage, when drawn maximally into the angle between the cone axis and its deflexed megasporophyll blade, the system is fully tensioned. The quartet of white megaspores now largely revealed by progressive retreat of their megasporangium walls have, in life, a smoothly glossy appearance and an apparently moist surface, and look like a clutch of four, shining, wet eggs, held in a tight and shrinking nest!

Sudden total release of the accumulated tension occurs when the four mature megaspores are shot forcefully from the plant, in two rapidly successive (i.e. split-second) volleys, the first pair by compression-ejection, the second pair by slingshot-ejection (Page, 1989*b* – see diagram opposite).

The resulting discharge trajectories achieve impressive megaspore dispersal distances for the size of the plant. Even in still air, these are mostly from 1.5 to 2.0 m from the parent cone, with a few spores occasionally being shot to horizontal distances of over 3 m. In the field, cone agitation by crosswind forces probably plays a further significant role not only by increasing the distance to which some megaspores may be finally hurled, but more especially by increasing the randomness of their direction of radial discharge.

Biologically, the significance of this process is presumably partly to remove the sites of new megaprothallial establishment (and hence the site of origin of new sporophytic progeny) further away from the parent plant than could be achieved by passive discharge alone. More poignantly, it also increases the opportunity for randomness of fertilisation of the resulting megaprothalli by the antherozoids of independent microprothalli derived from passively dispersed microspores originating from independent, differently placed, upwind sporophytes. For only when so removed, in large part, from settlement within the same downwind dispersal zone as that of a sporophyte's own microspores, do the biological benefits of megaspory in an airborne environment seem likely to become most fully realised.

The particularly wide edaphic range of *S. selaginoides* in these islands, especially in respect of soil moisture regime, compares favourably with that shown in more acidic habitats only by virtually the whole collected range of native species of Lycopodiaceae. Thus, the single species of *S.*

selaginoides parallels the entire family of Lycopodiaceae. It is therefore, in my view, a particularly good example of the expansion of the habitat range of a single plant into a wide range of otherwise relatively vacant available habitat niches in our species-undersaturated native pteridophyte flora.

Isoetes echinospora Durieu   Spring Quillwort

Preliminary recognition A usually fairly small, submerged aquatic plant, forming a spiky rosette of numerous, slenderly-tapering, bright-green, flexible, quill-like leaves.

Occurrence Sparsely distributed, although sometimes locally frequent, on the bottoms of clear-watered, upland lakes and tarns, restricted mainly to discontinuous mountainous northern and west-coast areas of Britain and Ireland.

Identification As with *I. lacustris* (q.v.), presence of this species in lakes can usually be detected by examination of drift material on downwind shorelines. On lake bottoms, plants occur as scattered individuals or gregariously. Each is of rosette-like habit, bearing numerous, slender, often spreading, fairly flexible, quill-like leaves, 3–15 cm long and 11–12 mm broad at the base, *which gradually taper throughout much of their length to end in slender, pointed apices.* Each leaf is of *much more flaccid and flexible texture* than that of *I. lacustris*, and if a washed-up plant is lifted out of the water, *the leaves mostly come together*, rather than remaining rigidly apart, *adhering by water surface-tension.*

The leaves arise from a slender corm, in a similar manner to those of *I. lacustris* (q.v.), but are of *a lighter, more grass-green colour*, with a larger proportion of the base very pale coloured. Their internal structure shows four lengthwise air chambers, as in *I. lacustris*, and they similarly have a translucent quality through which the numerous internal cross-walls can be seen.

Variation As with *I. lacustris* plants are variable in structure and in size, varying even in one lake with water depth (see 'Field notes'). They are, however, probably less variable than those of *I. lacustris*, and in the same environment, are usually the smaller of the two.

Possible confusion The altogether more gradually tapering, more slender, more flexible, more grass-green coloured leaves, and generally smaller size of plants, usually separate *I. echinospora* from *I. lacustris*. Like *I. lacustris, I. echinospora* can also be confused with a number of flowering plants of similar habit, but the features given under *I. lacustris* (q.v.) serve to distinguish both species of *Isoetes.*

Main southern boundary and scattered outlying stations of the sub-Atlantic range of *Isoetes echinospora* in Europe.

Technical confirmation Plants have $2n = c.$ 22 chromosomes. Megaspores are about 440–550 µm (0.44–0.55 mm) diameter, and are densely covered with long spines, which may be hooked at their tips. Their smaller size and more chalky-white colour (cf. greyish white in *I. lacustris*) can be distinctive even in the field.

Field notes *Isoetes echinospora* is a highly polymorphic species on a world scale, which in the broad sense occurs in Europe, with disjunctly separate populations in central and eastern Asia and in North America. Our species is probably entirely European in range, with its main centres of distribution throughout Fennoscandia, Iceland and western Britain, with numerous additional and often highly isolated stations in western Europe and as far south as the Pyrenees. To a large extent it is possible that especially the Icelandic, western Britain and Pyrenean stations follow the course of regular waterfowl (especially goose?) migration routes (see below).

Within Britain and Ireland, plants are predominantly west-coast in their range on both islands, most numerous in western Ireland, western Scotland and west Wales, with isolated stations also in south-west England. Plants are perennials, which occur scattered through northern and western Scotland and the Outer Isles, extending discontinuously down the west coasts of both Britain and Ireland, sporadically in lakes, pools, and upland tarns, up to about 610 m (*c.* 2000 ft) altitude, though they are more often at lower elevations and are less frequent than *I. lacustris*. They usually occur within or around the margins of still, clear, deep lakes, and seldom in flowing water. They seem restricted mainly to water of low calcium content, and of usually poor (oligotrophic) nutrient status.

Usually, *I. echinospora* occurs in more oligo-

Isoetes echinospora: a, Caernarvonshire; *b*, Wester Ross; *c*, Dorset; *d*, Barra, Outer Hebrides.

trophic waters than does *I. lacustris*, although there is a considerable overlap in the habitats of the two. Occasionally *I. echinospora* spreads into more mesotrophic waters, and sometimes occurs as a minor vegetational component amongst *I. lacustris*, where its presence can be easily overlooked. *Isoetes echinospora* seems to occur more often also in shallower water than *I. lacustris*, usually less than 4 m in depth and often under 2 m, hence is more likely to become exposed at the surface during periods of low water levels. It is, however, usually displaced from very shallow water when dense communities of flowering plants, such as Shore-weed (*Littorella uniflora*) are present. Plants at different depths are clearly differently exposed to water

turbulence and wave-action effects, and can show different morphologies. Those in the shallowest water usually adopt a stocky form of growth, with curved leaves under 5 cm long, whilst those at 1–2 m or more usually develop much longer, straighter, more ascending leaves over 10 and often 15 cm in length.

Isoetes echinospora occasionally forms dense turfy patches, but more often occurs either as scattered individuals, sometimes intermixed with other species, or, most often, as small, sporadic patches. The distribution of such patches within lakes is probably very largely determined by suitability of a range of physical factors, including clarity of water, water depth, degree of water movement, suitability of substrate and degree of silting. It seems confined to water that is clear, and where there is slight current flow and some wave-action. It seems to colonise mainly areas of coarse sandy or stony bottoms made of hard rocks such as granite, over which there is an accumulation of no more than about 1–2 cm of fine, peaty, silt. It occasionally spreads also to areas where the bottoms are entirely stony. In these areas, the thinness of the substrate and the degree of wave-action probably help to prevent long-stemmed submerged aquatics and reedswamp from developing (Seddon, 1965). It avoids areas of deep silt accumulation. Its probably inferior competitive ability in relation to that of the more vigorous flowering plant aquatics is probably the main factor limiting *I. echinospora* from success in more mesotrophic waters.

Plants of *Isoetes* can be confused at first sight with a number of flowering plants which have a similar rosette form and may be present in similar habitats. These include particularly Awlwort (*Subularia aquatica* – Cruciferae), Shore-weed (*Littorella uniflora* – Plantaginaceae), Water Lobelia (*Lobelia dortmanna* – Lobeliaceae), Pipewort (*Eriocaulon septangulare* – Eriocaulaceae) and eroded leafless small plants of Water Plantain (*Alisma plantago-aquatica* – Alismataceae). Other species which may be present with the Quillwort include Starwort (*Callitriche intermedia*), Alternate-flowered Water-milfoil (*Myriophyllum alternif-*

lorum), Pondweed (*Potamogeton perfoliatus*), Reddish Pondweed (*P. alpinus*), Floating Bur-reed (*Sparganium angustifolium*), Floating Scirpus (*Eleogiton fluitans*), and the Stonewort (*Chara vulgaris*).

The discontinuous distribution of *Isoetes echinospora* in upland lakes, its dominantly west-coast pattern, and its occasional sporadic appearance and rapid establishment in artificial reservoirs, raise obvious questions of biological interest about the methods of dispersal of this heterosporous plant. Liberated megaspores grow into female prothalli, microspores into small male prothalli, although little is known of the ecology of these. How both types of spore can become simultaneously dispersed over appreciable distances between isolated lakes requires further research. Whilst it is possible that microspores that dry at the water margin may become wind dispersed, other methods may be important for the large megaspores. Both spore types may be carried internally by birds such as swans and vegetarian ducks, as well as fish, which eat the foliage. Experimentally, the megaspores with rough protuberances have also been suggested to be capable of clinging externally to water-fowl feathers, particularly downy body feathers, and transport externally on such birds and dispersal along bird migration routes remains a possibility that needs further investigation (Stokoe, 1978, and personal communication).

Isoetes histrix Bory

Land Quillwort

Preliminary recognition An inconspicuous, very
small perennial plant, scarcely 2–4 cm high, with
short, pointed, linear, quill-like leaves, arising in a
spreading cluster from a small, knobbly, subterranean corm, the whole plant resembling a miniature
pineapple half-submerged in the ground.

Occurrence Very rare, in small patches of limited
extent, in winter-wet peaty hollows on downland
on the Lizard Peninsula of west Cornwall, and on
the Channel Islands of Alderney and Guernsey.

Identification Plants have small, slender, quill-like leaves, mostly 2–4 cm long (occasionally
more) and scarcely 1 mm wide, broadest at the base and tapering gradually to a pointed tip.
Each is triangular in section, with a flat surface uppermost, and contains a single lengthwise air
canal. The leaves are of soft, flexible texture and bright-green colour, pellucid towards the
base, and lack the visible transverse leaf septae of other quillwort species. Towards the corm,
the leaves broaden into *whitish, clasping bases*, some of which contain sporangia. The bases are
thickened along the midrib and edges and *these areas persist as dark-brown horny 'cusps'* when
the leaves decay.

The leaves arise in compact, spreading clusters, with the longest, oldest leaves forming a
spreading basal rosette, and the newest and shortest leaves standing erect in the centre. Plants
often have *all their leaves slightly curved around in one direction, giving the whole plant a
swirled-around appearance*, when viewed from above, which is quite distinctive.

The knobbly corms, from which the leaves arise, are 6–10 mm in diameter, and are buried
just below ground level. They are of a dirty, creamish-white colour, with numerous, fibrous,
spreading roots and, in their upper parts, may bear the old persistent bases of many previous
years' leaves.

Possible confusion Its diminutive size and amphibious or nearly terrestrial habit, make *Isoetes
histrix* totally distinctive from the other two quillwort species, which are much larger,
submerged, aquatics (q.v.). It is, however, much more likely to be confused with any of the
other narrow-leaved rosette-forming, small-flowering plants which grow in nearby sites,
forming the larger element in the surrounding turf – close examination of individual plants on
hands and knees is usually needed initially to distinguish the quillwort! Particularly similar are
young plants of Sea Thrift (*Armeria maritima*) and Buck's-horn Plantain (*Plantago coronopus*),
before the latter develop the serrations to their leaves, both of which may grow in quantity
nearby. Young plants of nearby Sand Sedge (*Carex arenaria*) can also look similar. In the turf
of the Lizard, Spring Squill (*Scilla verna*) and, in the Channel Islands, Autumnal Squill (*Scilla
autumnalis*) and Sand Crocus (*Romulea columnae*) can initially be confused with it. But only

Isoetes has the pineapple-like corm (the others having merely fibrous roots or smooth, fleshy corms or bulbs), but to avoid uprooting specimens to check, they can all be distinguished easily under a × 10 hand-lens by their thicker, mostly non-triangular leaves, whilst the *Isoetes* is also the only one with a single central canal, the others having solid leaf centres (the *Romulea* having 4 external furrows).

Variation Very little. Each patch presents individuals of very uniform appearance, except for seasonal changes.

Technical confirmation The presence of sporangia containing spores at the base of the leaves confirms *Isoetes*. Plants have $n = 10$, $2n = 20$ chromosomes. The dove-grey megaspores have short, blunt tubercles coalescing into ridges, but some differences in wall ornamentation pattern in Channel Island material have been reported by Jermy (see McClintock, 1975).

Field notes *Isoetes histrix* has a widespread but sparse and discontinuous range through southern, largely Mediterranean, Europe and north-west Africa, and overall from Cornwall and the Channel Islands and western France eastwards and western France eastwards to Turkey. In these islands it is known only from limited areas of the Lizard Peninsula of west Cornwall and to the Channel Islands of Guernsey and Alderney. It is unknown on the other Channel Islands.

Land Quillwort grows in sparsely vegetated hollows in dense turf vegetation, which are flooded or well irrigated in winter and in spring, but which partially or largely dry-out and bake in most summers. Most are on gentle southward or westward slopes, and occur over serpentine rock in the Lizard or over granitic rock in the Channel Islands, in soils which are particularly shallow (about 1–3 cm of fine, wet, friable, silty peat in winter, drying to much less in summer), but which warm rapidly in early spring sunshine.

Plants appear above ground early in the growing season, usually in March (but occasionally as early as the previous August; Frost, 1982: 14), and from mid-May to late June mostly wither back to the subterranean corm as the land dries out. In both Cornwall and the Channel Islands, plants are near enough to the sea to receive windblown seaspray, probably at all times of year. Their mature above-ground quills are typically somewhat swirled around the plant, usually in a clockwise direction when viewed from above, and Ryan (1990) records that in one luxuriant Guernsey site the leaves can reach 8–10 cm long.

Associated plants include some interesting species of restricted range, as well as several of similar, tufted habit. In the Channel Islands sites, these are mainly Buck's-horn Plantain (*Plantago coronopus*),

Autumnal Squill (*Scilla autumnalis*), Sand Crocus (*Romulea columnae*), and annual *Vulpia* grasses, whilst Sea Thrift (*Armeria maritima*) grows nearby. A fuller list of species recorded in Guernsey is given by Ryan (1990), who adds especially Chamomile (*Chamaemelum nobile*), Scarlet Pimpernel (*Anagallis arvensis*), various Trefoils including Birdsfoot Fenugreek (*Trifolium ornithopodioides*), Sweet Vernal grass (*Anthoxanthum odoratum*) and Annual Meadow grass (*Poa annua*) and points to the frequency of mosses, including *Acrocladium cuspidatum, Campulopus* sp., *Ceratodon purpureus, Eurhynchium praelongum, Hypnum cupressiforme* and *Pseudoscleropodium purum*. Other associates in the Lizard include Spring Squill (*Scilla verna*), Western Chives (*Allium schoenoprasum*), Dwarf Rush (*Juncus capitatus*), and *Festuca* grasses, as well as Twin-flowered Clover (*Trifolium bocconei*) and Upright Chickweed (*Moenchia erecta*) (Page, 1982a), to which Frost (1982) have added also Upright Clover (*Trifolium strictum*), Ciliate Rupturewort (*Herniaria ciliolata*), Hairy Birdsfoot-trefoil (*Lotus hispidus*), Vernal Sandwort (*Minuartia verna*) and Dwarf Rush (*Juncus capitatus*).

The methods of spore dispersal and re-establishment of the *Isoetes* remain to be explored, although the possibility of the involvement of insects of the order Collembola has been suggested. Its small, greatly disjunct, stations, have probably been long-isolated and each patch probably represents a small, inbreeding community.

The Cornish (Lizard) populations have been particularly well surveyed by Frost (1982) and those of Guernsey and Alderney by Ryan (1990). Nine populations are known on Guernsey near to the sea near the north and north-west coasts of the island and its immediate off-shore islets. On Alderney it is very scarce. In Cornwall,

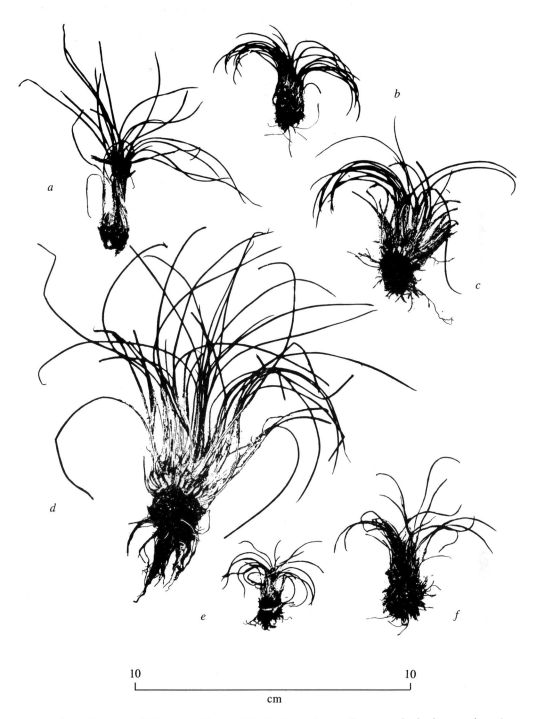

10   10
⊢————————————————————————————————————⊣
cm

Isoetes histrix: a–f, Guernsey, Channel Islands (from nineteenth century herbarium specimens).

over 30 populations have been recognised, growing mainly around rock outcrops in shallow soils with a cover of dwarf, herb-rich turf with some bare soil and situated mainly on slopes with a southerly component, but also with some patches in hollows in lightly trampled footpaths, and in coastal erosion pans (*sensu* Coombe & Frost, 1956).

The curious distribution of the plant in two widely separate stations on our extreme southern periphery parallels in type (as well as almost in habitat) that shown also by Least Adder's-tongue (*Ophioglossum lusticanicum*, q.v.), and the species could yet be found in other mild corners where appropriate habitats occur, such as the Scilly Isles or south-western Ireland.

Isoetes lacustris L.

Common Quillwort

Preliminary recognition A small to sometimes large-sized submerged aquatic plant, forming a spiky rosette of numerous, stiff, slender, dark-green, quill-like leaves.

Occurrence Not uncommon locally on the submerged bottoms of acidic, clear-watered, upland lakes and tarns, throughout most of the more mountainous regions of Britain and Ireland.

Identification Its presence can usually be best detected by examination of the downwind shorelines of upland lakes in late winter, for floating, broken-off leaves, often dislodged in numbers by winter storms. On lake bottoms, plants are usually highly gregarious. Each is of rosette-like form, bearing numerous, stiff, brittle, slender, cylindrical or bluntly squared, quill-like leaves, usually 2–3 mm broad and 8–25 cm long, although sometimes very much longer (see 'Variation'). The leaves all arise from a short, erect, corm-like base to each plant, which is partly immersed in lake-bottom silt, and from which arise numerous, thin, translucent pale-brown, dichotomously branched, roots. The leaves ascend to form a spiky rosette, and *are sufficiently stiff that, if a washed-up plant is lifted out of the water, the leaves mostly remain held rigidly apart* (cf. their immediate coalescence by water surface-tension in *I. echinospora*, q.v.). Each leaf is of *nearly constant thickness throughout the greater part of its length, and tapers only in the upper few centimetres to a rather abrupt and usually asymmetric point* – a feature that is usually less obvious in herbarium material than in the field.

The leaves, especially the older ones, are of *a deep-green colour*, and have an overall translucent quality. The numerous internal pale-coloured irregularly scattered cross-walls within the leaf can be seen conspicuously from the outside. Internally, they contain 4 lengthwise air chambers, readily visible with a hand lens.

The leaf bases are paler in colour than the rest of the leaf, and broaden out to membranous margins that clasp around the corm. On the lower inner surfaces of most leaves is a recess

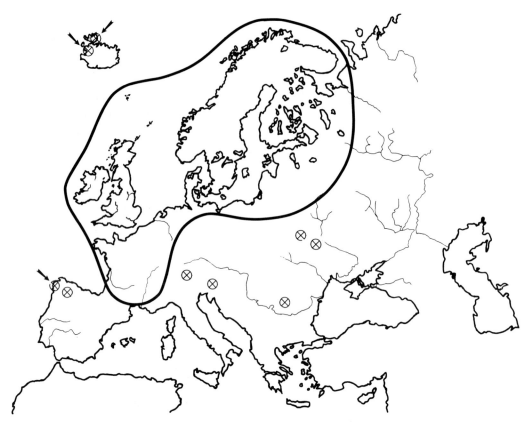

Main boundary and scattered outlying stations of the sub-Atlantic range of *Isoetes lacustris* in Europe.

covered by a membrane, within which the spores are produced. Large *megaspores*, each about the size of a grain of salt and hence sufficiently large to be seen individually with the naked eye, are produced earlier in the year, and thus occur in the outermost leaves of the summer rosette. Much smaller *microspores* occur in the sporangia of the leaves further into the rosette, and the most internal leaves are usually sterile. As leaves age, they gradually become pushed to the outside of the rosette, where they eventually break away, so rupturing the sporangia and releasing the spores.

Variation Plants vary both in structure and size. Plants with exceptionally long (30–60 cm), slender, tapering leaves occur in some Irish loughs (notably in Co. Wicklow), whilst other unusual, large or intermediate-sized forms are known in lakes in Wales and parts of western Scotland (Hyde *et al.*, 1978; Jermy & Crabbe, 1978). A viviparous form, in which vegetative buds replace sporangia, has been reported from Windermere (see Manton, 1950), and could occur elsewhere.

Possible confusion The more nearly parallel-sided leaves tapering only near their tips, darker green colour and much stiffer, less flexible texture, help to distinguish similar-sized plants of *I. lacustris* from *I. echinospora*. Both species of aquatic *Isoetes* are likely, however, to be readily confused with a number of flowering plants of similar habit, which also occur on lake bottoms

Isoetes lacustris: a, b–d, Angus; *e*, Co. Down; *f*, Mid Perthshire; *g*, Sutherland.

or around lake margins (see under *I. echinospora*). The *Isoetes* can be distinguished from these by the numerous, small transverse septae within the leaves (cf. non-septate in the flowering plants), four air channels within the leaves (contrasting with either one, two or a solidly spongy (parenchymatous) interior in the flowering plants), the presence of sporangia if the *Isoetes* is mature, and the brown-coloured roots that branch dichotomously branched (cf. white roots in all the flowering plants, which are not normally dichotomous).

Technical confirmation Plants have $n = 54$–56, $2n = > 100$ (probably 110) chromosomes. Megaspores are about 530–700 μm (0.53–0.70 mm) in diameter and are covered with only short blunt tubercles, looking rather like brain coral, or are almost smooth. Their large size and yellowish-brown or greyish-white colour (cf. more chalky-white in *I. echinospora*) can be distinctive even in the field.

Field notes *Isoetes lacustris* belongs to a highly polymorphic species complex which has a fragmented and disjunct distribution on a world scale, of which the European members are generally regarded as a distinct species of Europe and adjacent northern Asia. Our species has its main centres of distribution throughout Fennoscandia and in Britain and Ireland, with outlying populations in Iceland and numerous, often highly isolated stations in central and western Europe as far south as the Pyrenees and north-west Iberian Peninsula, and as far east as the Urals. As with *I. echinospora* (q.v.), to a large extent it is possible that especially the Icelandic, western British and Pyrenean stations follow the course of regular waterfowl (especially goose?) migration routes.

Within Britain and Ireland, plants are predominantly west-coast and northern in their range on both islands, most numerous in western Ireland, western Scotland, west Wales, and the Lake District, with isolated stations also in south-west England.

Plants of Common Quillwort are perennials that occur widely scattered in the clear, quiet waters of stream-fed upland lakes, mainly in northern Britain and the Scottish Islands, but extending southwards through the mountainous areas of western Britain and western Ireland, in a manner similar to, although more extensive than, *I. echinospora*. It thrives in large, deep-water, glacial lakes and tarns, especially in larger ones than *I. echinospora*, and those at generally higher elevations, ascending to about 825 m (*c*. 2700 ft). It also occurs more often in deeper water, and although the two species have a wide range of ecological overlap, *I. lacustris* is generally the more abundant. It grows in water depths of 1–6 m and sometimes forms submerged, meadow-like swards in the deeper bays of upland lakes, usually as pure stands.

It grows chiefly in water of a slightly higher base-content than *I. echinospora*, and most frequently colonises unsilted coarse to fine, consolidated sandy or gravelly lake bottoms and generally much more stony habitats than *I. echinospora*, but occasionally too on deep, nekron mud. On coarsely boulder-strewn lake bottoms, its leaves may emerge from between large boulders, accompanied by Floating Bur-reed (*Sparganium angustifolium*), Alternate-flowered Water-milfoil (*Myriophyllum alterniflorum*) and the Stonewort (*Nitella opaca*). It commonly extends into more mesotrophic waters than does *I. echinospora* (cf. Jermy & Crabbe, 1978), presumably through a greater competitive ability against flowering-plant growth, and its advantage in size. Plants can also grow well in eutrophic waters, but are probably naturally excluded from these by the density of competitors. Its preference for bays and backwaters of larger upland lakes, and its general absence from smaller ones, may be related to physical or biological factors such as the larger fetch and hence greater wave action and water movement in larger ones or the presence of large populations of water fowl or fish. At its preferred greater depths, the wave action is less, and it is possible that plants may depend more heavily on depradations of ducks or fish to remove old leaves, thereby releasing the spores.

The great size differences between plants from different lakes may well reflect underlying genetic differences between populations, those of different lakes possibly representing isolated inbreeding communities (Manton, 1950). The existence of late-glacial and early post-glacial subfossil records of *I. lacustris* from Irish Lake sediments in Co. Cavan and Co. Kerry emphasises that this species has probably had a long history in these islands, with adequate time for locally different forms to have evolved.

Horsetails

Equisetum arvense L.

Common Horsetail

Preliminary recognition Usually an upright-growing, medium-sized, colony-forming horsetail, with green stems and branches, without cones on top of the green shoots. Emerging shoots in late spring look like a forest of miniature spruce trees. By late summer, colonies often have a sprawling, untidy appearance.

Occurrence Widespread and common throughout Britain and Ireland, beside roads and footpaths, in waste places, on rough industrial sites and on margins of agricultural land, often forming dense stands.

Identification Plants have thin but tough, winter-deciduous, pale-green vegetative shoots furnished with dense regular whorls of *dull pale-green* branches. In early summer, after these shoots first emerge, the branches are short and ascending. But they rapidly lengthen to become *long and spreading, giving the shoot a broad outline and bushy appearance.* By late summer, the branches often become tangled and their shoots decumbent or leaning on other vegetation. If surrounded by dense vegetation, the lower nodes eventually become nearly white with deep-brown patches, and their nodes bent and distinctly inflated, giving the *unbranched lower part of the shoot a knobbly and arthritic appearance,* seen only in this species.

The main shoot sheaths have short, dark, triangular teeth, which *often adhere together in 2s and 3s.* The main internodes are 3–4 mm or more diameter, with about 6–18 shallow furrows, and have a *central canal about 2/3 to 3/4 the diameter of the shoot* (see diagram of transverse sections, p. 438). They are smooth to touch, but *when lightly squeezed, remain firm and unyielding* (cf. those of *E.* × *litorale*, q.v.).

The branches are *always fairly thick (1–2 mm)* (about twice as thick as those of *E. pratense*). The first *branch internodes exceed the length of the adjacent stem sheath* (often greatly so in the lower whorls of the shoot). The branch internodes have a very characteristic external structure, with 3–4 (mostly 4), *very prominent, almost knife-edged* ridges, separated by *deep, flat-sided,* V-*shaped furrows,* each of which has *a further groove along its very bottom* readily visible with a × 10 hand lens. At the branch nodes, the sheaths are *green throughout,* and their acutely pointed teeth *spread outwards, often very conspicuously* and are visible even without a lens. *These branch characters are constant, no matter how variable the form of the shoot, and the environment of the plant.*

Jan.	Feb.	Mar.	Apr.	May	Jun.	Jly	Aug.	Sep.	Oct.	Nov.	Dec.

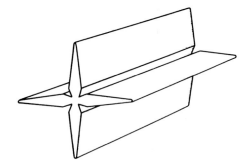

Equisetum arvense, block diagram of branch internode geometry.

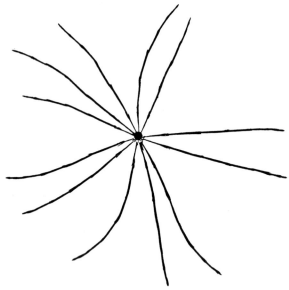

Equisetum arvense, branch whorl architecture.

Plants produce separate, short-lived, unbranched, ivory-white or pinkish, cone-bearing shoots, thicker than the vegetative ones, with larger, chaffier, pale-brown sheaths, which, at low altitude in the north of England, *emerge in the first week of April* (earlier further south), discharge large quantities of green spores, and wither after about 10 days.

Variation Of extremely variable appearance in the field, but almost all of the variation in Britain and Ireland seems environmentally induced. Plants in exposed situations (cliff-tops, fixed dunes) can be completely prostrate, those in the most sheltered situations tall and lax. Occasional woodland shoots may have drooping and branched branches. Some shoots occasionally bear terminal cones.

Possible confusion The constantly thick branches separate all specimens of *E. arvense* from all of those of the rare *E. pratense*, which has constantly thin ones. The branch internode structure described above, as well as the strongly dimorphic shoot habit, immediately separate all forms of *E. arvense* from all those of *E. palustre* (q.v.), the species with which it is very frequently confused. The bushier shoot, usual lack of terminal cones and especially the firm internodes

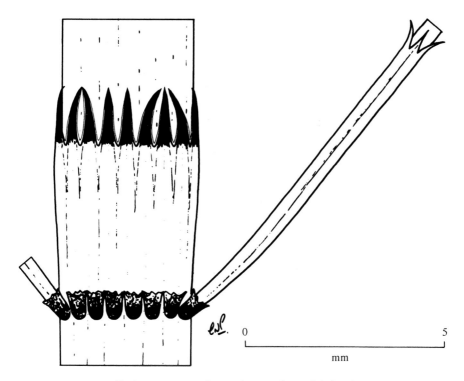

Equisetum arvense, lower shoot and a nodal sheath.

when squeezed, readily separate all fresh specimens of *E. arvense* from all those of *E. *×* litorale*, even in the absence of a cone. A valuable diagnostic character that helps to separate vegetative specimens of *E. arvense* from those of *E. palustre* is the tooth-habit on the branches. In *E. arvense* (left in diagram on p. 438) the teeth are usually green throughout and spread away from the internode above. In *E. palustre* (right), the teeth are usually blackish-tipped and clasped tightly around the internode above. These characters remain fairly constant even in extreme forms of each species.

Technical confirmation Little is needed. Plants have $n = 108$, $2n = 216$ chromosomes. Cross-sections of the branch internode confirm its structure: these appear like a Maltese cross, with pointed arms. Under the scanning electron microscope, the stomatal cell areas of the branch internodes lack the many rows of finger-like pilulae of *E. palustre* and *E. telmateia* (q.v.), and in this, are similar to those of *E. fluviatile*, to which, for this and other reasons (Page, 1972a), *E. arvense* is believed closely related.

Field notes *Equisetum arvense* is an extremely widespread species of horsetail, with a geographic range in southern Greenland and Iceland, through nearly the whole of Europe, across virtually the whole of northern Asia, with southerly enclaves as far south as the Tien Shan and south China, and throughout much of the North American continent.

In Britain and Ireland, Common Horsetail is extremely widespread through virtually every county, thinning mainly in south-western ones of both islands and in the far north of Scotland. It is a vigorous species, which is long lived and of very adaptable ecology. It is the least moisture-demanding of all species of horsetail, and is normally the only one that becomes a weed when established in cultivated ground. It is a plant of wide pH tolerance and succeeds on a variety of deep, freely drained mineral soils, especially those developed over shaly,

Equisetum arvense, transverse sections of main branch and shoot internodes.

Equisetum arvense (*left*) and *Equisetum palustre* (*right*), comparison of tooth habit on branches.

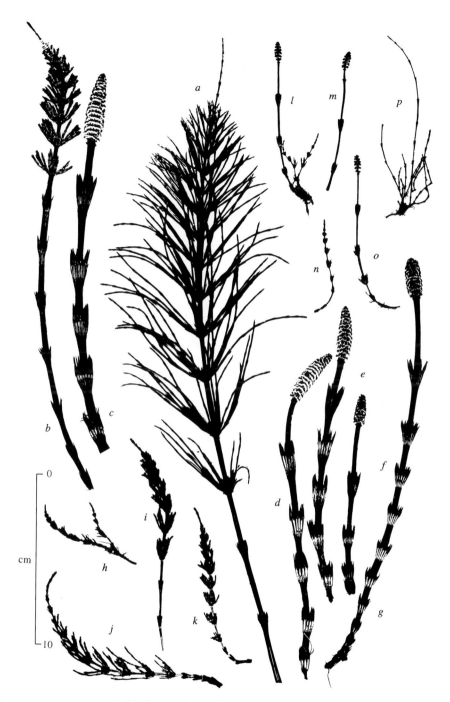

Equisetum arvense: a, typical hedgerow form, Kirkcudbrightshire; *b–g*, trackside and industrial
waste form, showing particularly free production of cone-bearing shoots, City of Newcastle-upon-Tyne;
h–k, typical sand-dune form, Isle of Rhum, Inner Hebrides; *l–p*, mountain form, Inverness-shire.
Note – In conditions of uniform cultivation, all turn into virtually identical plants.

Equisetum arvense, fertile shoots in early emergence, prior to spore liberation: *all* Morpeth, Northumberland, early April.

Equisetum arvense, fertile shoots in late stage of emergence, with spore discharge nearly complete: Newburn, Newcastle upon Tyne, mid-April.

sandy or gritty rocks, or over clays or gravels, wherever such ground has a moderate base status. It seems least frequent both in limestone districts and where there is an extensive depth of peat, in both cases perhaps due to lack of siliceous content, which seems to be required by all species of horsetail.

In least-disturbed situations, Common Horsetail is a plant of sandy river banks, fixed dune pastures, cliff-tops and other sandy grasslands, but seldom in great abundance. It also ascends on streambanks and in flush-margins to well over 915 m (3000 ft) on mountains. With human activity, it has become very much more frequent in habitats such as roadside verges and gravel dumps, along footpaths and canal banks, on old quarry scree and disused

mining spoil tips, on derelict ground around old industrial sites, on railway embankments and in the ballast of lightly used old railway tracks and sidings. In the latter habitats, it can be particularly dominant, where its abundance may be due in considerable part to its substantial herbicide resistance.

Establishment of new plants by spores may be an infrequent occurrence, as this requires the availability of moist, completely uncolonised, virgin habitats (Page, 1967). But colonisation of any such habitats available during the early spring spore-shedding time, once begun, can be very rapid, with young sporophyte horsetails appearing from prothalli by early summer. These form bushy plants by

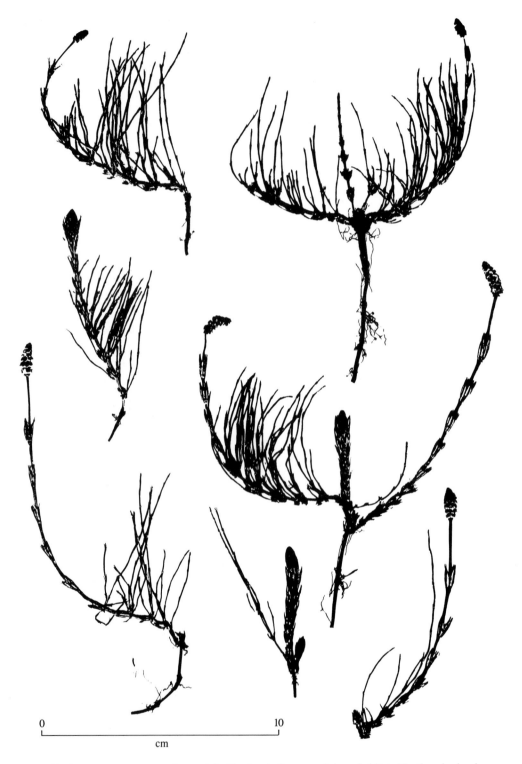

Equisetum arvense, vegetative and fertile shoots from sand-dune habitat, Northumberland.

the end of their first season, with up to two dozen shoots and several tuber-bearing rhizomes. Once colonies, are well established, the slender, black, felted, tuber-bearing, running rhizomes may extend down to depths of 2 m or very much more, and extend horizontally for greater distances.

The success of this horsetail in sites disturbed by humans is usually the result of freedom from competition coupled with lack of tree shade. In natural situations plants often play a pioneering role in early stages of vegetation succession. Where such successions eventually lead to the establishment of a dense herb layer shaded by developing tree growth, the horsetail eventually becomes a relatively subordinate component or dies out.

Equisetum arvense seems considerably tolerant of mineralogical contents of spoil heaps from mining operations, which are toxic to most other vegetation. It also survives, and sometimes thrives, in industrial ash and clinker, and in larger industrial cities seems considerably tolerant of associated atmospheric pollution. Large colonies sometimes succeed on pit heaps, where their colonising ability, allowing rudimentary soils to begin to form, and the stabilisation of the surface by the interwoven underground rhizomes, may be considerably beneficial towards the eventual establishment of vegetation on such inclement ground.

Equisetum fluviatile L.

Water Horsetail

Preliminary recognition A medium to large-sized, colony-forming, semi-aquatic horsetail, with slender, upright, hollow reed-like stems, frequently bearing only sparse branch whorls, and terminal blackish cones.

Occurrence Very widely distributed throughout Britain and Ireland, becoming most abundant in the north, sometimes forming extensive stands in swamps and carrs, and fringing the margins of still waters and lakes, extending into water 1.5 m in depth.

Identification A very distinctive species, producing winter-deciduous *tall, slender, pencil-thick reed-like stems*, up to 100–150 cm tall or occasionally more, *which snap readily if sharply bent*, revealing a *very large central hollow, 3/4 to 9/10 the diameter of the shoot* (see transverse section diagram on p. 444). Vallecular canals are present only in some specimens, and they are usually developed mainly in the lower parts of the stem. The internodes are a glossy yellow-green colour, with 10–30 *scarcely perceptible, very shallow ridges and grooves*, and very smooth to the touch. *When squeezed lightly between finger and thumb, the internodes collapse*, reflecting their pipe-like structure. The main stem sheaths are tight, and their green is frequently tinged with

Jan.	Feb.	Mar.	Apr.	May	Jun.	Jly	Aug.	Sep.	Oct.	Nov.	Dec.

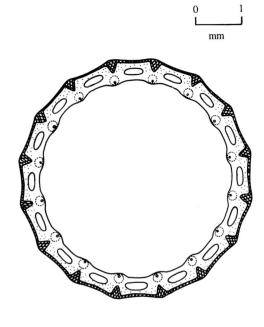

0   1

mm

Equisetum fluviatile, transverse sections of main shoot internodes.

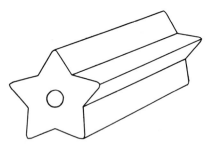

Equisetum fluviatile, block diagram of branch node geometry.

orange, a colour characteristic amongst the *deciduous* horsetails of only this species and its hybrids. The sheaths have relatively short teeth, which are dark and triangular with inconspicuous narrow pale margins, and rather similar to those of *E. arvense*. In exposed situations, whorled branches are completely suppressed, but with increasing shelter and shade, shoots are progressively better furnished with more complete whorls of fairly short, straight, slender branches, in spreading or slightly ascending nodal whorls, through about the middle 1/3 of the stem. *A dense stand of such shoots in heavy shade can thus have a strongly verticillate appearance.*

In all habitats, but especially in more exposed ones, a proportion of the shoots are shorter than others, though otherwise similar, and end in small, *barrel-shaped* greenish or blackish cones, borne out of the uppermost leaf-sheath on short, thick colourless stalks (peduncles).

Variation Extremely variable in size and degree of branching. Linnaeus regarded the unbranched forms as a separate species, *E. limosum*, and this treatment has been followed in many older British County Floras. Transplant experiments have, however, shown that all

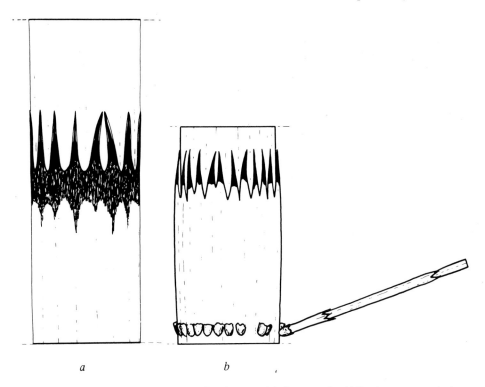

a *b*

Equisetum fluviatile, architecture of nodal sheaths: *a* and *b*, lower and middle shoot, respectively.

forms revert to virtually identical plants when cultivated alongside, the size and branching differences being entirely environmentally induced; shade and shelter are the most potent factors.

Possible confusion Branched shoots are unlikely to be confused with any other horsetail except possibly with some forms of *E.* *ˣlitorale* (q.v.). However, a simple squeeze of a stem internode immediately separates them in the field – those of *E.* *ˣlitorale* yielding only slightly, rather than totally collapsing. Unbranched shoots might be confused with *E. hyemale*, but the latter are of blue-green colour, firm and very rough to the touch, have no teeth to the sheaths and do not die down in winter.

Technical confirmation Abundant cones containing entirely good spores distinguish *E. fluviatile* from *E.* *ˣlitorale*. Plants have $n = 108$, $2n = 216$ chromosomes. Under the scanning electron microscope, the surface ornamentation pattern and stomatal cell microstructure are extremely similar to those of *E. arvense*, and this, with other evidence, suggest the existence of a close interrelationship between these two ecologically differently adapted species (Page, 1972*a*).

Field notes *Equisetum fluviatile* is an extremely widespread species of horsetail, with a geographic range in extreme southern Greenland, Iceland, through nearly the whole of Europe south to Spain and Turkey, across virtually the whole of northern Asia to Japan and Kamchatka, and throughout much of the North American continent.

In Britain and Ireland, Water Horsetail is extremely widespread through nearly every county, and most abundant in those areas with the largest shorelines of freshwater ponds and lakes. It occurs as a member of pond, ditch, Willow (*Salix*) and Alder (*Alnus*) carr vegetation, but usually is most extensive in shallow water of lakeside communities

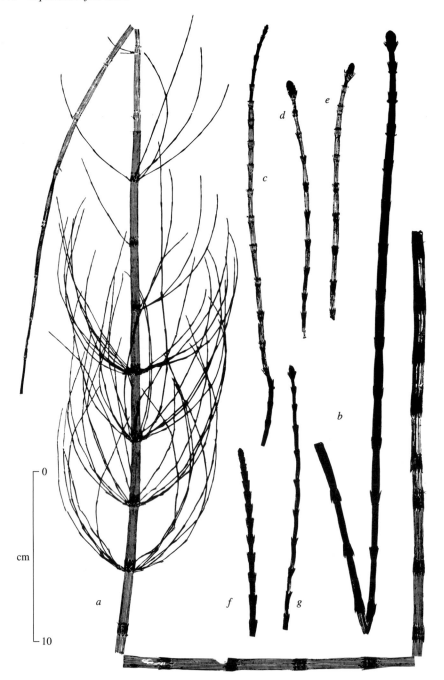

Equisetum fluviatile: a, sheltered pond form, Derbyshire; *b*, open, lake-margin form, Northumberland; *c–g*, exposed, moorland pool form, South Uist, Outer Hebrides. Note – In conditions of uniform cultivation, all turn into virtually identical plants.

in deep silty muds, especially in quiet backwaters. In such habitats, it may form extensive reedswamps, in up to 1.5 m water depth, or extend over much deeper water in mats of floating reedswamp vegetation.

Like Common Horsetail, Water Horsetail is a plant of extremely wide pH tolerance. It will survive in muds of poor, relatively acidic, oligotrophic upland peaty pools and lake waters, probably because of the lack of competition for such habitats. But it undoubtedly becomes much more strongly growing and vigorous in basic waters. It succeeds best where there is extensive accumulation of deep, silty, humus-rich mineral mud, in waters which are still, and suffer from little wave action or scouring, or which show little fluctuation in depth.

Where such suitable conditions exist, Water Horsetail can become an invasive species, often forming a conspicuous early pioneer stage of successions in freshwater hydrosere vegetation. Once well established, colonies are probably very long-lived, but eventually decline in abundance as the vegetation closes, only to become more frequent again in later successional stages of fen and carr establishment, where tree overgrowth again reduces the vigour of competing vegetation. Underground, it undoubtedly forms very extensive masses of rhizome growth, at a level perhaps deeper than that of most flowering-plant root competition. Although the aerial parts of the shoots are winter-deciduous, the relatively succulent lower portions (especially those immersed in considerable depth of water), often remain over much of the winter, an important storage adaptation perhaps in a species which, unusually amongst the deciduous horsetails, lacks tubers on its rhizomes. It is the only species of deciduous horsetail in which the central cavity of the aerial shoot persists in the rhizome, and this correlates with its unique ability amongst the horsetails successfully to extend its underground parts through anaerobic mud conditions. In this connection, the persistence of the lower parts of stems for much of the winter, when connecting with the water surface, may also be an important ventilating adaptation to enable passage of air to the rhizomes to continue out-of-season.

With its wide pH tolerance and long seral persistence, Water Horsetail occurs in association with a very great number of other aquatic or semi-aquatic plant species. In lakes, on the outer side of the reed zone which it pioneers, in water more than 1.5 m deep, the floating leaves of White Waterlily (*Nymphaea alba*), or Broad-leaved Pondweed (*Potamogeton natans*) usually fringe it. On the shore side of such vegetation, where the water is less than about 50 cm deep, dense growths of Common Spike-rush (*Eleocharis palustris*) or Bogbean (*Menyanthes trifoliata*) often largely replace it. Amongst the other most common herbaceous associates in lake-margin, fen or carr vegetation, are Flote-grass (*Glyceria fluitans*), Lesser Spearwort (*Ranunculus flammula*), Mare's-tail (*Hippuris vulgaris*), Marsh Cinquefoil (*Potentilla palustris*), Marsh Marigold (*Caltha palustris*), Water Avens (*Geum rivale*), Water Mint (*Mentha aquatica*), Marsh Hawk's-beard (*Crepis paludosa*), Ragged Robin (*Lychnis flos-cuculi*) and Meadow Sweet (*Filipendula ulmaria*).

Around lake shores, plants may become propagated from washed-up stem fragments (which float particularly readily). Ducks especially, have been seen deliberately to fragment shoots into numerous small (2–3 internodal) sections, thereby releasing abundant potential free-floating cuttings. Nevertheless, the species seems to have decreased steadily in abundance over the last 100 years, particularly in the south of England, due mainly to drainage of wet areas and filling-in of ponds.

Equisetum hyemale L.

Dutch Rush

Preliminary recognition A tall, evergreen horsetail, forming dense clumps of strongly erect, unbalanced, stout, rush-like shoots, which are blue-green in colour and totally lack teeth to the sheaths.

Occurrence Widely scattered throughout the whole of Britain and Ireland, everywhere scarce but slightly less so in the north. It is a very local, predominantly lowland species, of sheltered, shady, sandy or deep clay riverbanks, occurring occasionally at higher altitude.

Identification Plants produce pencil-thick (or thicker), *rough, tough evergreen shoots*, usually 70–100 cm tall. Once fully matured, shoots are of a dark blue-green colour, but young, emerging shoots are at first creamish-pink throughout before turning a bright green and then gradually darkening. Their internodes have about 12–24 fine, but conspicuous, longitudinal ridges and furrows *are minutely rough to touch, especially to the tip of a fingernail drawn along them*, and become *gradually inflated as they age*. Consequently, *when the length of an old shoot is run gently between finger and thumb, the internodes can be felt to bulge, and the nodes to be constricted.*

The toothless sheaths are about as long as broad, and are *close-fitting around the stem on young shoots, but with age often become eroded and split vertically into more of a funnel shape* – a shape which often becomes particularly exaggerated in herbarium material. They are at first green, but soon develop a black band around their base and a narrower black band around their upper margin, which is *minutely scalloped at the points where the missing teeth would attach*. With time, the black bands broaden (particularly the lowest ones). At some of the lower nodes of the shoot the whole sheath may become black, but, in the most visible upper ones, the *central part turns a very characteristic slightly pinkish ashen-grey colour.*

Scattered shoots, otherwise similar to the vegetative ones, terminate in small barrel-shaped cones, each about the size of a pea, which are capped by *a hard, dark-coloured, bluntly pointed tip* (apiculum). The cones are either green near the base and black above, or black throughout, and *seldom emerge far out of the uppermost sheath.*

Shoots first begin emergence in late spring, expand during the summer and overwinter in young, fully developed form. Those bearing cones retain these in an unopen condition until

Jan. Feb. Mar.	Apr.	May	Jun.	Jly	Aug.	Sep.	Oct. Nov. Dec.

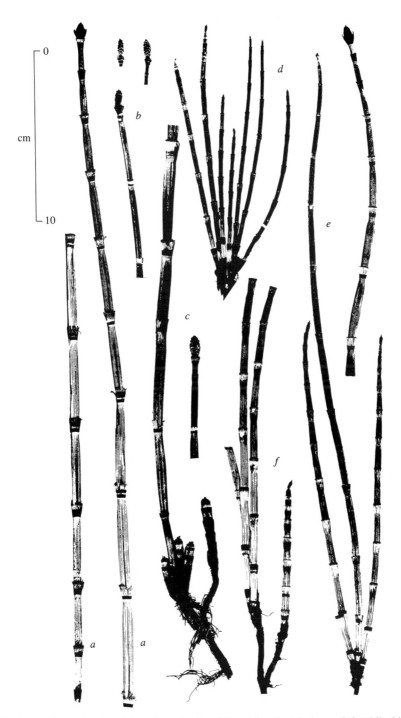

Equisetum hyemale: a/a, Orkney Islands; *b–c,* Weardale, Co. Durham; *d–f,* Midlothian.

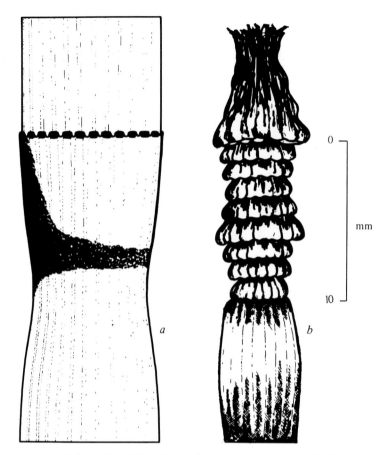

Equisetum hyemale, nodal sheath and tip of expanded shoot showing pagoda-like cap (see the text).

early the following spring, whence spores are shed. Some further stem elongation often takes place during the spring of the second year.

During initial expansion of the shoots, the teeth are lost from the sheaths in a manner unique to this species. As each internode expands and thrusts through the whorl of teeth capping the sheath in which it arises, so the *whole tooth whorl, intact, becomes torn away from the sheath* (rather than remaining attached to the sheath and each tooth splitting vertically apart from the others). This happens for each sheath; the intact tooth whorls, each like a Chinese hat in shape, *are carried up as a complete stack on the tip of the shoot, where they remain on the fully expanded shoots for some time, the whole like a miniature pagoda.* Lack of appreciation of what these objects are and how they originate have previously led to inaccurate reports that the teeth are merely shed from the sheaths, and of description of these dried-up looking objects on the shoot tips as abortive cones. In the diagram above, (*a*) shows architecture of the nodal sheath in a lower shoot and (*b*) the tip of a shoot with the accumulating pagoda-like structure of stacked and interlocked teeth torn from their sheaths as the shoot has expanded.

Variation This is not a very variable species. In juvenile plants and in very stunted colonies in exposed places, much thinner shoots can occur, which sporadically retain their narrow, black teeth on their sheaths.

Possible confusion Shoots of mature colonies can resemble ones of *E. fluviatile* (q.v.) in their size and lack of branching, but the rough-textured, tough construction, evergreen habit and lack of teeth to the sheaths clearly distinguish *E. hyemale*. It is distinguished with more difficulty from larger shoots of *E.* $^{\times}$*trachyodon* (q.v.), of which *E. hyemale* is one parent, but the lack of teeth to the sheaths on expanding shoots, inflated internodes on old stems, and sheaths about as long as broad, distinguish *E. hyemale*.

Technical confirmation Plants have $n = 108$, $2n = 216$ chromosomes. The stem internodes are each narrowly bi-angulate and armed with conspicuous angular, colourless siliceous tubercles.

Field notes *Equisetum hyemale* is a widespread species of horsetail, with a geographic range from southern Greenland, Iceland, Britain and Ireland through nearly the whole of northern and central Europe south to Spain, Greece and Turkey, and across very much of northern Asia to Japan and Kamchatka. A related, but larger, form occurs in North America.

In these islands, *E. hyemale* is predominantly a valley side and riverbank species of mainly low to sometimes moderate altitudes. Even once well established, it is very slow growing, with a long period of shoot emergence, and slow rate of spread. This may possibly be linked with the very large quantities of silica that are laid down in its structure, displayed when a shoot is burned by the amount of ash which remains, and which maintains much of the form of the original shoot.

Dutch Rush consequently appears to require highly mineral and silica-rich substrates, and thrives in deep sandy-clayey soils, which are kept permanently moist, at least beneath the surface, by moving groundwater, which flushes them with moderate quantities of bases. In sheltered localities on sloping riverbanks, it has the potential eventually to form fairly extensive colonies, almost totally excluding other vegetation.

In riverbank habitats, it succeeds best where there is a light tree overgrowth of such species as Ash (*Fraxinus excelsior*), Wych Elm (*Ulmus glabra*), Aspen (*Populus tremula*), Blackthorn (*Prunus spinosa*) or Elder (*Sambucus nigra*). Growing near or around the *E. hyemale* stands are usually species indicative of good, if moist and heavy, soils. Associates commonly include Lesser Celandine

(*Ranunculus ficaria*), Creeping Buttercup (*R. repens*), Water Avens (*Geum rivale*), Bugle (*Ajuga reptans*), Meadowsweet (*Filipendula ulmaria*), Dog's Mercury (*Mercurialis perennis*), Wild Angelica (*Angelica sylvestris*), Wild Strawberry (*Fragaria vesca*), Moschatel (*Adoxa moschatellina*), Wood Sedge (*Carex sylvatica*), Slender False-brome (*Brachypodium sylvaticum*) and Wood Melick (*Melica uniflora*). In its northern English and southern Scottish stations, *E. hyemale* occurs in valleys in which Great Horsetail (*Equisetum telmateia*) also forms extensive colonies, in slightly wetter ground.

Today most colonies of this particularly interesting horsetail are of small size. In former times, bundles of their shoots were collected for sale in major cities, where, because of their robust construction, high silica content and asperous surface, they were widely used for scouring cooking pots and for finely burnishing brass. Much of this material was probably imported from Holland, but doubtless, some commercial collection of British material may have taken a toll and had its effect on reducing native populations. Yet, long after its commercial use has ceased, Dutch Rush seems to be showing a continuing trend of rapid decrease even in the last half century or so and is now rare in many former sites. Drainage of originally moist sites may have contributed partially to this. But increased access to riverside pastures by cattle, which rapidly eliminate it by grazing and trampling, has probably been a major factor in its widespread disappearance from many former lowland riverside stations, as its habitats are often indicative of good pasture land.

Equisetum palustre L.

Marsh Horsetail

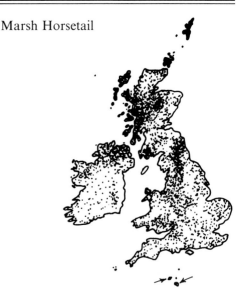

Preliminary recognition A winter-deciduous, colony-forming, medium-sized, horsetail, with upright-growing shoots, bearing regular whorls of rather stout, upcurving branching, with many of the main shoots ending in elongate greenish or blackish cones.

Occurrence Present throughout most of Britain and Ireland, becoming most frequent in the north, in marshes, damp pastures, ditches and occasionally sand-dune slacks in the lowlands, beside streams and in mountain flushes in the uplands, where there is a supply of at least some basic minerals.

Identification Plants often form colonies amongst other marshy vegetation, giving rise to *large numbers of thin, flexible, strongly erect, winter-deciduous shoots,* mostly 15–50 cm tall, *with 5–9 ridges,* mostly bearing sparse but regular whorls of *fairly thick branches, which themselves nearly always adopt a markedly upcurving and ascending habit,* and give the whole shoot a slenderly ovoid outline. Such branches typically arise throughout the middle part of each shoot, the upper part of each shoot terminating either in a cone or in a long, erect, unbranched, whip-like portion. In more exposed habitats, the branches may become more confined to the lower nodes of the shoot, and themselves become tall, whip-like and ascending. In very exposed terrain, branches can be absent altogether.

On the main shoots, the sheaths are relatively long (*at least twice as long as broad*), and end in *narrow, triangular, black teeth, each with conspicuous, broad, white margins which surround the central dark triangle in the shape of a gothic arch.* The first internode of the branches is *always much shorter than the length of the adjacent stem sheath,* and their basal scales (ochreolae) around the main stem are *always dark coloured.* On the internodes of the branches, the teeth are *clasping* and green *with minute black tips (never spreading and green throughout as in* E. arvense*).* The internodes of both the main shoot and branches each have only a *very shallowly rounded* ridge-and-furrow structure, are of a mid-green colour, with a slightly shining surface and feel very smooth to the touch. The diagram opposite represents the architecture of nodal sheaths on lower (*a*) and middle (*b*) shoots, respectively. Within the main shoot internodes, the central canal is the smallest of all species of horsetail, *usually about 1/4 or less of the diameter of the shoot,* and surrounded by a ring of other (vallecular) canals, *almost equalling it in size.* (See diagram on p. 454.)

Jan. Feb. Mar.	Apr.	May	Jun.	Jly	Aug.	Sep.	Oct. Nov. Dec.

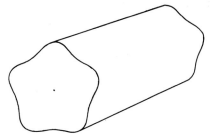

Equisetum palustre, block diagram of branch internode geometry.

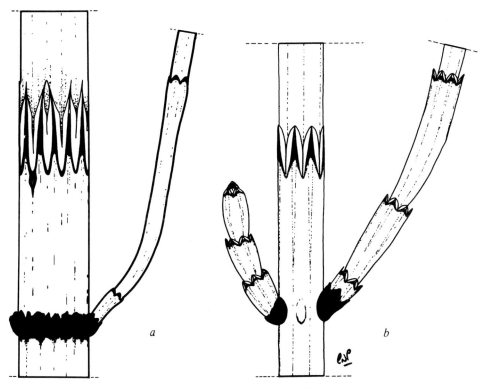

Equisetum palustre, nodal sheaths, lower and middle shoot respectively.

The cones are of *markedly cylindric, parallel-sided outline*, rounded at both ends, 2.0–3.5 cm or more long, and *raised above the uppermost sheath at maturity on long, slender stalks* (pedicels), usually half or more the length of the cone. Most colonies typically produce cones abundantly.

Variation *E. palustre* shows extensive environmentally induced variation in size and plant habit. Freely branched shoots are associated with greater shelter, more simple shoots with more exposed conditions.

Many individual colonies of *E. palustre* show particularly well the variation that can be observed between shoots of a numerical reciprocal relationship between the number of sporangiophore whorls in the cone and the number of nodes on the subtending vegetative parts of the same individual shoot (see diagram on p.456). This relationship probably

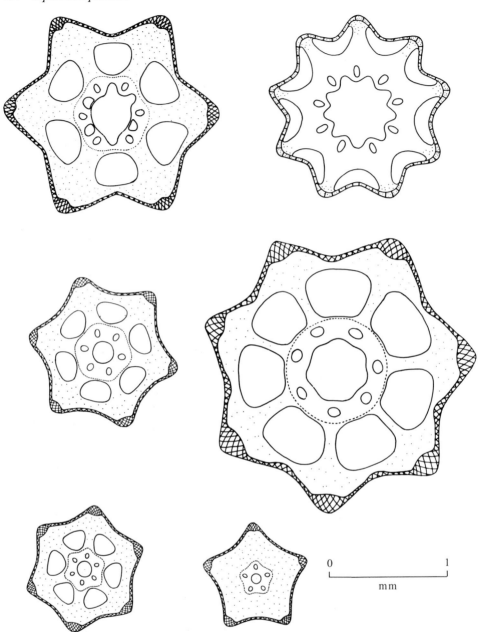

Equisetum palustre, transverse section of main shoot and branch internodes.

approximates in all species of *Equisetum*, but is usually especially clearly demonstrated in *E. palustre* because of the particularly wide inherent variation in cone size between individual shoots within single stands of this species. This reciprocity provides one strong line of evidence for the conclusion that the vegetative and fertile units of the shoot in the genus *Equisetum* are homologous with one another in their ultimate evolutionary origins (Page, 1972*b*).

Possible confusion Small, scarcely branched forms of *E. palustre* in coastal dune slacks or in

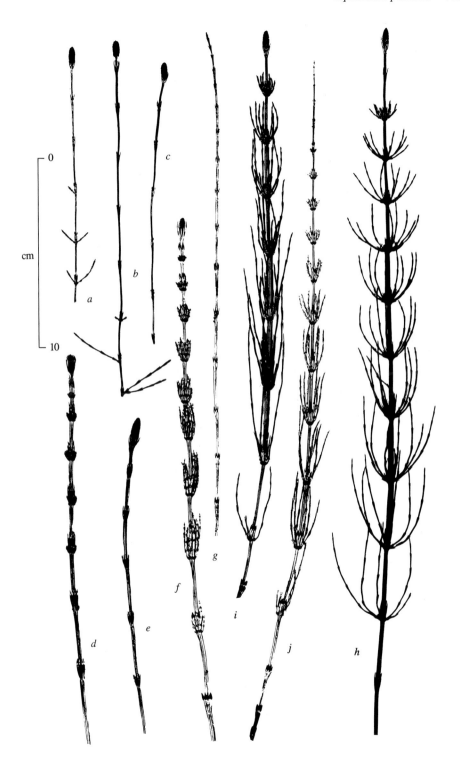

Equisetum palustre: a–c, upland form, Breadalbane mountains, Mid Perthshire; *d–e*, Forest of Dean, West Gloucestershire; *f–h*, Isle of Skye, Inverness-shire; *i–j*, Peeblesshire.

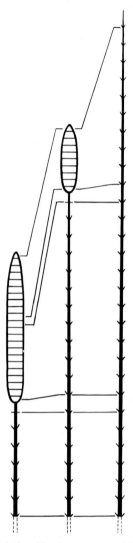

Equisetum palustre, statistical interrelationships between vegetative node numbers and cone sporangiophore whorl numbers present between typical shoots in an individual colony (after Page, 1972*b*).

gravels of mountain streams are very often taken for the much rarer *E. variegatum* (q.v.). Shoots of *E. palustre* are of smooth surface texture and under a hand lens their internodes lack the angular ridges with prominent tubercles and regular vertical rows of whitish stomatal pores seen in *E. variegatum*. The small central canal to the main stem internodes, the rounded ridge-and-furrow structure of the branch internodes, the clasping, black-tipped teeth to the branch nodes, and the very short first branch internode, provide good characters for separation of *E. palustre* from all forms of *E. arvense*.

Technical confirmation Plants have $n = 108$, $2n = 216$ chromosomes. Under the scanning electron microscope, the stomatal cell areas of the branch internodes are armed with regular ranks of elongated, finger-like projections (pilulae) that can move back from or interlock across the stomatal pores. This highly specialised structure is found elsewhere only in *E. telmateia*, and, with other evidence (Page, 1972a), suggests a close inter-relationship between these two species.

Field notes *Equisetum palustre* is a widespread species of horsetail, with a geographic range in extreme southern Greenland, Iceland, through nearly the whole of Europe south to Spain, Italy, Greece and Turkey, across very much of northern Asia to Japan and Kamchatka, and throughout much of the North American continent.

In Britain and Ireland, Marsh Horsetail is extremely widespread through nearly every county, and most abundant in those areas with the largest areas of streamsides and damp ground over appropriately mixed siliceous and calcareous mineral soils. It is a generally adaptable species which grows in a very wide range of deep, marshy, wet, sandy or peaty soils that are adequately lime rich or flushed by base-rich irrigation. In such habitats, it is much more tolerant of stagnation of the water supply than is Great Horsetail (*Equisetum telmateia*, q.v.), and also tolerant of a greater range of altitude and more acidic conditions, although it is often present in habitats where Great Horsetail grows too.

The most luxuriant lowland habitats are in base-rich water meadows and pastures, on deep, fertile soils, where the plant often spreads to densely occupy edges of drainage ditches. In more upland areas, extensive colonies may occur in areas where deep, peaty soils are flushed by basic mineral-rich groundwater movement, and it is a regular component of high-level basic flushes on mountains.

Marsh Horsetail associates with a wide range of other herbaceous plants of base-rich marshes and flushes. Notable associates in wet meadow communities commonly include Marsh Marigold (*Caltha palustris*), Meadowsweet (*Filipendula ulmaria*), Ragged Robin (*Lychnis flos-cuculi*), Meadow Buttercup (*Ranunculus acris*), Marsh Hawk's-beard (*Crepis paludosa*), Marsh Cinquefoil (*Potentilla palustris*), Water Mint (*Mentha aquatica*), Red-rattle (*Pedicularis palustris*), Northern Fen and Early Marsh Orchids (*Dactylorchis purpurella* and *D. incarnata*), and several Rushes (e.g. *Juncus acutiflorus*) and Sedges (e.g. *Carex nigra, C. lepidocarpa, C. dioica* and *C. hostiana*). In wet, calcareous, montane vegetation, Marsh Horsetail is often accompanied by Grass of Parnassus (*Parnassia palustris*), Lesser Spearwort (*Ranunculus flammula*), Autumnal Hawkbit (*Leontodon autumnalis*), Marsh Tormentil (*Potentilla palustris*), Eyebright (*Euphrasia officinalis* agg.) and Variegated Horsetail (*Equisetum variegatum*).

In suitable soils, rhizomes of *Equisetum palustre* usually penetrate to a depth of 60–90 cm or more. Here they develop very robust, long-running rhizomes, which are commonly 8–12 mm diameter with internodes 8–10 cm in length, spreading great distances through the ground. These rhizomes, which are usually five-sided, are extraordinarily large for the small shoot diameter of *E. palustre*, and in their construction and the tubers they bear in whorls at the nodes are almost indistinguishable from those of *E. telmateia*. The presence of such massive, deep rhizomes, confers on this species considerable abilities of both clonal spread and vegetative persistence.

Equisetum pratense Ehrh.

Shade Horsetail

Preliminary recognition A small to medium-sized, upright-growing, winter-deciduous horsetail, densely clothed with spreading branches as slender as those of Wood Horsetail, but not themselves branched.

Occurrence A northern species, with scattered stations in the northern Pennines of England and in Northern Ireland, but much more widely spread in Scotland. Seldom in great abundance, it occurs mainly as small patches on undisturbed, sandy or clayey river and streambanks, especially in moist, lightly shaded situations, chiefly in the lower valleys of mountain districts. It is sometimes in more open situations at higher altitudes in species-rich mountain grassland, usually with *Equisetum sylvaticum.*

Identification Plants form small, diffuse, colonies, with strongly erect, slender, *always freely branched, winter-deciduous* shoots, mostly 10–30 cm tall (but sometimes as little as 2–5 cm in mountain grassland). They are furnished throughout all but the lowermost nodes with abundant, regular whorls of usually long, *very slender,* decidedly spread, and sometimes drooping, *pale-green* branches, giving the whole shoot a *delicate* and markedly verticillate appearance. The diagram on p. 461 shows the characteristic branch whorl architecture, and typical asymmetry with branches curving downslope. In outline, shoots are approximately parallel sided, or broaden upwards, and well-developed shoots *always have markedly blunt tops.* The main shoot internodes are *slender,* usually little more than 2 mm diameter, but feel *firm, rigid and rough to the touch* when a fingernail tip is stroked lightly down one. They have *a large number (often 12–20) of ridges for their size,* and these are angular. The main shoot sheaths are *markedly urn shaped, and bear fine, dark, straight, unfused teeth,* which are also *very numerous* for the slender diameter of the shoot, and look rather like those of *Equisetum telmateia* in miniature (see diagram on p. 462). In the main shoot, the central hollow is approximately half the stem diameter (see diagram on p. 464).

The branches are themselves almost always completely unbranched, *very slender* (scarcely more than 0.5–1.0 mm diameter), and *regualrly have only three-angled internodes.* They are typically of a *very slightly glaucous green* colour, similar to those of *E. arvense,* presenting a surface which is poorly wettable – specimens in the field often glistening with tiny, separate, water droplets.

If present (see 'Field notes'), cones are borne on separate shoots, which arise simultaneously with, or slightly ahead of, the emergence of the purely vegetative shoots, and are paler in

Jan. Feb. Mar.	Apr.	May	Jun.	Jly	Aug.	Sep.	Oct. Nov. Dec.

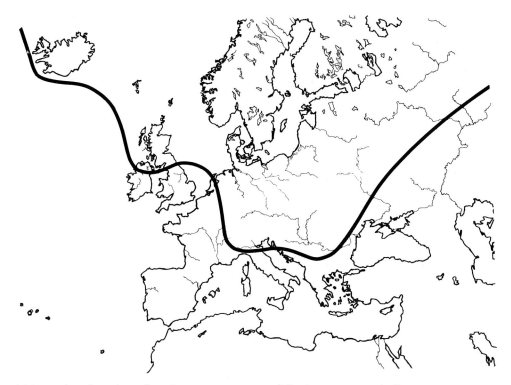

Main southern boundary of northern-montane range of *Equisetum pratense* in Europe.

colour, slightly thicker in diameter and have somewhat larger sheaths. They appear at first unbranched, but rapidly develop branch buds as the cone begins to mature. The branches expand as the cone withers, the whole shoot thereafter functioning as a vegetative one. The cone is somewhat spindle-shaped, and pale pinkish-green just before opening.

Variation Shoots show much variation in size according to degree of shelter of the habitat, and smaller shoots look like diminutive, although more sparsely branched, adults.

Possible confusion Very diminutive shoots from mountain habitats can be extremely similar to nearby diminutive shoots of *E. sylvaticum*, but can be separated by close examination of the main shoot sheath teeth, which are always finer in *E. pratense*, and never chaffy. Luxuriant plants might be confused with *E. arvense*, but the branches of *E. pratense* always remain very much more slender (about 1/2 the diameter of those of *E. arvense*), no matter how large the plant of *E. pratense*, or how small that of *E. arvense*, enabling shoots of the two species, even when growing together, to be separated at a glance. Shoots of *E. pratense* also have much larger numbers of teeth to the sheaths for their size than do those of *E. arvense*.

Technical confirmation Plants have $n = 108$, $2n = 216$ chromosomes. The *internodes of the main shoot* are armed with very fine, colourless, siliceous spines, which stand out at right-angles from the surface, and are clearly visible (especially just below each node) with a × 10 hand lens. Only *E. sylvaticum*, distinguished on other characters, shares this feature.

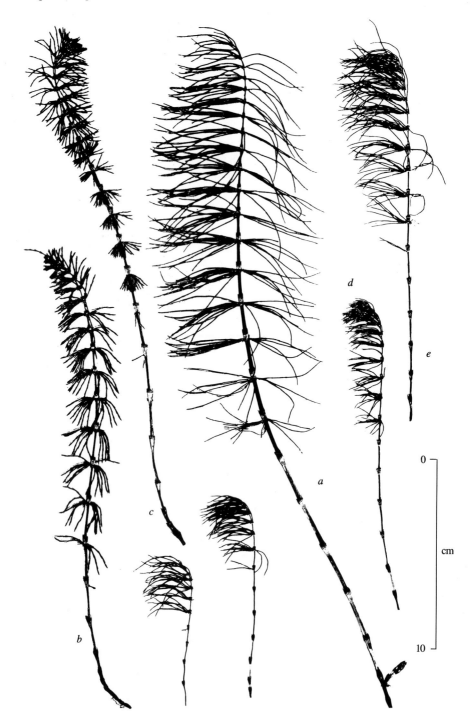

Equisetum pratense: a, sheltered river-bank form, Mid Perthshire; *b–c,* Upper Teesdale, Co. Durham; *d–g,* exposed upland form, Isle of Skye, Inverness-shire. Note – In conditions of uniform cultivation, all turn into virtually identical plants.

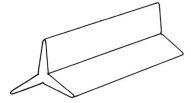

Equisetum pratense, block diagram of branch internode geometry.

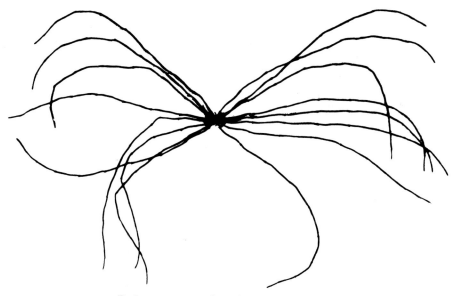

Equisetum pratense, branch whorl architecture.

Field notes *Equisetum pratense* is a widespread species of horsetail, with a geographic range from Iceland, Britain and Ireland through nearly the whole of northern and central Europe south to the Alps, across very much of northern Asia to Japan and Kamchatka, and throughout much of the higher latitudes of the North American continent from Alaska to Labrador. It is, however, somewhat sparser in density of stations as well as rather more northern than is the otherwise rather similar range of *E. sylvaticum* (q.v.).

In Britain and Ireland, Shade Horsetail is also rarer than *E. sylvaticum*, and is mainly confined to northern districts of both islands, where it thrives particularly on deep, moist, freely draining banks of essentially sandy soils, but chiefly where these have a relatively high base-status, through slight mineral flushing. Plants also seem to demand permanent soil moisture and high air humidity. They develop best where there is light shade from summer desiccation, but in cloudy regions also extend in open habitats, especially on mountains.

Shade Horsetail is essentially a pioneer species, growing best where slight soil instability or stream-side erosion maintains open ground surface, or where there is deep accumulation of sandy alluvium beside upland streams. Plants in more closed turf vegetation often occur as sparse, diminutive shoots only, and probably represent relatively late seral stages in vegetation in which the horsetail is steadily succumbing to higher plant competition.

In its most luxuriant habitats in a few upland valleys of the northern Pennines and in mountain glens of central Scotland, *E. pratense* occurs mostly in sloping banks above rivers and streams in semi-open sites or in thin-canopied, herb-rich, Birch–Hazel–Alder–Willow (*Betula–Corylus–Alnus–Salix*) woodland, with a great variety of scattered ground herbs, mostly characteristic of damp, base-flushed soils. These commonly include

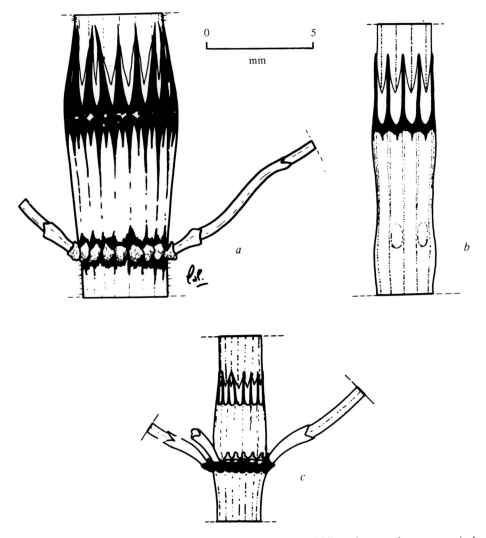

Equisetum pratense, architecture of nodal sheaths: *a–c*, lower, middle and upper shoot, respectively.

Meadosweet (*Filipendula ulmaria*), Water Avens (*Geum rivale*), Globe Flower (*Trollius europaeus*), Marsh Marigold (*Caltha palustris*), Marsh Hawk's-beard (*Crepis paludosa*), Angelica (*Angelica sylvestris*), Wood Cranesbill (*Geranium sylvaticum*), Cowslip (*Primula veris*), Primrose (*P. vulgaris*), Sweet Woodruff (*Galium odoratum*), Mountain Melick (*Melica nutans*), Beech Fern (*Phegopteris connectilis*) and Oak Fern (*Gymnocarpium dryopteris*). Herb Paris (*Paris quadrifolia*) may also occur nearby. In rather more exposed, more upland, habitats (but also descending nearly to sea-level in such habitats in western Scotland), associated species almost invariably include Primrose (*Primula*

vulgaris) plus a low turf-like growth of other species, amongst which are commonly Yarrow (*Achillea millefolium*), Purging Flax (*Linum catharticum*), Buck's-horn Plantain (*Plantago coronopus*), Eyebrights (*Euphrasia* spp.), Clovers (*Trifolium* spp.), Yorkshire Fog (*Holcus lanatus*), Beech Fern (*Phegopteris connectilis*) and Wood Horsetail (*Equisetum sylvaticum*).

Shade Horsetail is a circumboreal and arctic-alpine species, on the very southern and oceanic fringe of the Eurasian part of its range in Britain and Ireland. Although the plant succeeds vegetatively in our northern areas, its reproductive capacity seems in a delicate state of balance here. It is clear

Equisetum pratense, young shoots in life: West Perthshire, June.

from nineteenth century herbarium specimens that it then coned abundantly and probably regularly in sites as far south as Teesdale on the Durham/North Yorkshire border. Subsequent herbarium specimens this century very rarely include cone-bearing shoots, and this situations is still reflected in the field today. Cone production is extremely poor and sporadic in most colonies, the majority of which seem not to cone at all. Exceptions to this general trend seem to be virtually confined to the far north-east of Scotland, where at least one site (Morayshire) still cones well. Most of its colonies seem relatively undisturbed by man, and this decline in its reproductive capacity seems most

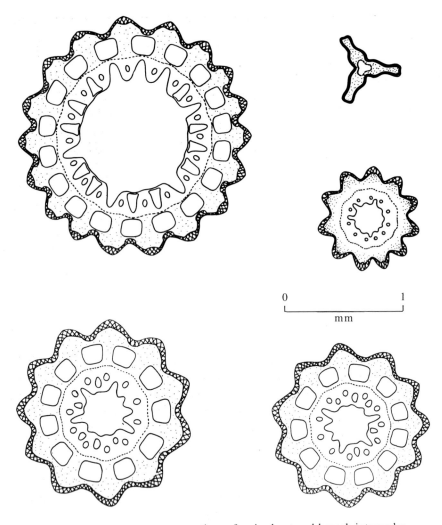

0   1
mm

Equisetum pratense, transverse sections of main shoot and branch internodes.

likely to be a response to natural climatic change, and perhaps correlates with a period of more generally sustained warmth with mostly milder winters this century than last (Lamb, 1972). Fortunately, the plant has considerable vegetative longevity. Most of the undisturbed colonies known a century or so ago still survive today, although some of these, at least, appear now to be approaching later stages of seral succession. In Britain and Ireland at present, *E. pratense* thus seems to be slowly declining.

Equisetum sylvaticum L.

Wood Horsetail

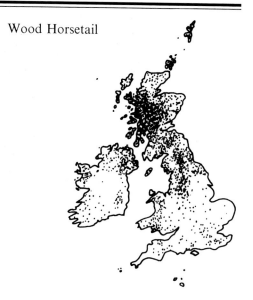

Preliminary recognition A colony-forming horsetail, with medium-sized, upright shoots, bearing abundant dense whorls of extremely fine, branched branches, gracefully drooping towards their tips, each shoot like a miniature spruce tree but *en masse* so fine that from a short distance a luxuriant colony appears to cover the ground like a fine, green mist.

Occurrence Widely spread over the whole of Britain and Ireland, but scarce in the south and most abundant in the valleys of northern mountain districts, where locally extensive colonies occur in wet woodlands, extending in a more stunted form onto mountain slopes.

Identification Colonies produce winter-deciduous, narrow (3–6 mm diameter) erect stems, mostly 10–50 cm in height, but occasionally to 80 cm or more, which are *densely furnished with nodal whorls of long slender spreading branches*, themselves giving rise to *numerous further similar branchlets in the same, horizontal plane. Towards their tips, the branches and branchlets all become drooping, and when fully developed the tip of the shoot also nods*, giving the whole plant its characteristically delicate and graceful appearance (see diagram on p. 466).

The main stem internodes are pale green, distinctly furrowed, and usually *quite rough to the touch*, with a central cavity internally about 1/4 to 1/3 the diameter of the shoot (see diagram on p. 467). The sheaths at the nodes are relatively long (usually 10–15 mm sometimes 20 mm) and *fit quite loosely around the shoot, with a short, green-ribbed, rather urn-shaped fused basal portion* and long, very distinctive teeth (see diagram on p. 468). The teeth have a slender central portion surrounded by *very broad, russet-brown papery membranous margins, with most teeth adhering together by their edges in 2s and 3s to form a small number of very broad lobes around each node.*

The branches are of pale-yellow-green colour, deeper green in more densely shaded conditions, with 3 or 4 (the ultimate always 3) acute angles and deep furrows, and small acutely point-tipped sheaths.

Cones are borne on separate shoots which arise approximately simultaneously with the emergence of the purely vegetative ones, and are of paler colour and slightly thicker diameter, with larger, chaffier and even more membranous sheaths. They appear at first unbranched, but rapidly develop young branches as the cone begins to mature. The cone is usually rather small for the size of the shoot, slenderly ovoid or spindle shaped, pale pinkish-green just before

Jan.	Feb.	Mar.	Apr.	May	Jun.	Jly	Aug.	Sep.	Oct.	Nov.	Dec.

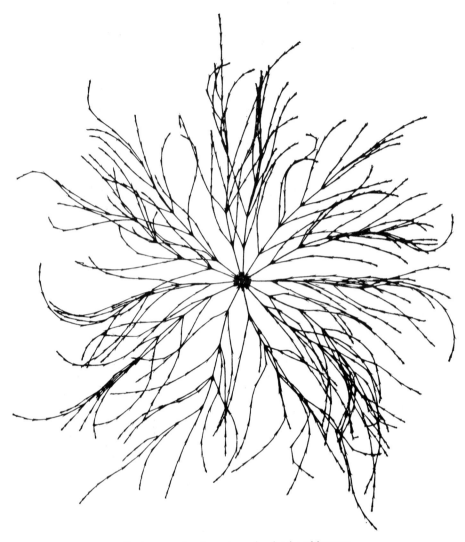

Equisetum sylvaticum, branch whorl architecture.

opening and *borne on an extremely slender stalk* (pedicel), which at maturity is usually longer than the length of the cone. As the cone withers, its shoot continues to develop the typical nodal whorls of branched branches, the whole shoot thereafter functioning as a vegetative one, often with the withered cone still adhering.

Variation There is considerable variation in size and luxuriance of growth, depending on the degree of shelter of the habitat. Exposed montane forms can be very stunted; sheltered woodland ones tall and luxuriant. 'Var. *capillare* Hoffm.' is a slender environmentally induced woodland form. In occasional rare specimens (usually only in particular colonies) short branch whorls or part-whorls arise from the annulus at the base of the cone (see diagram on p. 471). This observation provides evidence that the annulus, while adapted to protection of the underside of the lowermost sporangiophore whorl of the cone, represents a unit of the

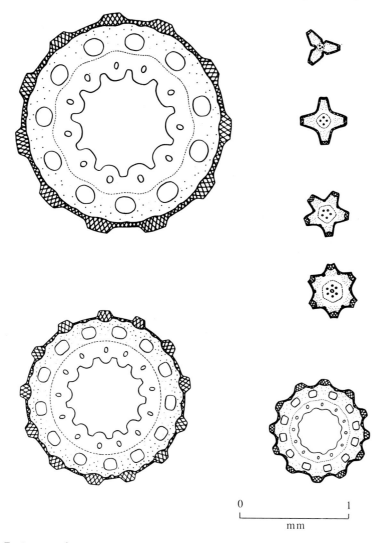

0 _____ 1

mm

Equisetum sylvaticum, transverse sections of main shoot and branch internodes.

shoot which is homologous with the vegetative nodes in its ultimate evolutionary origins (Page, 1972*b*).

Possible confusion This is the only horsetail that regularly has branched branches. The slenderness of the branches always distinguishes *E. sylvaticum* from occasional luxuriant woodland forms of *E. arvense*. The branched branches and broad russet-brown membranous margins to the rather few teeth of main stem sheaths, clearly separate even quite depauperate specimens of *E. sylvaticum* from those of *E. pratense* (q.v.).

Technical confirmation Plants have $n = 108$, $2n = 216$ chromosomes. The internodes of the main shoot are armed with very fine, colourless, siliceous spines, which stand out at right-angles from the surface, and are clearly visible (especially just below each node) with a $\times 10$ hand lens. Only *E. pratense*, distinguished from *E. sylvaticum* on other characters, shares this feature.

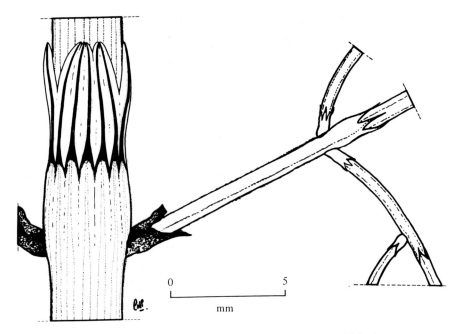

Equisetum sylvaticum, architecture of nodal sheath, middle shoot.

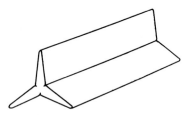

Equisetum sylvaticum, block diagram of branch internode geometry.

Field notes *Equisetum sylvaticum* is a widespread species of horsetail, with a geographic range mainly from Britain and Ireland through nearly the whole of northern and central Europe south to northern Spain, the Alps, northern Greece and Turkey, across very much of northern Asia to Japan and Kamchatka, and throughout much of the higher latitudes of the North American continent from Alaska to Labrador.

In Britain and Ireland, Wood Horsetail is widespread in both islands but most frequent in north-western districts, where it thrives through a wide range of altitudes, from sea-level to over 400 m (around 1200 ft)). Although a moderate calcifuge species, Wood Horsetail grows mainly where deep, acidic, humus-rich (often peaty) soils are flushed with more base-rich percolating ground-

water, keeping them permanently damp, and often wet. Gregarious stands of shoots often occur along lower slopes of mountain valleys, on steep, sheltered banks along streams and upland rivers, in flushed slopes of damp oak woodland, and beside lakes in carr of Alder (*Alnus*) or Willow (*Salix*). It reaches its most luxuriant development where conditions of high, but moving, soil moisture combine with good tree cover, giving high humidity, shade and shelter.

Cones occur quite abundantly in most colonies. The largest cones normally occur on shorter, stockier shoots, with progressively smaller cones borne on progressively taller shoots.

In colonies growing on slopes, each shoot of Wood Horsetail is typically developed asymmetrically, with more extensive branch development in

Equisetum sylvaticum, spring cone shoots in life: West Lothian, mid-May.

each whorl on one side than the other, with the densely branched side of each shoot as well as the nodding tip, always pointing downslope. Under misty conditions, as well as after dew, the finely dissected foliage is usually spangled with very numerous, extremely fine, water droplets (not only from condensation but also from internal secretion by the plant). The numerous, fine, very acutely angled branch sheath-teeth tips, the capillary effect of the fine deeply grooved branch internodes, and the numerous drooping branch tips, probably all combine in providing devices for shedding surface water in droplet form from a plant growing under very humid, sheltered conditions.

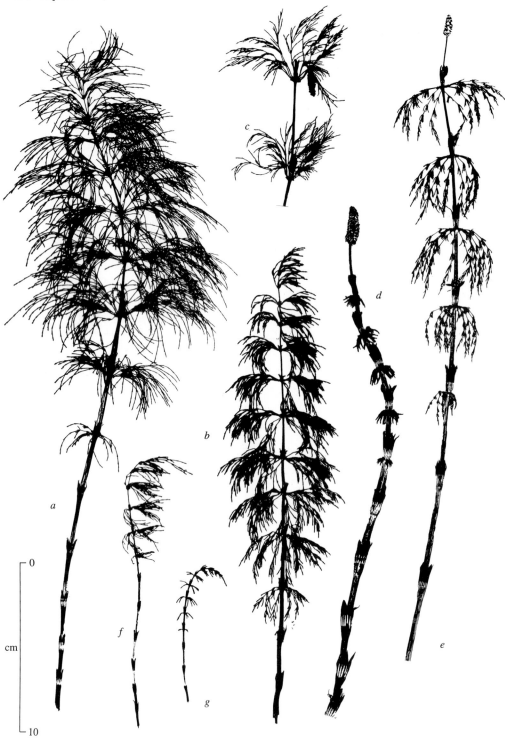

Equisetum sylvaticum: a, sheltered woodland form, Isle of Skye, Inverness-shire; *b*, West Argyll; *c*, with old cone still adhering, Mid Perthshire; *d–e*, Co. Durham; *f–g*, exposed-upland form, Peebles-shire.

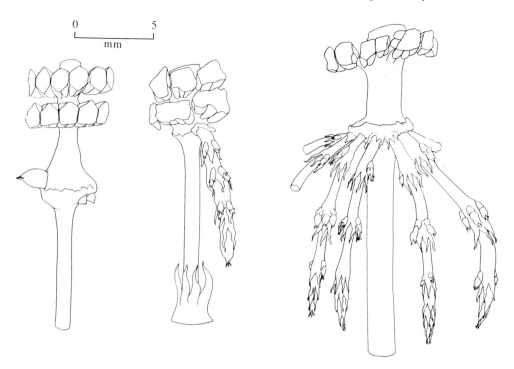

Equisetum sylvaticum, in rare specimens, occasional short branches may arise from the annulus at the base of the cone (after Page, 1972*b*).

Equisetum telmateia Ehrh.

Great Horsetail

Preliminary recognition A strongly upright-growing, large-sized, colony-forming horsetail, with stout, succulent, ivory-white stems bearing dense whorls of bright-green branches. Undoubtedly, a very impressive species, with shoots often forming extensive colonies, their dense, shoulder-high stands like miniature coal forests.

Occurrence Throughout Britain and Ireland, but often local, becoming more frequent southwards, mainly over the sedimentary rocks of central and southern England, especially on springlines and damp banks along shallow river-valley sides, where there is permanent seepage of lime-rich water over deep clay substrates.

Identification The tall, conspicuous, *ivory-white*, winter-deciduous aerial shoots, *densely furnished with nodal whorls of bright green branches*, are, without doubt, its most conspicuous feature. The shoots are usually 1–1.5 m or more in height, with internodes 10–15 cm in length and up to 3 cm or more diameter (see figure on p. 474). *The stems are succulent and smooth to touch, scarcely ridged, brittle and snap readily*, revealing an internal cavity about 1/2 to 1/3 the diameter of the shoot (see figure on p. 476). On the stems, only the sheaths, which surround the nodes quite tightly, are green, and these usually a greenish-grey through only their fused portion. *Their teeth are long and deep brown, with buff-brown, narrow, margins near their base, and end in long, slender, straight but flexible points, occasionally adhering together in 1s and 2s, but mostly free.*

The branches arise out of a ring of small dark scales (ochreolae) at the base of each sheath. The branches are at first straight and rather stiffly ascending but as the season progresses, mostly become long and arching. Expanding shoots are at first of *narrowly spindle-shaped outline*, with dense short branches like bottle-brushes, but when fully expanded, the shoots become more parallel sided and broaden upwards. Branches are absent from the lower few nodes of the shoot, and also cease abruptly near its tip, giving the upper part of fully expanded shoots a markedly flat-topped, squared-off, appearance (although on young shoots, there is usually a single filamentous tip, no thicker than a branch, which is rapidly lost).

The branches themselves have a very characteristic structure. The internodes are 4–5 angled, and *each angle is two-ridged and armed with minute, pointed, saw-like teeth*, set outwards from the stem along each internode. Each pair of ridges enters the next nodal sheath above it and

Jan. Feb. Mar.	Apr.	May	Jun.	Jly	Aug.	Sep.	Oct. Nov. Dec.

Equisetum telmateia: a, Cotswold Hills, East Gloucestershire; *b–d*, Northumberland, showing surface detail.

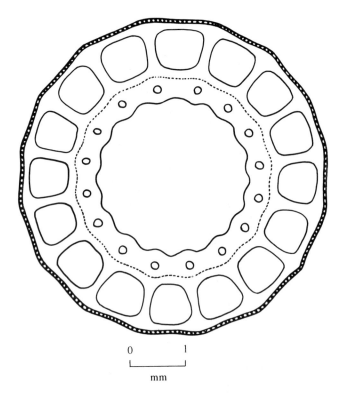

0   1
⎣_____⎦
mm

Equisetum telmateia, transverse section of main shoot internode.

continues through the sheath to give each tooth *a distinctly two-keeled appearance*. These features are clearly visible with a × 10 hand lens.

Plants normally produce separate, short-lived, unbranched fertile shoots. These are thick, ivory white and succulent, about 15–25 cm in height, with large, rather loose sheaths and long, brown, chaffy teeth, each shoot with a large ellipsoidal, terminal cone, 1.5–2.0 cm in diameter and 5.0–8.0 cm in length. The cone is greenish-cream coloured, with a brown tip. In northern England, these shoots emerge *during the second week of April* (earlier further south), discharge large quantities of green spores, and wither after about 2 weeks.

Variation Apart from the differences in size, the form of shoots does not normally vary very widely. Occasionally specimens with cones borne on the tops of otherwise vegetative shoots (so-called 'var. *serotinum* Milde') sometimes occur. Extraordinary freak shoots, which completely switch from whorled to spiral growth, have occasionally been found (see p. 481).

In some British and Irish colonies of *E. telmateia* shoots regularly occur in which either the cones have annuli which are part leaf-like or part sporangiophore-like on different radii (in some colonies), or which have serotinous cones on the otherwise vegetative shoots (in other colonies). Such serotinous cones can regularly bear organs which are variously intermediate between fertile and vegetative appendages (see diagrams on pp. 478–479). Even in normal cones, individual sporangiophores show wide variation in the arrangement of the vascular strands supplying them, although throughout, these are composed of an essentially dichotomous branching series in constant planes of successive orientation (see illustration on p. 480).

Equisetum telmateia, mature vegetative shoot outline: East Gloucestershire.

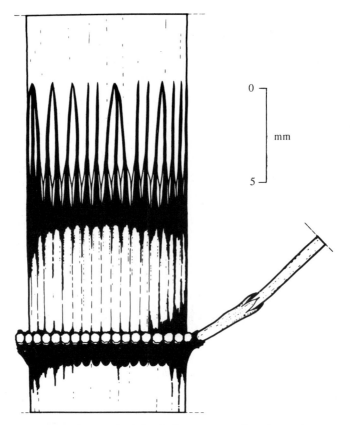

Equisetum telmateia, leaf sheath from mature plant, lower shoot.

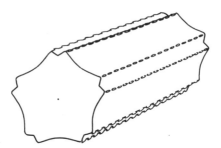

Equisetum telmateia, block diagram of branch internode geometry.

Detailed study of these standard and non-standard structures provides a variety of strong lines of evidence that the annulus, the positions of insertion of the sporangiophore whorls in the cone and the nodes of the vegetative portions of the plant, while adapted to specific modern roles, all represent units of the shoot that are essentially homologous with one another in their ultimate evolutionary origins. Further, the detailed anatomy of their appendicular organs still retains traces of a fundamentally dichotomous primitive vascular structure (Page, 1972*b*).

Technical confirmation Little needed. Plants have $n = 108$, $2n = 216$ chromosomes. Under the scanning electron microscope, the stomatal cell areas of the branch internodes have regular

Equisetum telmateia, emerging fertile shoots at the beginning of spore liberation: all Morpeth, Northumberland, mid-April.

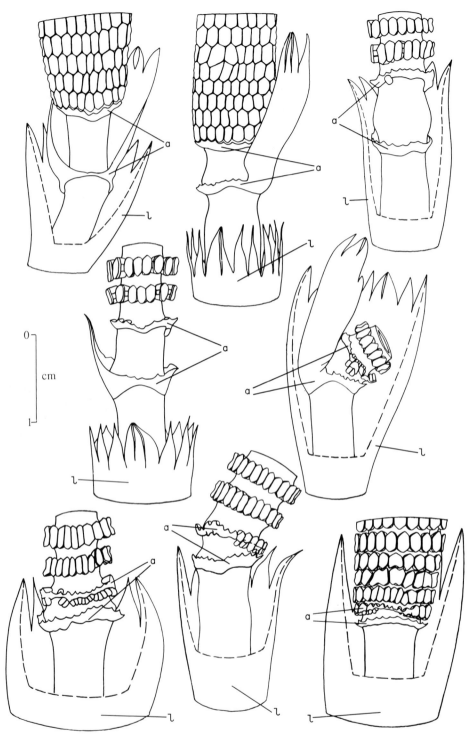

Equisetum telmateia, in occasional specimens of the cone-bearing shoots, multiple annuli (a) may occur, which are in part leaf-like, above the uppermost true leaf whorl (1), providing evidence that the true annulus is homologous with a normal vegetative node (after Page, 1972*b*).

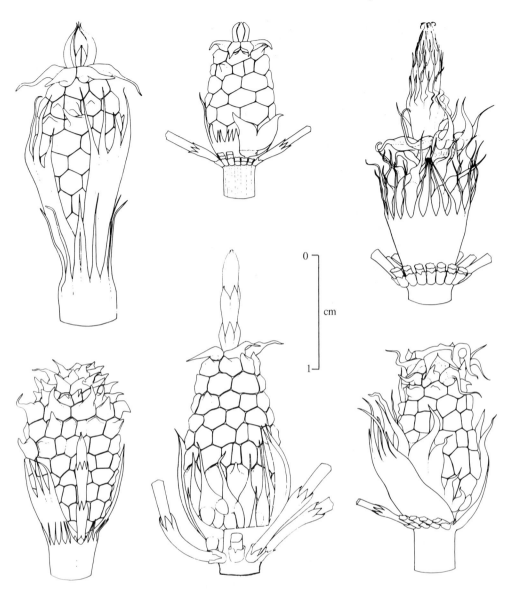

Equisetum telmateia, abnormal cones, when occasionally present on the tips of otherwise vegetative shoots, often show a tendency to reversion to vegetative axes, bearing numerous intermediate appendages. These are of use in the interpretation of the homologies between vegetative and fertile appendages (after Page, 1972*b*).

longitudinal ranks of elongated, finger-like projections (pilulae), very similar to those of *E. palustre* (q.v.), suggesting, with other evidence, existence of a close relationship between these two species.

Possible confusion The ivory-white main stem internodes and double-ridged (bi-angulate) branch internode angles are characters that separate this species and its only known hybrid

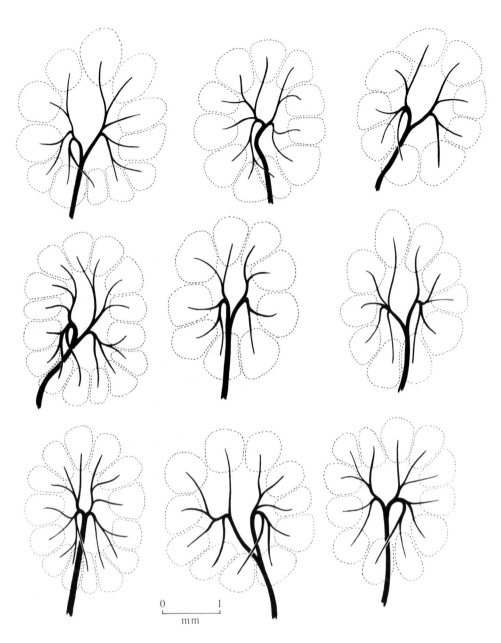

0 ___ 1
m m

Equisetum telmateia, arrangement of the vascular structure of the sporangiophores showing its essentially dichotomous structure (after Page, 1972*b*).

from other horsetails with regularly whorled branches. The larger size, entirely good spores in the cone and normally dimorphic shoot habit, separate *E. telmateia* from *E.* ×*font-queri* (q.v.), of which it is one parent. The fertile shoots differ from those of *E. arvense* in their much larger size and more massive cone.

The short-lived habit of these cone shoots, their whitish colour, succulent appearance,

Equisetum telmateia: *a–c*, occasional shoots occur in certain colonies in which the normal whorled growth pattern of the axis switches abruptly and irrevocably to a spiral construction: all East Gloucestershire (after Page, 1968).

curious shape, and rapid disappearance have undoubtedly led on numerous occasions to the presumption that they are some sort of fungus!

Field notes Great Horsetail is a widely distributed but sometimes local species which characteristically forms dense and extensive colonies, in which it is the vegetationally dominant plant. It occurs almost exclusively on deep clay ground which is more or less permanently irrigated and flushed with lime-rich seepage water. It is thus most abundant in areas of predominantly sedimentary rock, where seepage lines emerge between the junctions of limestones and natural underlying clay strata along the sides of shallow river valleys, its colonies often marking the lines of geological junctions of such strata. Particularly extensive colonies often mark the outcrops of some of the better-known clays, such as the Middle and Upper Lias of the Cotswold Scarp and the Yorkshire coast, the Kimmeridge clays of south-western England, London clays of the south-east and the Marl clays of the north-east Midlands. It also spreads to damp roadside and railway banks, where base-rich soils have been used, and rhizomes have probably been transplanted.

Equisetum telmateia is a warmth-loving species, restricted to Britain and Ireland to rather low-altitude habitats from sea-level to little over 250 m (about 800 ft). Extensive colonies sometimes occur on moist, often slumping, low coastal cliffs, especially where clay outcrops are common. But the tall, brittle shoots reach their maximum development where topographical features such as valleys and light tree overgrowth give better shelter and light shade. In such habitats, shoots nearly 2.5 m tall can occur.

Underground, there is an extensive development of running rhizomes, bearing whorls of pear-shaped, potato-like, storage tubers at the nodes. Such rhizomes can put on more than 2 m of horizontal growth through soft clay media in one season, and sometimes reach depths of over 4 m below the ground surface. Intimate contact with clays in lime-rich districts may well be essential not only for giving physical support to such large plants, but also in providing a source of silicates, which seem essential to the growth of all species of *Equisetum*, and on which the large annual aerial biomass of this species doubtless makes considerable demand.

The structure of the rhizomes, each about 1 cm in diameter, with internodes of 8–10 cm or more, and a ring of usually five large vallecular canals acting as air spaces, is extremely similar to that seen in *E. palustre*.

As a result of its preference for springline habitats, colonies of *E. telmateia* are nearly always on slopes. Towards the bottoms of such slopes, if water becomes sluggish or stagnates, *E. telmateia* dies out, sometimes becoming replaced by *E. palustre*, suggesting that Great Horsetail requires a much higher oxygen content in the groundwater than Marsh Horsetail. Towards the upper sides of such slopes, colonies are limited by increasing soil dryness. But it is usually along these drier upper margins that the short-lived cone-shoots appear in greatest numbers in spring.

The cone-shoots arise each from a specialised bud produced at the base of the stem of a previous year's vegetative shoot, and persist over winter, whilst new vegetative shoots arise in spring from fresh vertical rhizomes that push up from deeper, subterranean ones. Dates of emergence of the cone shoots seem remarkably constant in any one site from year to year, seldom varying by more than a few days from their time of emergence in previous seasons.

Other species associated with Great Horsetail commonly include Meadowsweet (*Filipendula ulmaria*), Lesser Celandine (*Ranunculus ficaria*), Bugle (*Ajuga reptans*), Primrose (*Primula vulgaris*), Wood Bitter-cress (*Cardamine flexuosa*), Marsh Thistle (*Cirsium palustre*), Wild Angelica (*Angelica sylvestris*), Hemlock Water Dropwort (*Oenanthe crocata*), Moschatel (*Adoxa moschatellina*), Ragged Robin (*Lychnis flos-cuculi*), Water Avens (*Geum rivale*), and sometimes other horsetails, particularly Marsh Horsetail (*Equisetum palustre*) in adjacent more stagnant pockets and Dutch Rush (*Equisetum hyemale*) on riverbanks in northern Britain. In addition, Hazel (*Corylus avellana*) is very frequently amongst the woody growth, providing light overhead shade. The edaphic preferences of Meadowsweet in particular, are, indeed, so similar to those of several of the more moisture-loving species of horsetail that in the field it often acts as a valuable botanical pointer to horsetail sites.

Equisetum variegatum
Schleich. ex Weber & Mohr

Variegated Horsetail

Preliminary recognition A small to medium-sized horsetail, with slender, often decumbent, filament-like unbranched evergreen stems bearing black-rimmed sheaths with conspicuous white-margined teeth.

Occurrence Widely spread through Britain and Ireland, never in abundance and usually as rather solitary clumps or scattered shoots in moist calcareous shingle, in a wide range of situations from coastal dune slacks to river and canalbanks, lake margins, mountain streamsides, erosion banks and base-rich alpine flushes.

Identification Plants have narrowly filamentous, evergreen shoots, up to about 4 mm diameter and 80 cm tall, though are very often very much smaller than this (see 'Variation'). They may be decumbent, but can be erect or largely prostrate, are of a dark *blue-green colour, evergreen, and feel rough to the touch*, especially if a fingernail tip is drawn gently along an internode. Often, some of the lowermost nodes adopt a *sinuously winding growth habit*. The green colour is broken at the nodes by sheaths, which carry a conspicuous black band and are topped by *short triangular teeth with narrow dark centres and very broad, papery-white margins*. These contrasting markings give the shoot a banded appearance that is characteristic even from some distance. *The white teeth are shaped like a broad gothic arch, and end in a* small, dark hair-like portion, which readily becomes eroded in the field and is usually lost in herbarium specimens. Sometimes, especially in plants growing near the sea, the nodes and internodes can become rather orange-tinged.

The tips of shoots end in small dark points or in very small cones, which are usually greenish in colour and black tipped, but can also be rather orange-tinged. The cones are slightly longer than broad, scarcely more than 3–4 mm diameter, and have a bluntly pointed tip (apiculum).

Variation Plants show considerable morphological plasticity, particularly in response to apical damage, when numerous lateral buds at or below ground level may develop into much finer filamentous shoots than are typical of the main stem. But in addition to this, there seem to be other differences between plants (especially in shoot size and erect versus decumbent habit) that are consistently maintained in experimental cultivation and show strong ecological and geographic correlations. These seem good ecotypes, and the majority of populations differ from one another at least in very small ways in such characters. There are at least two

Jan.	Feb.	Mar.	Apr.	May	Jun.	Jly	Aug.	Sep.	Oct.	Nov.	Dec.

Equisetum variegatum, var. *variegatum: a–b*, typical upland streamside/flush form, showing erect, ± filamentous growth, Upper Teesdale, Co. Durham; *c*, coastal sand-dune slack form, showing semi-decumbent growth, with very numerous, contorted branches arising from the base of each strong but easily damaged main shoot, North Devon. *Equisetum variegatum* var. *wilsoni: d*, typical deep-water, erect lake plant with extremely smooth internodes, Co. Kerry (courtesy A. Willmot). *Equisetum variegatum* var. *majus: e–f*, typical tall, erect lowland streamside plant, Co. Clare. Uniquely amongst native horsetail species, all these growth forms persist or are even enhanced in conditions of uniform cultivation, and are thus genetically determined (see the text).

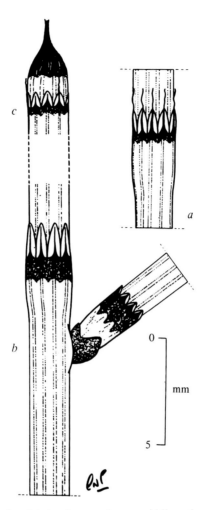

Equisetum variegatum, architecture of nodal sheaths: *a–c*, lower, middle and upper shoot, respectively.

well-marked groups of such ecotypes which are widespread, and a third, equally distinct and very local one, still known as herbarium material and each subsequently re-found in Ireland, all of which deserve to be treated as traditional taxonomic varieties. This is the only species of *Equisetum* in the British and Irish flora in which this seems to be the case.

Var. *variegatum* is the usual plant over much of Scotland, England and Wales, and probably occurs in some Irish stations too. Most plants produce shoots seldom more than 15–25 cm tall and about 2 mm or less diameter. Larger and smaller ecotypic forms occur within this range, some with a highly decumbent habit, others more erect. It includes the very prostrate-shooted forms of coastal sandhills often referred to as 'var. *arenarium*' in older County Floras. Var. *variegatum* occurs predominantly in dune slacks in coastal situations, and again in mountain river shingle and cut-offs, especially in the northern Pennines and Scottish Highlands, often in sites where there are concentrations of rare 'arctic-alpine' species. In addition, a few inland lowland stations occur in southern England.

Var. *majus* Syme is the more usual plant in many Irish habitats, with probably only a

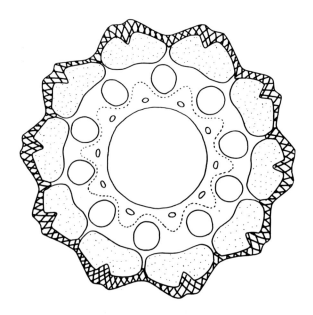

Equisetum variegatum, transverse section of main shoot internode.

handful of stations elsewhere. On the whole it is a much less variable, more vigorous, more robust plant, with shoots frequently over 40 cm high, some reaching 80 cm high and 4 mm in diameter, which grow strongly erect. These differences are maintained or even enhanced when transplanted via cuttings into experimental cultivation. It is predominantly a lowland plant of lake margins, river and streambanks, establishing also in banks of Irish canals, and surviving in more closed vegetation than is typical for var. *variegatum*.

Var. *wilsoni* Newman is a unique plant, described in the mid-nineteenth century from Co. Kerry, south-west Ireland, differing from all others known in having completely smooth stem internodes, which are also much less prominently angled: features which were described last century by Newman and which can be still confirmed from the harbarium specimens.

Possible confusion Small, unbranched forms of *E. palustre* (q.v.), which can look superficially similar to *E. variegatum*, are often taken for this species. The rough surface texture, and evergreen shoot as well as stomatal characters (see 'Technical confirmation') separate *E. variegatum* from all specimens of *E. palustre*. Smaller shoots of *E.* ˣ*trachyodon*, can approach this parent very closely indeed, and examination of young, emerging shoots in spring to inspect the characters of the fresh sheaths and teeth helps to distinguish them, whilst the presence of only good spores in the cones confirms *E. variegatum*.

Technical confirmation Plants have $n = 108$, $2n = 216$ chromosomes. The central canal is usually about 3/4 to 1/3 of the width of the stem. The ridges are broadly biangulate in section. On the surface, examination under a $\times 10$ hand lens shows that the stomata (visible as white dots) are in regular single vertical files (distinguishing them from the broad bands of irregularly scattered stomata of *E. palustre*, q.v.).

Field notes *Equisetum variegatum* is a widespread species of horsetail, with a geographic range from southern Greenland, Iceland, Britain and Ireland through nearly the whole of northern and central Europe south to the Alps, Pyrenees and mountains of Turkey across very much of northern Asia to Japan and Kamchatka, and throughout much of the higher latitudes of the North American continent from Alaska to Labrador.

In Europe, its main concentrations of stations have a distinctly arctic-alpine distribution of the various high mountain forms of this variable plant, with a scattered distribution of more lowland forms between. In Britain and Ireland, each of these forms are also present, with further additional ones, and the whole species range is consequently similarly mostly montane and especially in northern districts of both islands, with more scattered lowland and coastal populations elsewhere.

In these islands, *E. variegatum* (and especially var. *variegatum*) is a rather slow-growing horsetail, establishing in base-rich sites where moist shingle surfaces are regularly water-scoured and kept free from competing vegetation. Its thin, wiry, shallowly creeping rhizomes can spread considerable distances under and around boulders, doubtless helping plants to retain a hold in mobile sandy situations. Var. *majus* appears to be restricted to the margins of fast-flowing lowland streams, especially in Ireland, and var. *wilsoni* is entirely a lake-margin taxon, restricted to south-west Ireland. Var. *variegatum* occurs typically in a wide range of open, fairly basic, moist sandy sites, especially in gravel fans beside mountain streams, as well as at low altitude on a few sandy heaths in southern England and in sand dune slacks widely scattered around the coasts of Britain and Ireland. In Ireland in particular, it also occurs on lake shores and along canalbanks. It grows mostly in sands with a pH range of about 6.5–7.3 or higher, and where vegetation competition is not dense, due to winter flooding or mobility of the sand surface. Its habitats characteristically have much bare sand, and both in mountain habitats and in calcareous coastal dune slacks, similar species associates are often present. These commonly include scattered plants of Ribwort Plantain (*Plantago lanceolata*), Autumnal Hawkbit (*Leontodon autumnalis*), Meadowsweet (*Filipendula ulmaria*), Self-heal (*Prunella vulgaris*), Bird's-foot Trefoil (*Lotus corniculatus*), Common Dog-violet (*Viola riviniana*), White Clover (*Trifolium repens*), Daisy (*Bellis perennis*), Red Fescue (*Festuca rubra*), Jointed Rush (*Juncus articulatus*), Sedges such as *Carex flacca, C. lepidocarpa* and *C. capillaris*, Marsh Horsetail (*Equisetum palustre*). Lesser Clubmoss (*Selaginella selaginoides*) and the moss *Acrocladium cuspidatum*. More unusual species present may include Marsh Helleborine (*Epipactis palustris*) and orchids such as *Dactylorchis incarnata, D. fuchsii* and *D. purpurella*.

In upland northern Pennine sites of England and Highland sites of Scotland, Variegated Horsetail is also a frequent member of flush vegetation associated with base-rich springs, where there is a low percentage plant cover amongst bare gravelly stones, and a high pH of usually about 6.5–7.3. Associated species in such habitats commonly include dominant Yellow Saxifrage (*Saxifraga aizoides*), with various sedges, such as *Carex panicea, C. flacca, C. pulicaris* and *C. nigra*, Bog-rush (*Schoenus nigricans*), Lesser Clubmoss (*Selaginella selaginoides*, q.v.) and a large number of mosses, including *Drepanocladus revolvens, Ctenidium molluscum, Cratoneuron commutatum, Bryum pseudotriquetrum, Philonotis calcarea, Campylium stellatum* and *Scorpidium scorpioides*.

The disjunct geography of the native highland and lowland stations for *E. variegatum* presumably results in considerable genetic isolation. Elsewhere in more northerly latitudes, *E. variegatum* has been reported to be amongst the pioneer species of shingly conditions in areas of recent glacial retreat. Its ecology suggests that in Britain and Ireland it may have been amongst the earliest species to re-establish following post-glacial ice retreat (see p. 32). Its present ecotypic and regional differentiation could well be the result of a post-glacial history involving restriction to progressively more isolated open habitats as forest vegetation developed, with the present surviving populations genetically long-isolated from one another. The survival of predominantly different plants in Ireland from those of the arctic-alpine relict habitats of Scotland and northern England as well as the very variable but distinctive dune form may well be significant in this perspective. The plant, however, seems to be becoming increasingly rare in many sites today, and its future conservation status may need examination in both islands.

Equisetum [×]bowmanii C. N. Page   Bowman's Horsetail

(*Equisetum sylvaticum* L. [×]*E. telmateia* Ehrh.)

Preliminary recognition A medium to large-sized horsetail of bushy habit, looking like a more robust and more profusely branched form of *E. sylvaticum*, but with usually thicker shoots, paler main stem internodes and looser pale-green sheaths with longer, darker teeth.

Occurrence Plants of this parentage were described for the first time in 1988 from the New Forest in Hampshire (Page, 1988*a*). The hybrid is clearly rare, although is now known from more than one local station. It could presumably occur virtually anywhere within these islands within the sympatric portions of the ranges of the two parent species, although the habitats are limited in which the two species occur together.

Identification Shoots of *Equisetum* [×]*bowmanii* are usually 30–50 cm tall, erect, 2.5–4.0 cm diameter, succulent, mostly very pale green becoming nearly white in the lower part of the stem, *with a minutely rough surface* almost throughout, and a slender, usually nodding tip to each shoot. Internodes have usually about 8–14 fairly deep grooves, with sheaths (excluding teeth) 5–7 mm long, pale greenish-grey in colour, sometimes blackish above and below; with *teeth 4–6 mm long, slender, acute, deep brown with pale brown scarious margins, 2-ribbed, and frequently adhering in 2s–4s.* Internally internodes have a central hollow about 1/3 to 1/2 of the diameter of stem. Abundant branches are present throughout the upper 3/4 of most stems, in regular whorls, numerous, very slender, *spreading and drooping at their tips, mostly 4-angled, bright pale green in colour and bearing abundant further lateral branches of similar form and colour* in a more or less horizontal plane. Branch sheaths are pale, with dark-tipped triangular-acuminate teeth, and the branch internodes are flat-topped to subbi-angulate in section.

Cone-bearing shoots have not, as yet, been found at its site.

Overall, vegetative shoots of *Equisetum* [×]*bowmanii* are strikingly intermediate in size and morphology between those of *E. telmateia* (q.v.) and *E. sylvaticum* (q.v.) and present an initial appearance of a rather large and brighter-green version of *E. sylvaticum*. They resemble *E. telmateia* particularly in their large size, stem thickness and succulence, paleness of internode colour (especially on the lower parts of the stem), greenish-grey sheath colour (more deeply coloured than the adjacent internodes), long, narrow teeth and bright pale-green colour of the branches, forming the dominant colour of the plant. They resemble *E. sylvaticum* closely in the general habit of the plant, and especially in the long nodding tip to the main stem, the slenderness of the branches, which are themselves copiously branched, the minute roughness of the main stem internodes and the depth of the grooves, and the brown scarious margins to the main stem teeth, which are broad and tend to adhere loosely between many of the teeth. The stem in section has a central hollow in proportion to the stem diameter and carinal and vallecular canal number (about 8–14) approximately intermediate between those of both parents.

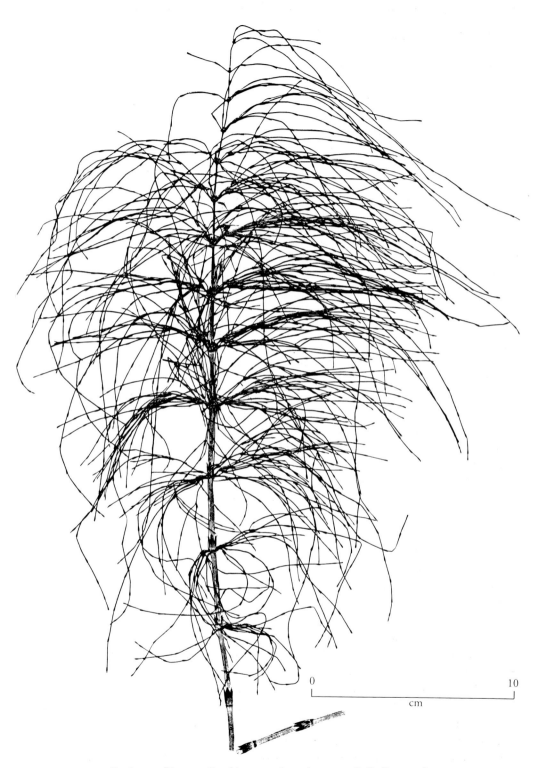

Equisetum ×*bowmanii*, mid-season shoot (courtesy R. P. Bowman).

Equisetum ×*bowmanii*, early season shoot.

Variation Mostly that of the size of the shoots, though in the wild, frequently damage to growing leaders through depradations of ponies results in many short broken stems of eventually somewhat bushier habit.

Possible confusion Overall, the succulent, pale-coloured shoots, main stem sheath characteristics and bright grass-green branch colour characteristic of *E. telmateia* combined with slender branches that are themselves regularly and copiously branched, produce a horsetail of quite unusual and particularly spectacular appearance. The combination of these characteristics

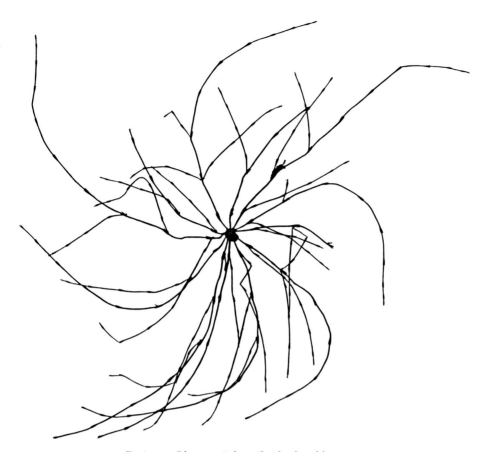

Equisetum ^x*bowmanii*, branch whorl architecture.

readily distinguishes this hybrid from both parents, as well as from other hybrids in *Equisetum* (Page, 1973*a*, 1982*a*).

Technical confirmation The combination of the above characters, and especially the details of the main stem sheaths together with branch angles, rough and pale-coloured main stem internodes should provide a reliable diagnosis. Spores are unknown.

Field notes *Equisetum* ^x*bowmanii* was described fairly recently (Page, 1988*a*) from material found by R. P. Bowman in July 1986 at Minstead in the New Forest, South Hampshire (*R. P. Bowman s.n,* holotype E). It was found in September 1986 as a single colony spreading over heathland road banks, grass verges and in adjacent short turf on wet clay soil. Plants occur in open sites, or amongst low *Rubus fruticosus, Ulex, Prunus spinosa* and *Pteridium aquilinum,* with some *Quercus robur* seedlings, in association with *Mentha aquatica, Cirsium palustre, Senecio aquaticus, Betonica officinalis, Prunella vulgare, Galium palustre, Lysimachia nemorum* and *Juncus conglomeratus. Equisetum arvense* occurs

nearby on the same roadbanks. The altitude of the site is approximately 35 m. The clay exposure is that of the Barton Clay of Eocene age.

The habitat of the plant in many ways represents an amalgamation of two types of habitat more typical of the two parent species: *E. telmateia* on damp, usually seepage-flushed, sloping clay banks, and *E. sylvaticum* in more acidic sites amongst heathland. Both parents are, indeed, present in several localities in the general area.

The bulk of the habitat of *E.* ^x*bowmanii* occurs along a section of road which was constructed as part of a road realignment about 22 years previously (*c.* 1974). It is not known whether the hybrid was present before this event, but it is possible that its occurrence here may be a direct result of the soil disturbance resulting from the activity, providing a site where gametophytes of both parents could establish nearby one another, in habitats initially free from competition of an established vegetation.

Equisetum ×dycei C. N. Page

Hebridean Horsetail

(*E. fluviatile* × *E. palustre*)

Preliminary recognition A small horsetail, with slender, unbranched or sparsely branched shoots and long, branchless terminal segments, looking like a smaller, weaker form of *E.* [×]*litorale.*

Occurrence Known from a small number of scattered stations in Scotland, in the Outer Hebrides (Harris and Lewis) and in Perthshire, but perhaps overlooked as a weak form of either parent or of *E.* [×]*litorale.*

Identification Plants have thin shoots, usually not much more than 2–3 mm in diameter and about 15–45 cm tall, which are of erect or semi-decumbent habit, and may have somewhat sinuous lower nodes. The shoots are either entirely unbranched throughout, or have a few, slender, whip-like, ascending branches from some of the nodes in the lower third of the shoot, above which the shoot continues as a long, slender, unbranched, tail-like portion. They thus somewhat resemble poorly branched specimens of *E. palustre*, and are of a similar green colour, with a small (about 9) number of shallowly rounded ridges and furrows. As in *E. palustre*, the sheaths have dark-coloured ochreolae around their bases when branches are present. The sheaths are, however, shorter in relation to their width than those of *E. palustre* (more like *E. fluviatile*), and the teeth have much narrower white margins, than those of *E. palustre*. The first internode of the branches is about the same length as the sheath, the branches 4–5 angled, and slender, similar to those of *E. fluviatile*. In section, the shoot internodes have a fairly small central hollow, about 1/3 to 1/2 of the diameter of the shoot, and the vascular bundles have individual endodermises (see 'Technical confirmation'). On old shoots, the sheaths often develop an orange tinge, characteristic amongst the deciduous horsetails of only *E. fluviatile* and its hybrids.

A few shoots occasionally bear cones on their tips. These cones are very small, dark coloured, and appear not be open fully.

Variation Even individual shoots on a single plant vary widely in size, habit and degree of branching, and seem clearly of highly plastic morphology, as are both parents.

Possible confusion The much smaller stem hollow distinguishes *E.* [×]*dycei* plants readily from *E. fluviatile*, whilst this feature, their much weaker stems and the sparse, thin, ascending

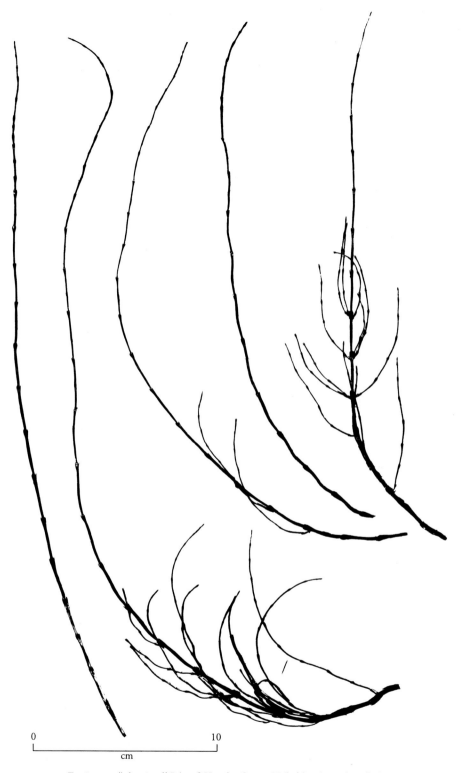

0   10
                     cm

Equisetum ×*dycei: all* Isle of Harris, Outer Hebrides (type locality).

branches, occurring only at a few of the very lower nodes, help to distinguish them from *E.* ×*litorale* (q.v.). They are most likely to be taken for a weak form of *E. palustre* (q.v.), from which the slightly larger stem, hollow, shorter sheaths and individual endodermises to the vascular bundles (see 'Technical confirmation') help to separate them.

Technical confirmation This is difficult without, from a fresh (not pressed) specimen, evidence of the small size of the central hollow and the endodermal pattern. The vascular bundles of the stem internodes are each surrounded by their own, individual endodermis (rather than a common one around the whole ring of vascular bundles). In the deciduous horsetails (subgenus *Equisetum*), this condition is unique to *E. fluviatile* and its hybrids.

Field notes *Equisetum* ×*dycei* is an elusive and possibly short-lived plant, first found on the Isle of Harris, Outer Hebrides, in 1962 (Page, 1963), where it formed a small colony for a metre or so along a single roadside ditch, on bare sandy mud. The colony grew about 100 m from a larger one of *E. palustre*, which is frequent in ditches in the vicinity. The hybrid colony was, by comparison, a weak one, and when revisited 5 years later, had disappeared completely following a minor road-widening scheme, which a more vigorous horsetail might well have survived. A second locality for this hybrid was, however, found, near the west coast of Lewis, Outer Hebrides, in a roadside ditch adjacent to a marsh. Here also occurred, partly intergrown with it, much more vigorous shoots of both parents as well as *E.* ×*litorale*, and on adjacent roadside banks, also *E. arvense*. When this site was visited again in 1979, the ditch had become choked with a dense growth of vigorous marsh vegetation, including all four other horsetails present, but there remained no sign of shoots of *E.* ×*dycei*. Other finds of shoots answering this description have, however, also been made by Duckett elsewhere in Scotland (see Duckett, 1979), in areas where both parents occur.

As with certain plants of the related *Equisetum* ×*rothmaleri* (q.v.), which shares the *E. palustre* parent, vegetatively more extensive (though seldom very large or vigorous) colonies of *E.* ×*dycei* occur in more constantly surface-scoured sites, which are also more or less frequently inundated with moving fresh water. These sites include the gravelly shores of various loughs and the margins of streambanks especially through the west of Ireland (where I have seen it occurring as thin colonies in such sites from Sligo to Kerry) and in usually lower mountainside flushes in upland areas, especially in Highland Scotland and notably in some of the partly ultrabasic flushes of the Cabrach area of Aberdeenshire. In the former it is fluctuations in stream and lake levels and the waves generated that serve to scour the semi-mobile gravelly surfaces; in the latter, it is the semi-toxic nature of the terrain that presumably itself serves to reduce long-term plant competition and thereby promote the relatively long-term persistence of this undoubtedly weak hybrid. In each of these areas, scattered shoots of *E.* ×*dycei* and *E.* ×*rothmaleri* may be co-associated, and occasionally (notably in western Ireland) survive (probably briefly) into denser vegetation, sometimes including nearby much more extensive and vigorous colonies of *E.* ×*litorale*, which are probably their main lake-shore as well as occasional ditch-line competitor.

The number of stations for this hybrid seem extremely few in contrast to the abundance and widespread occurrence of the two parents. Plants have been synthesised in artificial culture by Duckett (see Duckett & Page, 1975), but those in wild habitats at least, seem weak growers that are poorly competitive, and probably confined to a relatively brief existence mainly in habitats opened by human beings. They may, therefore, have considerable difficulty in persisting in the wild, apparently rapidly succumbing to encroachment of other vegetation, including that of their parents, and this may account for the very few records of this hybrid, between two relatively common species.

Equisetum ×font-queri Rothm.   Skye Horsetail

(*E. palustre × E. telmateia*)

Preliminary recognition A medium- to large-sized, vigorous and freely branched, colony-forming horsetail, with spindle-shaped shoots, resembling overgrown plants of *E. palustre*, with the ivory-white internodes and long sheath-teeth of *E. telmateia*.

Occurrence First recognised as a British plant in 1968 from a single vast colony on the Isle of Skye, western Scotland, occupying a wide range of wet, flushed terrain (Page, 1973*a*) and subsequently in a number of scattered localities within the ranges of both parents (see map). It could yet be found elsewhere.

Identification The pencil-thick shoots grow to 65 cm or more in height, and have *conspicuous, rather succulent-looking, ivory-white internodes*, with a moderate number (about 8–12) of inconspicuous, shallowly rounded ridges and furrows. The shoots always feel firm and smooth to the touch, and they are *distinctly intermediate in appearance between their very different-looking parents* (q.v.).

They resemble *E. telmateia* in their large size, fairly thick stems, ivory-white stem internodes and vivid green branch colour, and in having *bi-angulate branch ridges, stem and branch sheaths with two-ribbed teeth, and relatively long, somewhat flexuous teeth on the main stem sheaths*. The diagram shows the characteristic branch whorl architecture of the plant.

They resemble *E. palustre*, however, in their comparatively slender, spindle-shaped, overall shoot outline, long terminal branchless portion to the shoot, spreading to subascending branches, which are *more robust and rounded than those of* E. telmateia, stems with a relatively small central cavity, later coning season and monomorphic shoots. Most cone-bearing shoots thus closely resemble the vegetative ones, with profuse whorls of green branches. The cones mature in July and are ovoid in shape, slimmer than those of *E. telmateia* and pale green in colour, with chestnut-brown tips.

Variation Most shoots show only minor variation, mainly in size and degree of branching. Occasional cone-bearing shoots occur which are tall and slender, but have few or no branches. Between the two known British stations, there is only minor variation in detail between the plants, those from Worcestershire having even whiter internodes, longer free teeth to the stem

Jan.	Feb.	Mar.	Apr.	May	Jun.	Jly	Aug.	Sep.	Oct.	Nov.	Dec.

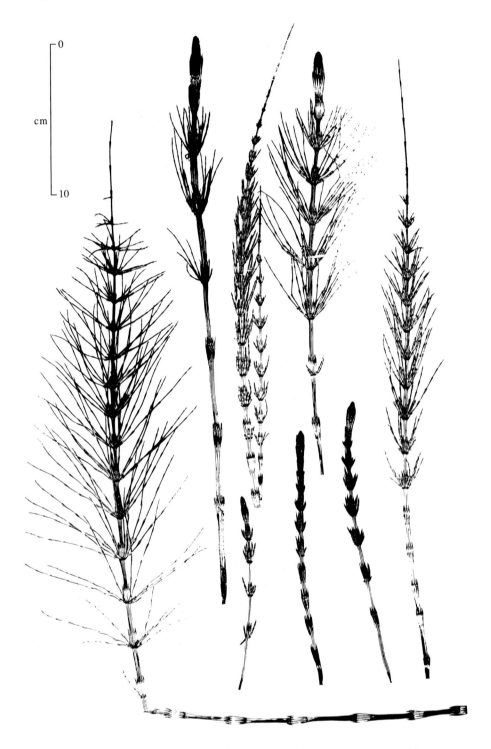

Equisetum ˣ*font-queri:* all Trotternish Peninsula, Isle of Skye, Inverness-shire.

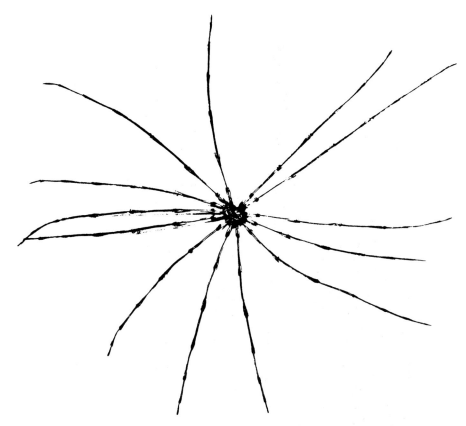

Equisetum ×*font-queri*, branch whorl architecture.

internodes and less bi-angulate branch internodes (Roberts & Page, 1979). This seems true also in comparison with plants in the other known overseas stations (see 'Field notes').

Possible confusion The combination of above characters distinguishes *E.* ×*font-queri* from either parent. The robustness of the shoots, ivory-white internodes, typically more abundant, more robust, branches, bi-angulate branch internodes and two-ribbed teeth to the stem and branch internodes, distinguish it from the other known *E. palustre* hybrids (*E.* ×*rothmaleri* and *E.* ×*dycei*) as well as from the unrelated *E.* ×*litorale* (q.v.).

Technical confirmation Stomata are present on the internodes of the main stem (absent here in *E. telmateia*). Under the scanning electron microscope, the sculpturing of the branch internodes shows stomatal cell areas bearing regular ranks of long, interlocking pilulae, virtually indistinguishable from the unique appearance of both parents in this character. All cones of the hybrid appear well filled (exceptional amongst hybrid horsetails). Some sporangia and cones contain only spores which are misshapen, lack good elaters, are colourless, small and obviously abortive. Others contain a proportion (varying from 5% to 50% in different sporangia and different cones) which are relatively well formed, are larger, are green in colour, and possess well-developed elaters.

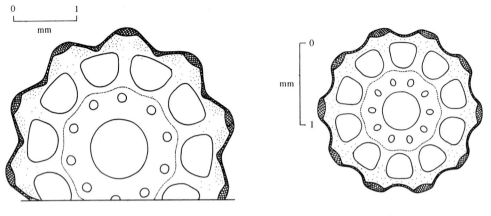

Equisetum [×]*font-queri*, transverse section of main shoot internodes.

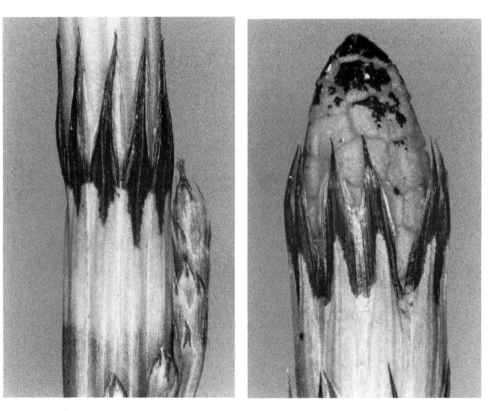

Equisetum [×]*font-queri*, mid shoot and cone sheath detail: Skye, June.

Field notes In its extensive Skye habitat, *E.* [×]*font-queri* is the dominant plant where base- and mineral-flushed peaty ground occurs over Jurassic limestones, in a region of high and frequent rainfall. Here, it is clearly extremely successful, occurring over 5 km² of terrain as a nearly continuous stand, becoming particularly abundant in most damp depressions, irrigated slopes, flushes, seepage lines, screebank, drainage channels, ditches and stream-banks, and vigorously colonising roadside verges and rubble along most of a 3 km stretch of road. Its ecological aggressiveness and adaptability are impressive, the plant appearing to have successfully displaced both parents throughout its area (they are otherwise common nearby), and itself occupying not only all intermediate situations but also virtually the whole range of habitats of both.

In one of its other British stations, the hybrid occurs on the site of a disused Worcestershire railway, forming a dense stand for about 75 m along both sides of the top of an artificial embankment, which is capped by a layer of coarse limestone chippings. Neither parent is present nearby, and because of the unlikely suitability of this habitat for them, it seems possible that the hybrid established from fragments originally brought in from elsewhere in Worcestershire, or adjoining counties, during the nineteenth century construction of the railway (Roberts & Page, 1979). It may yet still occur elsewhere.

In the Skye colony, the frequency of production of cone shoots has been found to differ widely from year to year, varying from a few some years to up to half the shoots in other years. The occurrence of well-filled cones so regularly in a colony of hybrid *Equisetum*, and, furthermore, the occurrence within them of a proportion of spores which appear to be well filled and possess regularly formed elaters (see 'Technical confirmation'), are unique amongst all known *Equisetum* hybrids. Whether such spores are viable and have played any role in the spread of the plant, or whether its area has been established entirely through vegetative spread, remains to be studied.

The success of the plant in all British habitats underlines its considerable adaptability. It is of interest to note that the same hybrid has recently been recognised in at least three French localities where it is also vigorous, and in one of which, it also occurs on a railway embankment near to a river (Badré & Prelli, 1980). The number of known stations for *E.* ˣ*font-queri* make it the second most common hybrid amongst the deciduous horsetails, after *E.* ˣ*litorale*. Significantly perhaps, the parents of both crosses are in each case plants that have been suggested, on morphological evidence (Page, 1972*a*), to be closely interrelated species pairs.

Equisetum ˣlitorale
Kühlew. ex Rupr.

Shore Horsetail

(*E. arvense* × *E. fluviatile*)

Preliminary recognition A medium- to large-sized, vigorous-growing, colony-forming winter-deciduous horsetail, with upright green stems bearing regular whorls of fairly short branches throughout approximately the middle half of the stem.

Occurrence Widely scattered throughout Britain and Ireland in well over 150 known stations, especially in high-rainfall districts near coasts. It grows in damp water-edge habitats mainly around ponds, lakes, and canals, and on coastal foreshores around marshy ground, along ditches, and where streams meet the sea, almost always in sites which have been disturbed or created by humans, and mostly at low altitude.

Identification Plants produce *thickets of tall, slender stems*, up to 80–100 cm or more in height, slightly slimmer stemmed and tougher than those of *E. fluviatile*. They are also *more densely furnished with branches through much of the central portion* of the shoot, although the *uppermost portion of the shoot always remains long and unbranched*. The internodes are smooth to the touch, and *when squeezed gently between finger and thumb, yield slightly to pressure but*

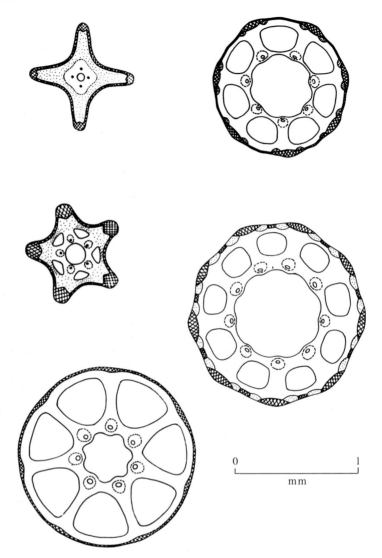

Equisetum litorale, transverse sections of branch and main shoot internodes.

are elastic, reflecting the size of their central hollows (about half the stem diameter), and contrasting with that of both parents.

Shoots are of a glossy yellow-green colour, and *often inherit the orange-tinge of* E. fluviatile around the nodes. The sheaths of the branch teeth usually inherit the *very spreading-tipped tooth character of the* E. arvense *parent*, and the channelled furrows to the branch internodes. All shoots of this hybrid are typically strong and vigorous. Those which bear cones are similar to the vegetative ones, but bear on their tips small, blackish, barrel-shaped cones, which seldom arise far out of the uppermost leaf sheath. These cones never completely open, and contain only abortive spores.

Variation Shoots are of extremely variable appearance, depending much on habitat. Those in drier situations are often shorter and more bushily branched, superficially approaching *E.*

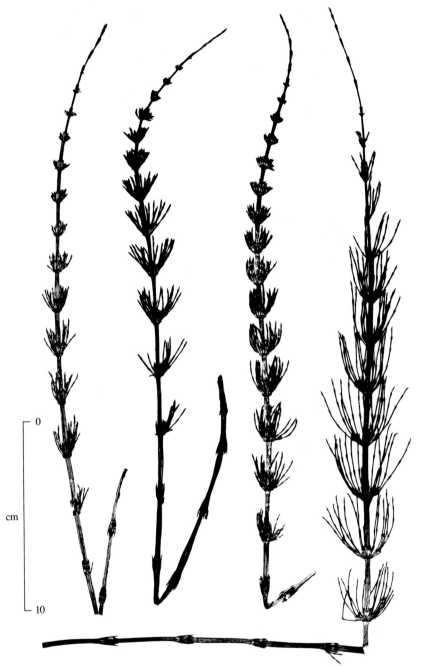

Equisetum *litorale: a*, South Uist; *b*, Isle of Harris; *c*, Isle of Lewis, Outer Hebrides; *d*, Kirkcudbrightshire.

arvense in form, those in wetter situations taller and more slender with shorter branches, more similar to *E. fluviatile*. Transplant experiments, however, show that all forms revert to almost identical hybrid plants, virtually intermediate between the parents, when cultivated alongside, the field differences being entirely environmentally induced.

Possible confusion Extreme forms are very likely to be taken for either parent, but the intermediate size of the central hollow and the resulting characteristic elastic feel of the internode when squeezed are absolutely diagnostic of all specimens. This feature also separates them readily from luxuriant shoots of *E. palustre*, which can sometimes look quite similar. The size and vigour of the shoots distinguishes them readily from those of *E. ×rothmaleri* (q.v.), which shares the same *E. arvense* parent, and from *E. ×dycei* (q.v.), which shares the same *E. fluviatile* parent. The green internodes separate *E. ×litorale* from *E. ×font-queri* (q.v.), which is superficially similar in size and vigour but has neither parent in common.

Technical confirmation Plants have $n = 108$, $2n = 216$ chromosomes, show irregular meiotic chromosome pairing, and have entirely abortive spores (small, misshapen and lacking elaters and chlorophyll). Internodal vascular bundles in the shoot have individual endodermises (in contrast to the *E. arvense* parent which has a single common endodermis), but always have tubers on the rhizomes (in contrast to *E. fluviatile*).

Field notes Although the apparent absence of *E. ×litorale* in many parts of Britain and Ireland may be due to lack of recognition, it is undoubtedly the most common *Equisetum* hybrid, and seems to occur with greatest concentration in areas of high and frequent rainfall.

It is a vigorous hybrid, which normally colonises ground which is too wet for *E. arvense* and too dry for *E. fluviatile*, although there is some overlap in habitats, particularly with the latter. Colonies may occur in either the presence or the absence of one or both parents nearby, and grow mainly in unsheltered lowland conditions, often initially pioneering ones that are edaphically open.

It occurs mainly around the edges of lakes and reservoirs, in river shingle, in roadside ditches, and in damp coastal slacks. It thrives where there is deep, moist, mineral-rich (often base-flushed) muddy or sandy ground, and in many stations occurs where the habitat is, or has been, in some way disturbed. In Ireland, many of its sites are lake-margin ones. In the Scottish Inner and Outer Hebrides and Shetlands, where this hybrid occurs frequently, its habitats are mostly either in roadside ditches, coastal marshes, or in damp, flushed patches around stream mouths in coastal machair vegetation, where irrigated, lime-rich, wind-blown sand lies over deep, moist peat. This hybrid seems to be especially frequent in its number of sites around lake margins in central and western Ireland and in some parts of the Lake District.

Although in many ditch and lakeside habitats, *E. ×litorale* may grow as a pioneer species with few associates, in coastal marshes and flushes in machair vegetation it can survive amongst considerable flowering plant competition. In such habitats, commonly associated species include Yellow Flag (*Iris pseudacorus*), which is often the dominant plant, Lesser Spearwort (*Ranunculus flammula*), Marsh Bedstraw (*Galium palustre*), Water Forget-me-not (*Myosotis scorpioides*), Ragged Robin (*Lychnis flos-cuculi*), Common Spotted-orchid (*Dactylorchis fuchsii*), Sharp-flowered Rush (*Juncus acutiflorus*), and sometimes Marsh and Water Horsetails (*Equisetum palustre* and *E. fluviatile*.

Plants have occasionally been seen to establish in lake-margin shingle from washed-up shoot fragments, probably originally broken up by water fowl. Experiments in cultivation indicate that small stem fragments of this hybrid root with greater facility on damp sand than do those of either parent, suggesting that dispersal of such fragments may be an important method of local secondary establishment of colonies around still waters or along streamcourses in the wild.

Overall, however, *E. ×litorale* is the most frequent and widespread horsetail hybrid known, and the distances between very many of its stations probably indicate that it has also re-formed anew very many times. This abundance and its general vigour everywhere seem significant in relation to the view that its two rather different-looking and ecologically dissimilar parents appear, on grounds of the micromorphological structure of the sporophytes (Page, 1972*a*), to be a taxonomically closely interrelated species pair, and supports the view proposed by Page & Barker (1985) that the most vigorous and most ecologically successful hybrids in *Equisetum* are those between taxonomically close but ecologically divergent pairs of parents.

Equisetum ˣmildeanum Rothm.   Milde's Horsetail

(*Equisetum pratense* Ehrh. ˣ *E. sylvaticum* L.)

Preliminary recognition A medium-sized rather weak and slender horsetail with a nodding tip and sparsely branched horizontally spreading branches drooping at the tips, the whole often somewhat bilaterally compressed appearance looking like a more profusely branched form of *E. pratense*, with sheath teeth closely intermediate in form and number between those of the parents.

Occurrence Plants of this parentage were recognised for the first time only recently in the central and eastern Scottish Highlands (Page, 1988*a*). The hybrid is clearly rare, but could occur anywhere within the mainly Scottish sympatric portions of the ranges of the two mainly northern parent species.

Identification Shoots of *Equisetum* ˣ*mildeanum* are mostly 10–35 cm high, erect, slender, 0.8–1.5 mm diameter, the tip usually nodding; with main stem internodes very pale green and minutely rough with perpendicular setose spicules. Internodes of all shoots have mostly 8–15 fairly deep grooves and angular ridges, and main stem nodes with sheaths (excluding teeth) which are 3.0–4.5 mm long, somewhat loose and usually greenish-grey, becoming pale brown to blackish above, and teeth which are 2–3 mm long, 1-ribbed, slender, acute, mostly straight, their central portions deep brown above, grey-green at the base, their margins scarious, or occasionally slightly flexuous, narrow near the tips of the teeth but broadening rapidly downward. The teeth frequently adhere in 2s and 3s for their full length, especially in the lower parts of most specimens.

The central hollow of each main stem internode is about 1/2 of the diameter of the stem. Whorled branches numerous, up to 5.5 cm long, throughout much of the central portion of most shoots and arising from chaffy, pale-brown basal ochreolae, whorled but somewhat asymmetrically arranged to give the shoot *an often somewhat bilaterally compressed appearance*, spreading horizontally and drooping at the tips, very slender, all 3-angled, with acute uni-angulate ridges and deep, flat-sided, V-shaped furrows each with a narrow lengthwise basal channel, mid-green throughout, each usually bearing a small number of regular short, spreading secondary branches of similar form and colour, the branch sheaths pale green, their teeth triangular-acuminate, green throughout.

Fertile shoots have not been found and are possibly normally absent in most colonies.

Overall, shoots of *E.* ˣ*mildeanum* are strikingly intermediate in morphology between those of *E. pratense* and *E. sylvaticum*, and resemble a rather more profusely branched, though weaker, version of *E. pratense*.

Variation Shoots in each known locality vary mainly in their profusion and length of branches, degree of secondary branching, and the slenderness of the main shoot. All, however, maintain their distinctive interspecies intermediacy.

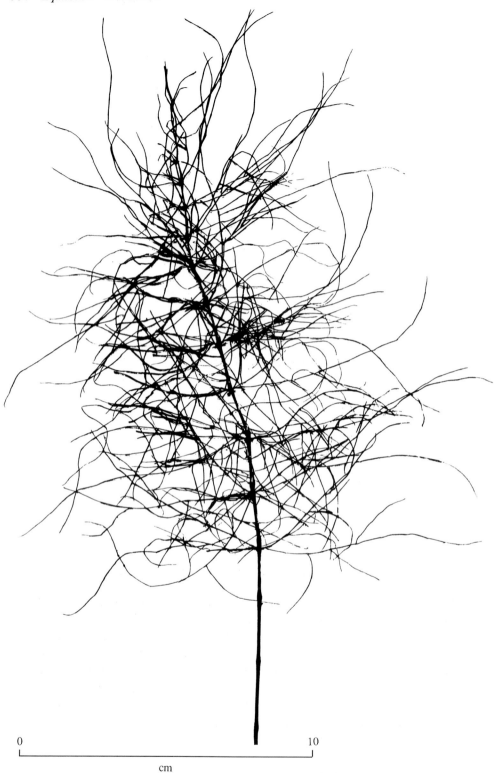

0   10

cm

Equisetum [×]*mildeanum*, mid-season shoot: Mid Perthshire.

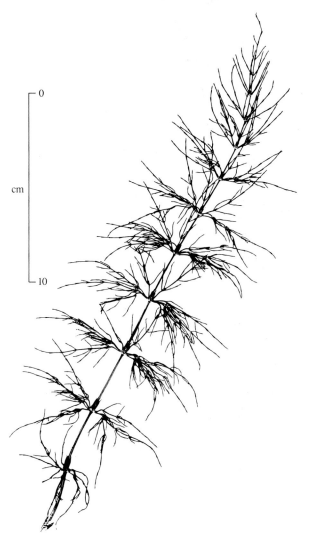

Equisetum [×]*mildeanum*, early-season shoot: East Perthshire.

Possible confusion Shoots differ most conspicuously from *E. pratense* in having secondarily branched branches, somewhat looser main stem sheaths which bear fewer teeth with broader paler margins and which also typically adhere together. They differ from *E. sylvaticum* in the more bilaterally compressed (and hence not quite radially symmetric) habit of the overall shoot and in its shorter, somewhat fewer, branchlets to the branches, the paler ochreolae to the branchlet bases, concolorous branch node teeth, and straighter darker-centred more acute teeth to the main stem sheaths, which are also less cup-shaped, especially in the lower stem.

The combination of these characters also distinguishes *E.* [×]*mildeanum* from *E.* [×]*bowmanii* (q.v.), which shares the *E. sylvaticum* parent, as well as from other hybrids in *Equisetum*.

Technical confirmation The combination of the above characters and additionally the spiculate structure of the main stem internode surface is unique to this hybrid.

Field notes Although long-known in the mountains of northern and central Europe (e.g. Rothmaler, 1944, who reported it from 'Hassia, Holsatia, Saxonia, Prussia, Silesia, Rossia'), *E.* ×*mildeanum* was found only recently to be a member of the British Flora (Page, 1988*a*). I have confirmed the identity of Rothmaler's type (at Jena, J), and the match of these British specimens to it. Throughout northern and central Europe and Asia the two parents of this hybrid have very wide sympatric ranges, and their ecology is such that the two species frequently meet. Here, so far as is known, it is restricted in occurrence to the Scottish Highlands, where three native stations are known with certainty in central and eastern Perthshire between 625 and 860 m (2000 and 2900 ft). All these localities occur also within the sympatric portions of the parental ranges within these islands, and others may well occur at lower altitude in western Scottish sites, where both parents descend nearly to sea-level.

All three confirmed locations are in montane sites where both parents occur (*E. sylvaticum* frequently and *E. pratense* sometimes in reasonable abundance) in the general vicinity or in the immediate neighbourhood of the hybrids. All are in fairly base-flushed (and in one site wet) habitats over mica schist rocks, and two of the sites are amongst boulders or scree. In one locality, plants grow with *Carex nigra, C. echinata* and *Sphagnum recurveum*. In the other two localities, plants grow in sites where, in addition to both parents, other pteridophytes of upland basic screes are frequent, including *Dryopteris expansa, Polystichum aculeatum* and *P. lonchitis*.

The occurrence of each of these hybrids in upland areas in sites well away from river banks suggests that translocation of vegetative material from one locality to another has, in this case, been unlikely, and that each of its sites of occurrence marks a site of independent formation of this hybrid *de novo*.

Although the differences observed in the field are of virtually exact intermediacy between *E. pratense* and *E. sylvaticum*, it could be argued that, because of the environmental plasticity of all horsetails, these could have been dismissed as merely a growth form of especially the latter. However, small rhizome fragments sampled non-destructively from the wild have established successful populations in experimental cultivation which, after 8 years, still clearly maintain the original morphology and its interspecies intermediacy. It is thus clear that these wild native Scottish plants of *E.* ×*mildeanum* are indeed genetically determined individuals of genuine hybrid origin between these two distinctive northern horsetail parents.

Only one other hybrid involving *E. pratense* in its parentage is known in the literature. This is *E.* ×*montellii* Hiitonen (*E. arvense* L. ×*E. pratense* Ehrh.), reported from Finland, Sweden and the Canadian Arctic (Duckett & Page, 1975), and which could also yet be found in the British Isles.

Equisetum *moorei* Newman   Moore's Horsetail

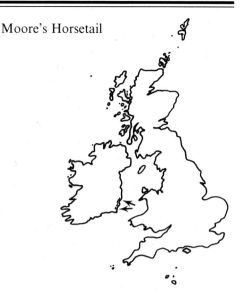

(*E. hyemale* × *E. ramosissimum*)

Preliminary recognition A fairly tall, upright-growing, unbranched horsetail, looking like a much
more slender, more yellow-green form of *E. hyemale*, but of less rough texture and with longer
sheaths.

Occurrence Confined to Co. Wicklow and Co.
Wexford in south-east Ireland, where it occurs in
scattered stations along about 50 km of coast from
Ardmore Point to Wexford Harbour. Here, it
grows in the immediate neighbourhood of the sea,
on low sandy and clayey banks, descending nearly
to the high-tide mark.

Identification Plants form scattered clumps, which give rise to 40–60 cm tall, *quill-thick,
yellow-green stems, which are normally completely unbranched* and arise usually in fairly close
groups of 3–4 together. They die down either partially or completely (depending on exposure)
in winter. The internodes are *only slightly rough* to touch, notably less so than those of *E.
hyemale*, and are *not inflated.* They have a *slightly translucent quality when freshly emerged.* The
sheaths are notably long – about twice as long as broad – and are more loosely fitting and
funnel shaped than those of *E. hyemale.* They end in narrow dark-brown or black teeth, each
drawn into a *fine tapering, straight or flexuous tip.* The teeth *usually persist until the shoots are
fully expanded, but then frequently break away at their base* (usually so in herbarium material),
leaving a truncate, slightly crenate, upper margin to the sheath.

The sheaths are at first a uniform green, but develop a black band around the base of the
teeth, followed by one around the base of the sheath. In the upper nodes of the shoot, the
central portion usually turns white as the shoot ages, but in the lower nodes, the whole sheath
usually becomes black.

Small, black, oblong-ovoid cones are produced only occasionally, and contain only abortive
spores.

Variation Very little.

Possible confusion Likely to be confused only with *E. hyemale* (q.v.) or *E.* *trachyodon* (q.v.),
which shares the same *E. hyemale* parent, but the yellower-green shoot colour, longer sheaths,
and distinctly non-winter-hardy shoots distinguish *E.* *moorei* from both.

Jan. Feb. Mar.	Apr.	May	Jun.	Jly	Aug.	Sep.	Oct. Nov. Dec.

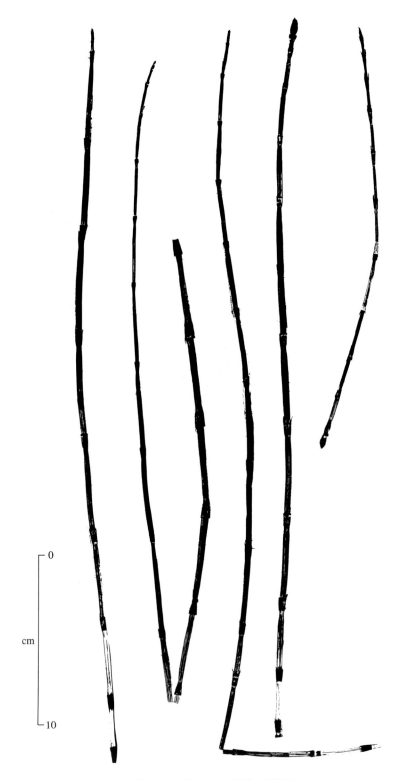

Equisetum ×*moorei: all* Co. Wicklow.

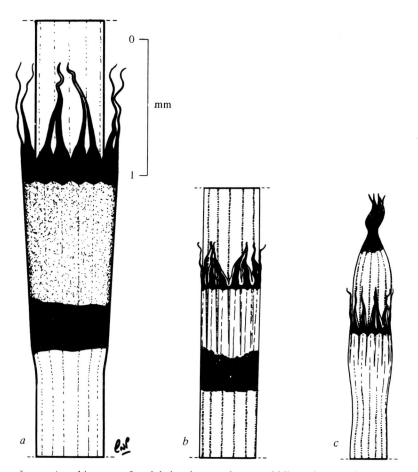

Equisetum ˣ*moorei*, architecture of nodal sheaths: *a–c*, lower, middle and upper shoot, respectively.

Technical confirmation Plants have $2n = 216$ chromosomes, showing irregular meiotic pairing. In cross-section of the stem, each ridge is broadly rounded, and non-biangulate, and bears a single row of rounded tubercles (not bi-angulate, with two tubercle rows as in *E.* ˣ*trachyodon*).

Field notes Plants mostly colonise bare moist, sandy and clayey banks, in full exposure, in places so close to the high-tide line, that they must receive appreciable salt spray. Their occurrence over so small an area and in such maritime habitats poses some interesting questions regarding their survival and origin. Transplants from cuttings have proved completely hardy out of doors as far north as Edinburgh, even through one particularly severe winter, unharmed. Even its least hardy parent, *E. ramosissimum*, has maintained itself as an introduced clone at a similar latitude in eastern England. It thus seems likely that restriction of this plant to a maritime environment is unlikely to be due to lack of hardiness. Gradual erosion of the ground surfaces by the sea, plus the limiting effects on potential competing species of salt-laden air, are, however, factors which probably have been of importance in its survival.

Experimenta have shown that shoot fragments can survive up to 10 days floating in seawater, without impairing their ability to root subsequently as cuttings on damp sand. Longshore vegetative dispersal of fragments by sea currents might thus have contributed towards the concentrations of stations for this hybrid along so limited an area of coast.

The present absence of the *E. ramosissimum*

parent in a totally native state from the British or Irish native floras seems indicative of a likely past wider range of this now southern central European species, as the more probable reason to account for the initial occurrence of the hybrid in Ireland, although long-distance transport and establishment cannot be ruled out. Very likely, the hybrid has outlasted its parent through some historic period of greater coldness, through its enhanced vigour and inheritance of increased hardiness from the very cold-hardy *E. hyemale* parent.

Equisetum ˣrothmaleri
C. N. Page

Ditch Horsetail

(E. arvense × E. palustre)

Preliminary recognition A small- to medium-sized, colony-forming horsetail, with slender, upright stems, looking like duller, paler-green shoots of *E. palustre*, with broader overall outlines, much fewer, smaller cones and conspicuously more angled branch internodes.

Occurrence Known only from the Isle of Skye, western Scotland, where a single extensive colony was discovered in roadside ditches and adjacent marshy fields in 1971 (Page, 1973*a*), and a second smaller colony in a similar habitat in 1979.

Identification Plants have shoots usually not more than 25–50 cm high and 2–3 mm broad, which look strongly intermediate in appearance between the parents.

They resemble *E. arvense* in having the teeth of the sheaths on the main shoot with *only narrow scarious margins*, the longest branches typically near the middle of the shoot, the first internode of the branch *about as long as or longer than* the adjacent stem-sheath, the branches with *prominent acute ridges separated by deep furrows* (usually not more than four), *spreading teeth to the branch-sheaths*, and dull, non-shining branch surfaces.

They resemble *E. palustre* in typically having a rather *long branchless terminal portion to the stem*, branches which are *mostly ascending and upward-curving, a gradually tapering outline* to the upper part of the shoot, *black-tipped teeth* on the branch sheaths, small vallecular canals and central cavity, and *cones always borne on tops of ordinary green vegetative shoots*.

The cones are small (4–9 mm long and 2–3 mm broad), ovoid-cylindrical, and black in colour.

Possible confusion The combination of the above characters distinguishes *E. ˣrothmaleri* from either parent. *Equisetum ˣrothmaleri* can be distinguished from *E. ˣlitorale* (*E. arvense × E. fluviatile*) by its smaller size, more ascending branches, darker ochreolae, and by the transverse section of the stem, which shows a much smaller central cavity (about 1/4 of the stem width) and a single common endodermis. *Equisetum ˣrothmaleri* is distinct from *E. ˣfont-queri* (*E. palustre × E. telmateia*) in its smaller size, green stem-internodes, stem and branches with simple ridges and single ribs to the teeth. It is less easy to distinguish from *E. ˣdycei* (*E. fluviatile × E palustre*) but it can be separated by its smaller size, more abundant and more

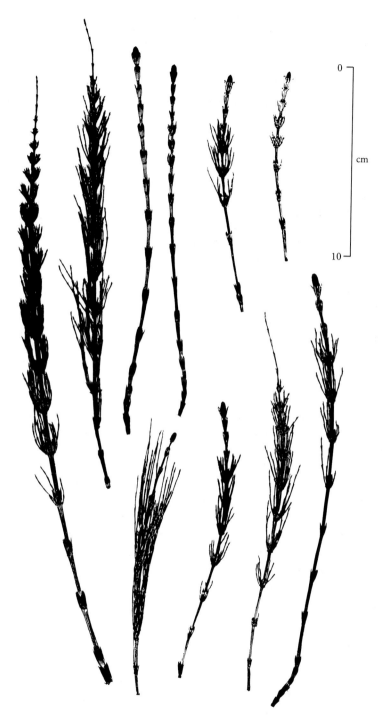

Equisetum ˣ*rothmaleri: all* Trotternish Peninsula, Isle of Skye (type locality).

regular branching, more deeply furrowed branches with more prominent ridges and a longer first internode, a single common endodermis in the stem, and the lack of any orange-brown tinge to the internodes or sheaths of the main stem (a characteristic of only *E. fluviatile* and its hybrids in the deciduous-stemmed horsetails, and *E. variegatum* and *E.* ×*trachyodon* in the evergreen ones).

Variation Shoots in the known localities all present a remarkably uniform appearance.

Technical confirmation The small cones and spores, which are entirely misshapen and colourless, lack all but the most rudimentary elaters and are obviously abortive, confirm its hybrid status. The single common endodermis in the stem internodes confirms it is not *E.* ×*litorale* or *E.* ×*dycei*.

Field notes So far as is known, *Equisetum* ×*rothmaleri* is entirely confined in Europe to Britain and Ireland. The two originally known nearby localities for this hybrid at the north end of the Trotternish Peninsula on the Isle of Skye occur in sites of slightly flushed, herb-rich, old marshy peaty pasture, in both sites predominating in (and perhaps having originated within) shallow, muddy roadside ditches. The habitats are in areas of high rainfall, more or less permanent ground moisture, near to the coast, and within a very few feet of sea-level. The larger colony also spreads a short distance into the surrounding marshy fields, occurring as rather sparsely scattered shoots amongst relatively dense, herb-rich marsh vegetation.

As with the related *Equisetum* ×*dycei* (q.v.), which shares the *E. palustre* parent, further exploration since the original discovery has shown *E.* ×*rothmaleri* to be rather more extensively widespread in these islands, but like *E.* ×*dycei*, to have a highly diffuse and strongly discontinuous overall range.

Vegetatively more extensive (though seldom very large or vigorous) colonies of *E.* ×*rothmaleri* occur in a number of more or less constantly surface-scoured sites, which are also ones which are frequently or regularly inundated with moving fresh water. These sites include the gravelly shores of various loughs and the margins of streambanks especially through the west of Ireland (where I have seen it occurring as thin colonies in such sites from Sligo to Kerry) and in usually lower mountainside flushes in upland areas, especially in Highland Scotland and notably in some of the partly ultrabasic flushes of the Cabrach area of Aberdeenshire. In the former it is fluctuations in stream and lake levels and the waves generated that serve to scour the semi-mobile gravelly surfaces; in the latter, it is the semi-toxic nature of the terrain which presumably itself serves to reduce long-term

plant competition and thereby promote the relatively long-term persistence of this undoubtedly weak hybrid.

In each of its known areas, scattered shoots of *E.* ×*rothmaleri* and *E.* ×*dycei* may be co-associated, and occasionally (notably in western Ireland) survive (probably briefly) into denser vegetation, sometimes including nearby more extensive and vigorous colonies of *E.* ×*litorale*, which are probably their main lake-shore as well as occasional ditch-line competitor.

As with both *E.* ×*litorale* and *P.* ×*rothmaleri*, in habitats such as lake margins, it is certainly possible that secondary vegetative spread of *E.* ×*dycei* may take place widely from local establishment of small vegetative stem and eroded rhizome fragments spread either by ducks or geese or by natural wave action and longshore drift. But, as too with both *E.* ×*litorale* and *E.* ×*rothmaleri*, the highly discontinuous overall distribution of the majority of known sites of *E.* ×*dycei* in both Britain and Ireland seems strongly indicative of a polytopic and polychronic origin, by *de novo* hybridisation between its parents, at most, if not all, of its known major sites.

The main associates of *Equisetum* ×*rothmaleri* in the two stations include Yellow Flag (*Iris pseudacorus*), Meadowsweet (*Filipendula ulmaria*), Lesser Spearwort (*Ranunculus flammula*), Marsh Marigold (*Caltha palustris*), Marsh Cinquefoil (*Potentilla palustris*), Ragged Robin (*Lychnis floscuculi*), Marsh Lousewort (*Pedicularis palustris*), Yellow Rattle (*Rhinanthus minor*), several orchid (*Dactylorchis*) and rush (*Juncus*) species, Bottle and Common sedges (*Carex rostrata* and *C. nigra*), and Marsh, Water and Shore Horsetails (*Equisetum palustre, E. fluviatile* and *E.* ×*litorale*), the former in particular quantity. Indeed, the apparent extraordinary suitability of the terrain for success of species of *Equisetum* locally is demonstrated not only by the frequency and extent of the above taxa, but also by the occurrence nearby of an unusually extensive stand of Great Horsetail (*E. telmateia*), with other stands of Shore Horsetail (*E. ×litorale*) and Skye Horsetail (*E. ×font-queri*) only a few kilometres away.

Equisetum [×]trachyodon A. Braun   Mackay's Horsetail

(*E. hyemale* × *E. variegatum*)

Preliminary recognition An upright growing or decumbent, infrequently branched, evergreen horsetail, looking like overgrown shoots of *E. variegatum* but with longer, darker sheaths and teeth.

Occurrence A local, but widely scattered hybrid. It is known in over two dozen stations throughout Ireland, mainly in wooded glens, and in one English and five Scottish stations, where it is mainly a plant of damp areas in calcareous coastal dunes and of flushed sandy, calcareous river banks. The English and most of the Scottish localities have been only recently reported, and the plant could well occur in similar habitats elsewhere.

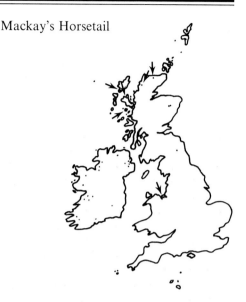

Identification Shoots arise singly or in clumps, are decumbent or erect and 30–60 cm (sometimes up to 90 cm) high and 3–5 mm wide, though are mostly much smaller than this in English and Scottish stations – see 'Variation'. The shoots are *dark green*, but sometimes have the orange tinge inherited from the *E. variegatum* parent (q.v.), as well as the tendency of some of the lower nodes to adopt a sinuously winding growth habit. *Shoots persist usually into a second season, but old internodes do not become inflated as do those of* E. hyemale. The internodes vary widely in roughness between localities, but have about 8–14 rather shallow furrows, separated by *acute-angled ridges, each of which is longitudinally grooved (bi-angulate)*, with a row of small tubercles arranged along each angle.

The sheaths and sheath-teeth are particularly characteristic (see diagram on p. 515). The sheaths are fairly close fitting, slightly funnel shaped and are *longer than broad* (more so than in either parent). They are at first *pale green with a narrow black band near the top, but during the course of their first season, become mostly black* (particularly on the side of the shoot most exposed to light). Unlike *E. hyemale*, *the teeth persist on the shoot after its initial expansion*, and mostly remain throughout the first season. *The teeth are long, black, and narrowly tapering. Near their junction with the sheath they usually spread outwards slightly before turning upwards for most of their length*, although their extreme tips may also become spreading. They have

Jan.	Feb.	Mar.	Apr.	May	Jun.	Jly	Aug.	Sep.	Oct.	Nov.	Dec.

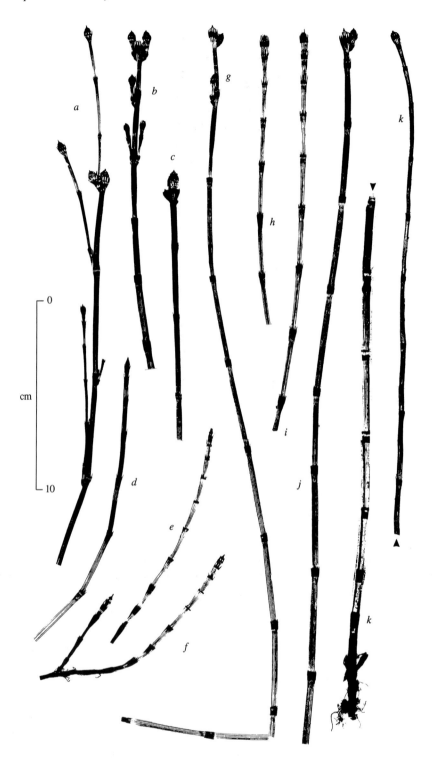

Equisetum ×*trachyodon: a, c*, West Donegal; *b*, West Sutherland; *d, e–f*, South Harris, Outer Hebrides; *g*, Wirral Peninsula, Cheshire; *h–j*, Isle of Skye, Inverness-shire; *k/k*, Co. Antrim.

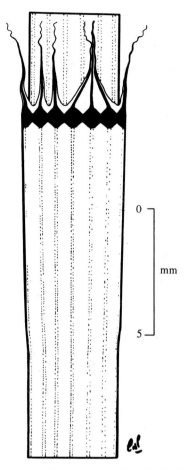

Equisetum [×]*trachyodon*, architecture of nodal sheath, lower shoot.

scarious margins of variable width, but these are usually conspicuous *only near the base of the tooth and are seldom as wide or as white as those of* E. variegatum.

After overwintering, most teeth become irregularly broken off, and are eventually completely shed from the top of the sheath. During their second year, the sheaths often develop a characteristic *ash-grey or sometimes whitish band around their tops, leaving only a narrow black apical band, and the sheaths often become vertically split (as in* E. hyemale*).*

Shoots are mostly unbranched during their first season, but *scattered branches often arise during their second year*, especially in shoots that have suffered apical damage during the first winter or which previously bore a cone. They thus often become more freely branched than do those of either parent. The branches are like diminutive shoots, and smaller ones can approach the appearance of *E. variegatum* quite closely.

Small, apiculate cones are borne singly on some of the stems, each cone about 4–5 mm long, black in colour or sometimes orange-tinged, usually remaining half within the uppermost sheath, and containing only highly misshapen, entirely abortive, spores.

Variation Most of the Irish material is more distinct from *E. variegatum* than it is elsewhere, with taller, more erect stems, longer black sheaths, and quite persistent teeth with scarcely any white margin. The Irish plants are fairly constant in these characters, and are quite easily recognised, whilst those elsewhere are mostly smaller and much more variable, and hence more often overlooked. These differences presumably reflect differences in parentage on the *E. variegatum* side (q.v.), the larger and more constant var. *majus* probably being most often involved in the Irish *E.* ˣ*trachyodon*, whilst various ecotypes of the smaller and more variable var. *variegatum* are probably involved in crosses arising in England and Scotland.

Possible confusion Extreme specimens can be confused with either parent, but, in addition to the abortive spores, the characters of the sheaths and teeth elaborated above are usually diagnostic. It has also been confused with *E.* ˣ*moorei*, which shares the common *E. hyemale* parent, but the darker green colour, bi-angulate ridges, more rapidly blackening sheaths and wintergreen habit, distinguish *E.* ˣ*trachyodon*.

Technical confirmation Plants have $2n = c.$ 216 chromosomes, showing irregular pairing at meioisis. The paired rows of tubercles (rather than a single broad row in *E.* ˣ*moorei*) confirms *E.* ˣ*trachyodon*.

Field notes Like many hybrid horsetails, *E.* ˣ*trachyodon* has a predominantly western distribution, occurring especially in areas of mild climate and frequent rainfall. It is a lime-loving plant, growing in England and Scotland near moist areas in lime-rich (often largely shell-sand) dunes, and in moist, base-flushed sandy river banks. It seems to inherit the high silica requirement of its *E. hyemale* parent (q.v.). In its most notable northern Irish stations, and in at least one Scottish station, it occurs in base-flushed areas on the sides of sheltered wooded glens. Its habitats are mostly at low altitude (up to 120 m in Scotland), and many are near the sea.

It thrives particularly in sites where competition is low, especially where this state is maintained by surface erosion or continual addition of wind-blown sand. But from such spots it can maintain itself, at least temporarily, as the habitat changes, or can spread into adjacent, more closed, turfy vegetation, albeit in more depauperate and scattered form. In its extensive and diffuse Cheshire sand-dune locality, it occurs in a community which includes Early Forget-me-not (*Myosotis ramosissima*), Lady's Bedstraw (*Galium verum*), and Creeping Bent-grass (*Agrostis stolonifera*), becoming slowly overwhelmed by a sand-accumulation community dominated by Marram Grass (*Ammophila arenaria*) and Creeping Fescue (*Festuca rubra*). It maintains itself for a while amongst the latter, but can come away vigorously again in damp patches which have been burned (Barker, 1979). In its Inner and Outer Hebridean localities, it occurs in moist areas of wind-blown shell-sand and persists in peaty turf in coastal machair and riverine localities, and moist-slack grassland. Associated species in these herb-rich turf communities commonly also include *Festuca rubra* and *Galium verum*, as well as several lime-indicating herbs, including Silverweed (*Potentilla anserina*), Birdsfoot-trefoil (*Lotus corniculatus*), Thyme (*Thymus drucei*), Purging Flax (*Linus catharticum*), Daisy (*Bellis perennis*), Primrose (*Primula vulgaris*), Kidney Vetch (*Anthyllis vulneraria*), Lesser Clubmoss (*Selaginella selaginoides*), and the lichen *Peltigera apthosa* (Page, 1979c). Its occurrence in Sutherland in a *Carex panicea-Campylium stellatum* flush and *Molinia-Myrica* mire (Ferreira, 1980), and on Skye on a collapsed clay riverbank with rich-meadow species such as Meadowsweet (*Filipendula ulmaria*), Meadow Buttercup (*Ranunculus acris*), Zigzag Clover (*Trifolium medium*), and Common Spotted Orchid (*Dactylorchis fuchsii*), underline this plant's wide ecological adaptability in rich, basic conditions.

Experiments have shown that small fragments of shoot can take root very readily when placed on damp sand, even after floating for as much as 10 days in either freshwater or seawater. There thus seems a considerable potential for local dispersal and re-establishment by such means in the wild. In this connection, it is interesting to note that in at least three of its localities, it may well have done so, occurring at several stations along the length of a single river in Skye, in similar unconnected habitats on both sides of a broad river estuary in the Outer Hebrides, and scattered along at least 1.3 km of coastline in Cheshire.

Equisetum [×]willmotii C. N. Page Willmot's Horsetail

(*Equisetum fluviatile* L. [×]*E. telmateia* Ehrh.)

Preliminary recognition A tall slender horsetail looking like a more robust and more profusely branched form of *E. fluviatile*, but with usually thicker shoots, paler main stem internodes and pale-green sheaths with long black teeth.

Occurrence Plants of this parentage were described for the first time only recently from Co. Cavan, Ireland (Page, 1995*a*). The hybrid is clearly rare, but could occur virtually anywhere within these islands within the sympatric portions of the ranges of the two widespread and locally frequent parent species.

Identification Shoots of *Equisetum* [×]*willmotii* are strong growing and semi-dimorphic, with those of both types tall and bearing whorled green branches, but those of the purely vegetative shoots with longer and more profuse branches and a slender, erect, unbranched tip. Overall, stems are mostly 65–125 cm high (vegetative mostly 75–125 cm, the fertile mostly 65–105 cm), erect, robust, succulent, 5–8 mm diameter, with main stem internodes ivory-white in colouration throughout the lowermost parts of most stems, pale green in the upper parts and usually becoming blackish towards the very base. Internodes of all shoots are smooth with numerous (mostly 16–20) very shallow grooves, and main stem nodes with sheaths (excluding teeth) which are 5.5–6.5 mm long, as wide or slightly wider than long, tightly fitting (becoming somewhat loose in dried material), usually greenish-grey throughout or with brown length-wise markings alternating with the position of the teeth which are 3–5 mm long, hence nearly as long as the sheaths, shallowly 2-ribbed and rather flat-topped in section at the base, acutely acuminate, straight or occasionally slightly flexuous, with a long, acutely tapering dark-brown central portion and narrower greenish-grey wings at the base broadening rapidly downwards. The teeth frequently adhere in pairs for their full length, especially in the lower parts of most specimens.

The central hollow of each main stem internode is very large, 3/4 to 9/10 the diameter of the stem. Whorled branches are very numerous, simple, straight, *c.* 6–14 cm long, and adopt a spreading-ascending angle at about 30–45° from horizontal (those on purely vegetative shoots often the longer and more ascending), forming strongly verticillate whorls throughout much of the central portion of most shoots (more extensive and densely so on purely vegetative shoots). All branches are very slender (*c.* 4–5 mm), bright-green throughout, 4–5 angled, the angles acute, simple to slightly flat-topped, ornamented with a conspicuous row of prominent, rounded, forward-pointing teeth, the furrows rounded or slightly grooved; the branch sheaths are long, their teeth long, usually straight or slightly flexuous, mostly green.

Fertile shoots are similar to, or slightly shorter than, those of the vegetative, but with usually shorter, sparser and more spreading branches, the shoot terminating in a short but broad

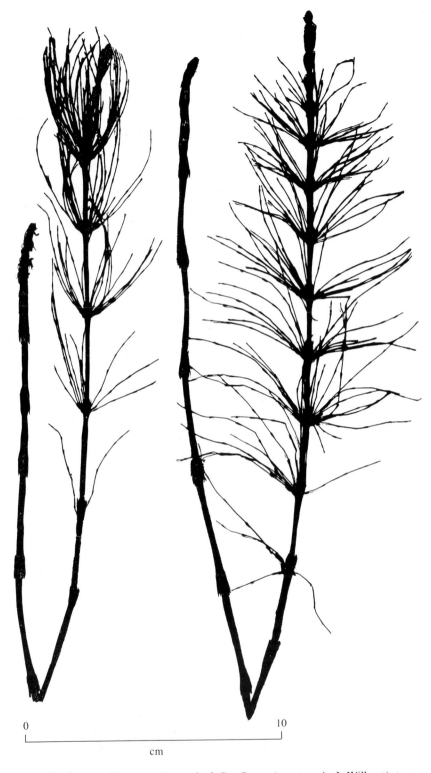

Equisetum ˣ*willmotii*, mid-season shoots: *both* Co. Cavan (courtesy A. J. Willmot): type.

unbranched portion of usually 3–4 pale nodes with large chaffy sheaths subtending an ovate cone up to 1.5 cm long. The spores are abortive.

Overall, shoots of *E.* ×*willmotii* are strikingly intermediate in morphology between those of *E. fluviatile* and *E. telmateia*, and resemble a rather taller and stouter version of *E.* ×*litorale* (*E. arvense* × *E. fluviatile*) with brighter-green branches which are more ascending, more abundant per whorl, more ridges to the main stem internodes, paler internode colouration and greyer-green sheaths and teeth of the *E. telmateia* type, and branch internode morphology approaching the bi-angulate type and conspicuous ridge ornamentation of *E. telmateia*.

Shoots of this hybrid thus differ conspicuously from *E. fluviatile* especially in the ivory-white colouration of the main stem internodes throughout the lowermost parts of most stems, the green-grey colouration of their sheaths, the conspicuous long dark teeth to the main stem sheaths, the prominent ridge ornamentation to all branch internodes, and the size and chaffiness of the sheaths subtending the cone. They differ from *E. telmateia* in their overall more slender and delicate main shoot with very much larger central hollow, their greener main stem internodes in the upper part of the stem, their less conspicuously 2-keeled teeth to the main stem and branch sheaths, less bi-angulate branch internode ridges with less acutely dentate ornamentation, and the presence of cones on the tops of otherwise tall, regular, green, branch-bearing shoots.

Variation Shoots in the single known locality vary mainly in their profusion and length of branches, both being greater on the vegetative than the cone-bearing shoots. There also seems to be notable variation in cone size between shoots.

Possible confusion The cones borne regularly on the tops of the vegetative shoots, and the large size of the cones plus their abortive spore content distinguishes *E.* ×*willmotii* from either parent. The combination of a more profuse branching habit plus ivory-white internodes (at least in the lower shoot) and branch internode ridge ornamentation characteristic of *E. telmateia* combined with a much more delicate shoot architecture and very large central hollow to the main stem internodes of the vegetative portions of the shoot characteristic of *E. fluviatile* are useful features in the separation of this hybrid from both *E.* ×*font-queri* and *E.* ×*bowmanii*, both of which share the *E. telmateia* parent, and from *E.* ×*litorale* and *E.* ×*dycei*, both of which share the *E. fluviatile* parent.

Technical confirmation Spores are abortive. The combination of the above characters and especially the *E. telmateia*-like sheaths combined with an anatomy showing large central hollow to the main stem internodes should provide a reliable diagnosis.

Field notes *Equisetum* ×*willmotii* was described only recently (Page, 1995a) from material found by A. J. Willmot in July 1984 near the Black River north of Dowra, Co. Cavan, Ireland (*A. J. Willmot* no. 1313 E & DBN). The habitat is a roadside bank of tall herbs with a few bushes, near to the river. Here, the hybrid is recorded to be locally abundant and grows with *E. palustre* and *E. arvense*.

Equisetum fluviatile is widespread in Ireland, and *E. telmateia* too, although more sparsely. Although the origin of *E.* ×*willmotii* as the cross between these two species at first sounds an unusual parentage, both are species known elsewhere to enter into hybridisations with other species of *Equisetum*, and thus that they should cross together is, in this respect at least, not totally unexpected. Indeed, the occurrence of *E.* ×*willmotii* in Ireland is the third hybrid known to have formed with *E. telmateia* as one parent. The other two are *E. palustre* × *E. telmateia* (*E.* ×*font-queri* Rothm.; Page, 1973a), and *E. sylvaticum* × *E. telmateia* (= *E.* ×*bowmanii* C. N. Page; Page, 1988a), It is also the third known hybrid with *E. fluviatile* as one parent, the others are *E. arvense* × *E. fluviatile* (= *E.* ×*litorale*) and *E. fluviatile* × *E. palustre* (= *E.* ×*dycei* C. N. Page), q.v.

As *E. fluviatile* and hybrids with it are known to be capable of spread along river banks by natural (or even artificial) translocation of vegetative material from one locality to another (e.g. Page & Barker, 1985), it might be of interest to search for the possibility of further stations for this so-far unique Co. Cavan plant further along the course of the same river.

Bibliography

Allen, D. E. (1969). *The Victorian Fern Craze.* London: Hutchinson.

Alston, A. H. G. (1940). Notes on the supposed hybridisation in the genus *Asplenium* found in Britain. *Proceedings of the Linnean Society of London*, **152**, 132–44.

Backhouse, I. (1853). *Pseudathyrium flexile. Phytologia*, **4**, 974–5.

Badré, F. & Prelli, R. (1980). New records of *Asplenium* and *Equisetum* hybrids in France. *Fern Gazette*, **12**, 115–17.

Balfour, J. H. (1902). Botanical excursion made by Professor J. H. Balfour – the years 1846–1878 inclusively. *Notes from the Royal Botanic Garden, Edinburgh*, **11**, 418–19.

Barker, M. A. (1979) *Equisetum* × *trachyodon* in Cheshire, new to the English flora. *Fern Gazette*, **12**, 59–60.

Baroutsis, J. S. & Gastony, G. J. (1978). Chromosome numbers in the fern genus *Anogramma*. II. *American Fern Journal*, **68**, 3–6.

Bell, P. R. (1985). The essential role of the Pteridophyta in the study of land plants. *Proceedings of the Royal Society of Edinburgh*, **86B**, 1–4.

Bennell, A. P. & Henderson, D. M. (1985). Rusts and other fungal parasites as aids to pteridophyte taxonomy. *Proceedings of the Royal Society of Edinburgh*, **86B**, 115–24.

Bennert, H. W. & Ficher, G. (1993). Biosystematics and evolution of the *Asplenium trichomanes* complex. *Webbia*, **48**, 743–60.

Bennert, H. W., Jager, W. & Theren, G. (1982). Spore characters of taxa within the *Asplenium adiantum-nigrum* complex and their systematic significance. *Berichte der Deutsche Botanisher Gesellschaft*, **95**, 297–312.

Benoit, P. M. (1964). The two types of *Asplenium trichomanes*. *Nature in Wales*, **9**, 75–9.

Benoit, P. M. (1966). Some recent work in Wales – the *Polypodium vulgare* aggregate. *British Fern Gazette*, **9**, 277–82.

Birks, H. J. B. (1973), *Past and Present Vegetation of the Isle of Skye. A Paleo-ecological Study.* Cambridge: Cambridge University Press.

Birks, H. J. B. (1976). The distribution of European pteridophytes: a numerical analysis. *New Phytologist*, **77**, 257–87.

Blasdell, R. F. (1963). A monographic study of the fern genus *Cystopteris. Memoirs of the Torrey Botanical Club*, **21**, 1–102.

Bonharmont, J. (1972). Origin de la polyploidie chez

Asplénium ruta-muraria L. *Bulletin et Transactions Botaniques de la Société Naturelle Belgique*, **42**, 375–83.

Boudrie, M., Garraud, L. & Rasbach, H. (1994). Discovery of *Dryopteris* × *brathaica* in France (Dryopteridaceae: Pteridophyta). *Fern Gazette*, **14**, 237–24.

Bower, F. O. (1908). *The Origin of a Land Flora.* London: Macmillan.

Bower, F. O. (1929) The evolutionary relations of the British Ferns. *London Naturalist*, 1929, 15–29.

Bower, F. O. (1935). *Primitive Land Plants.* London: Macmillan.

Brown, D. F. M. (1964). A monographic study of the fern genus *Woodsia. Beiheft zur Nova Hedwigia, Weinheim*, **16**, 1–154.

Brown, R. W. (1993). Ticks and Lyme disease. *Biologist*, **40**, 4.

Brown, R. W. (1994). Bracken and the ecology of Lyme disease. In *Bracken: An Environmental Issue*, ed. R. T. Smith & J. A. Taylor, pp. 116–19. Aberystwyth: International Bracken Group.

Butler, T., Baker, J. G. & Syme, J. T. B. (1881). *Asplenium germanicum* Weiss. *Report of the Botanical Society and Exchange Club of the British Isles*, **1**, 38–9.

Callaghan, T. V., Svensson, B. & Headley, A. (1986). The modular growth of *Lycopodium annotinum. Fern Gazette*, **13**, 65–6.

Camus, J. M. (ed.) (1991). *The History of British Pteridology 1891–1991.* London: British Pteridological Society.

Caulton, E., Keddie, S. & Dyer, A. F. (1994). The incidence of airborne spores of Bracken, *Pteridium aquilinum* (L.) Kuhn in the rooftop airstream over Edinburgh, Scotland, U.K. In *Bracken: An Environmental Issue*, ed. R. T. Smith & J. A. Taylor, pp. 82–9. Aberystwyth: International Bracken Group.

Clapham, A. R. (1969). *Flora of Derbyshire.* Derby: County Borough of Derby Museum and Art Gallery.

Clapham, A. R., Tutin, T. G. & Warburg, E. F. (1962). *Flora of the British Isles*, 2nd edn. Cambridge: Cambridge University Press.

Clausen, R. T. (1938). A monograph of Ophioglossaceae. *Memoirs of the Torrey Botanical Club*, **19**(2), 1–177.

Clowes, F. (1860). *Lastrea remota. Phytologist*, **4**, 227–9.

Connolly, A. P. & Dahl, E. (1970). Maximum summer temperature in relation to the modern and

Quaternary distributions of certain arctic-montane species in the British Isles (*Cryptogramma*). In *Studies in the Vegetational History of the British Isles*, ed. D. Walker & R. West, pp. 159–23. Cambridge: Cambridge University Press.

Coombe, D. E. & Frost (1956). The nature and origin of the soils over the Cornish serpentine. *Journal of Ecology*, **44**, 605–15.

Cooper-Driver, G. A. (1976). Chemotaxonomy and phytochemical ecology of Bracken. *Botanical Journal of the Linnean Society*, **73**, 35–46.

Cooper-Driver, G. A. (1980). The role of flavenoids and related compounds in fern systematics. *Bulletin of the Torrey Botanical Club*, **197**, 116–27.

Cooper-Driver, G. A. & Haufler, C. H. (1983). The changing role of chemistry in fern classification. *Fern Gazette*, **12**, 283–94.

Cooper-Driver, G. A. & Swain, A. (1976). Cyanogenic polymorphism in bracken in relation to herbivore predation. *Nature*, **260**, 604.

Corley, H. B. (1967). *Dryopteris filix-mas* agg. in Britain. *Proceedings of the Botanical Society of the British Isles*, **7**, 73–5.

Corley, H. V. & Gibby, M. (1981). *Dryopteris × sarvelae* in Scotland – a new hybrid for the British Isles. *Fern Gazette*, **12**, 178.

Corner, R. W. M. (1967). *Lycopodium* species in colliery debris in southern Scotland. *Transactions and Proceedings of the Botanical Society of Edinburgh*, **40**, 337–8.

Cousens, M. I., Lacey, D. G. & Kelly, E. M. (1985). Life-history studies of ferns: a consideration of perspective. *Proceedings of the Royal Society of Edinburgh*, **86B**, 371–80.

Crabbe, J. A., Jermy, A. C. & Walker, S. (1970). The distribution of *Dryopteris assimilis* S. Walker in Britain. *Watsonia*, **8**, 3–15.

Crabbe, J. A. & Shivas, M. G. (1975). *Polypodium*. In *Hybridisation and the Flora of the British Isles*, ed. C. A. Stace, pp. 121–2. London: Academic Press.

Cubas, P. & Sleep, A. (1994). *Asplenium × sarniense* (Aspleniaceae, Pteridophyta) from Guernsey (Channel Islands, U.K.): a cytological enigma. *Fern Gazette*, **14**, 269–88.

Dandy, J. E. (1969). *Watsonian Vice-counties of Great Britain*. London: Ray Society.

Darwin, F. (1877). On the nectar glands of the common Brake Fern (*Pteris aquilina*). *Journal of the Linnean Society (Botany)*, **15**, 406–9.

Derrick, L. N., Jermy, A. C. & Paul, A. M. (eds.) Checklist of European pteridophytes. *Sommerfeltia*, **6**, 1–94.

Dickie, G. (1860). *The Botanists Guide to the Counties of Aberdeen, Banff and Kincardineshire*. Aberdeen: Brown.

Dopp, W. (1932). Die Apogamie bei *Aspidium remotum* A. Br. *Planta*, **17**, 86–152.

Dostal, J. (1984). Genus *Diphasiastrum* Holub. In Kramer, K. (ed.) Pteridophyta, *Illustrierte Flora von Mittel-Europa*, 3rd edn, ed. G. Hegi, Teil I, Band I, *Pteridophyta*, ed. K. Kramer. Berlin: Parey.

Doyle, D. J. (1987). A new station for the Killarney Fern *Trichomanes speciosum* in Killarney oakwoods (Blechno-Quercetum). *Irish Naturalists Journal*, **22**, 353–6.

Druce, G. C. (1882). On *Lycopodium complanatum* L. as a British plant. *Journal of Botany, London*, **20**, 321–3.

Druce, G. C. (1892). Plants of Glen Spean, Westerness. *Annals of Scottish Natural History*, **1**, 178–85.

Druce, G. C. (1915). *Lycopodium complanatum* L. *Report of the Botanical Society and Exchange Club of the British Isles*, **3**, 219–23.

Druce, G. C. (1919). *Cystopteris regia* Presl. var. *Dickieana* (Sm.) Druce. *Report of the Botanical Society and Exchange Club of the British Isles*, **9**, 317–18.

Druery, C. T. (1911). Our Common ferns. The Lastreas (Nephrodiums). *British Fern Gazette*, **1**, 258–62.

Druery, C. T. (1912). *British Ferns and Their Varieties*. London: Routledge.

Duckett, J. G. (1970*a*). Spore size in the genus. *Equisetum. New Phytologist*, **69**, 333–46.

Duckett, J. G. (1970*b*). Coning behaviour of the genus *Equisetum* in Britain. *Fern Gazette*, **10**, 107–12.

Duckett, J. G. (1979). An experimental study of the reproductive biology of hybridisation in the European and North American species of the genus *Equisetum*. *Journal of the Linnean Society (Botany)*, **79**, 205–29.

Duckett, J. G. & Duckett, A. R. (1980). Reproductive biology and population dynamics of wild gametophytes of *Equisetum*. *Journal of the Linnean Society (Botany)*, **80**, 1–40.

Duckett, J. G. & Page, C. N. (1975). *Equisetum*. In *Hybridisation and the Flora of the British Isles*, ed. C. A. Stace, pp. 99–103. London: Academic Press.

Duerden, H. (1929). Variations in the megaspore number in *Selaginella*. *Annals of Botany*, **43**, 451–9.

Dyce, J. W. (1962). The British Pteridological Society Excursion to Dalmally. *British Fern Gazette*, **10**, 100–4.

Dyce, J. W. (1976). Growing ferns from spores. *Natural Science in Schools*, **14**, 43–5.

Dyce, J. W. (1979). British Pteridological Society Meetings 1979: Criccieth, North Wales. *British Pteridological Society Bulletin*, **2**, 13–18.

Dyer, A. F. (ed.) (1979) *The Experimental Biology of Ferns*. London: Academic Press.

Dyer, A. F. (1990). Does bracken spread by spores. In *Bracken Biology and Management*, ed. J. A. Thomson & R. T. Smith, pp. 35–42. Canberra: Australia Institute of Agricultural Science.

Dyer, A. F. & Hadfield, P. R. H. (1985). Polymorphism for cyanogenesis in British bracken (*Pteridium aquilinum* subsp. *aquilinum* var. *aquilinum*). *Proceedings of the Royal Society of Edinburgh*, **86B**, 462–4.

Dyer, A. F. & Lindsay, S. (1992). Soil spore banks of temperate ferns. *American Fern Journal*, **82**, 89–122.

Dyer, A. F. & Page, C. N. (eds.) (1985). *Biology of*

Pteridophytes. Edinburgh: Royal Society of Edinburgh.

Edees, E. S. (1972). *Flora of Staffordshire*. Newton Abbot: David & Charles.

Emmot, J. I. (1964). A cytogenetic investigation in the *Phyllitis–Asplenium* complex. *New Phytologist*, **63**, 306–18.

Evans, I. A. (1976). Relationship between Bracken and cancer. *Botanical Journal of the Linnean Society*, **73**, 105–12.

Evans, I. A. (1986). The carcinogenic, mutagenic and teratogenic toxicity of Bracken. In *Bracken. Ecology, Land Use and Control Technology*, ed. R. T. Smith & J. A. Taylor, pp. 139–46. Carnforth: Parthenon Press.

Evans, I. A. (1987). Bracken carcinogenicity. In *International Quarterly Scientific Reviews*, ed. G. V. James, pp. 161–99. Tel-Aviv: Fround Publishing.

Evans, I. A. & Galpin, O. P. (1990). Bracken and leukemia. *Lancet*, **335**, 736.

Evans, I. A. & Mason, J. (1965). Carcinogenic activity of Bracken. *Nature*, **208**, 913–14.

Evans, W. C. (1976). Bracken thiaminase-mediated neurotoxic syndromes. *Botanical Journal of the Linnean Society*, **73**, 113–31.

Evans, W. C. & Patel, M. C. (1982). Acute Bracken poisoning in homogastric and ruminant animals. *Proceedings of the Royal Society of Edinburgh*, **81B**, 29–64.

Farrar, D. R. (1967). Gametophytes of four tropical fern genera reproducing independently of their sporophytes in the southern Appalachians. *Science*, **155**, 1266–7.

Farrar, D. R. (1985). Independent fern gametophytes in the wild. *Proceedings of the Royal Society of Edinburgh*, **86B**, 361–9.

Farrar, D. R., Parks, J. C. & McAlpin, B. W. (1983). The fern genera *Vittaria* and *Trichomanes* in the northeastern United States. *Rhodora*, **85**, 83–92.

Fataliyev, R. A. (1960). [On the distribution of certain fern genera present as fossils in the USSR.] *Botanicheskii Zhurnal*, **45**, 1213–18. (See map 2, p. 1215.)

Ferguson, D. K. (1986). An additional site for *Polystichum × illyricum* (Borbas) Hahne in north-west Scotland, *Watsonia*, **16**, 331–2.

Ferrarini, E. (1977). Un antico relitto ai piedi des Alpi Apuane: *Trichomanes speciosum* Willd., entita nuova per las flora italiana. *Giornale di Botanica Italiana*, **111**, 171–7.

Ferreira, R. E. C. (1958). Scottish mountain vegetation in relation to geology. *Transactions and Proceedings of the Botanical Society of Edinburgh*, **37**, 229–50.

Ferreira, R. E. C. (1980). A new find of *Equisetum × trachyodon* in north-west Scotland. *Fern Gazette*, **12**, 113–14.

Fletcher, W. E. (1978). Bracken past and present. *Transactions and Proceedings of the Botanical Society of Edinburgh*, **42**, Suppl., 1–9.

Fraser-Jenkins, C. R. (1977). Three species in the *Dryopteris villarii* aggregate (Pteridophyta, Aspidaceae). *Candollea*, **32**, 305–9.

Fraser-Jenkins, C. R. (1979). A new name for a European *Dryopteris. Fern Gazette*, **12**, 56.

Fraser-Jenkins, C. R. (1980). *Dryopteris affinis*: a new treatment for a complex species in the European pteridophyte flora. *Willdenowia*, **10**, 207–15.

Fraser-Jenkins, C. R. (1982). *Dryopteris* in Spain, Portugal and Macaronesia. *Boletin de la Sociedad Broteriana*, ser. 2a, **55**, 175–335.

Fraser-Jenkins, C. R. (1984). *Dryopteris affinis*. In *Illustrierte Flora von Mitteleuropa*, 3rd edn, ed. G. Hegi, Band I, Teil I, *Pteridophyta*, ed. K. Kramer, pp. 146–8.

Fraser-Jenkins, C. R. (1987a). A new subspecies of *D. affinis*. In *Checklist of European Pteridophytes*, ed. L. N. Derrick, A. C. Jermy & A. M. Paul, *Sommerfeltia*, **6**, 1–94 (and see p. xi).

Fraser-Jenkins, C. R. (1987b). Hybrids between *D. affinis* and *D. filix-mas*. In *Checklist of European Pteridophytes*, ed. L. N. Derrick, A. C. Jermy & A. M. Paul. *Sommerfeltia* **6**, 1–94 (and see pp. xi–xiii).

Fraser-Jenkins, C. R. & Corley, H. V. (1972). *Dryopteris caucasica* – an ancestral diploid in the male fern aggregate. *British Fern Gazette*, **10**, 221–31.

Fraser-Jenkins, C. R. & Gibby, M. (1980). Two new hybrids in the *Dryopteris villarii* aggregate (Pteridophyta, Dryopteridaceae), and the origin of *D. submontana. Candollea*, **35**, 305–10.

Fraser-Jenkins, C. R. & Jermy, A. C. (1977a). Two hitherto unnamed hybrids of *Dryopteris expansa. Fern Gazette*, **11**, 338–9.

Fraser-Jenkins, C. R. & Jermy, A. C. (1977b). The tetraploid subspecies of *Dryopteris villarii. Fern Gazette*, **11**, 339–40.

Fraser-Jenkins, C. R. & Reichstein, T. (1977). *Dryopteris × brathaica* Fraser-Jenkins & Reichstein hybr. nov., the putative hybrid of *D. carthusiana × D. filix-mas. Fern Gazette*, **11**, 337.

Frost, G. (1982). Lizard Survey: *Isoetes histrix*. Unpublished report, University of Bristol.

Galpin, O. P., Whitaker, C. J., Whitaker, R. H. & Kassab, J. Y. (1990). Gastric cancer in Gwynedd. Possible links with Bracken. *British Journal of Cancer*, **61**, 737–40.

Gastony, G. J. & Gottleib, L. D. (1982). Evidence for genetic heterozygosity in a homosporous fern. *American Journal of Botany*, **69**, 634–7.

Gibby, M. (1991). The development of laboratory based studies in fern variation. *The History of British Pteridology*, ed. J. M. Camus, pp. 59–63. London: British Pteridological Society.

Gibby, M. & Fraser-Jenkins, C. R. (1985). Hybridisation and speciation in the genus *Dryopteris* in Pico, Azores. *Proceedings of the Royal Society of Edinburgh*, **86B**, 473–4.

Gibby, M. & Lovis, J. D. (1989). New ferns of Madeira. *Fern Gazette*, **13**, 285–90.

Gibby, M. & Walker, S. (1977). Further cytogenetic studies and a reappraisal of the diploid ancestry on the *Dryopteris carthusiana* complex. *Fern Gazette*, **11**, 315–24.

Gilbert, O. L. (1966). *Dryopteris villarii* in Britain. *British Fern Gazette*, **9**, 263–8.

Gilbert, O. L. (1970). Biological flora of the British

Isles: *Dryopteris villarii* (Bellardi) Woynar. *Journal of Ecology*, **58**, 301–13.

Girard, P. J. (1967). The discovery in Guernsey of an extremely rare fern ×*Asplenophyllitis microdon*, and the subsequent investigations carried out by J. D. Lovis at Leeds University. *Report and Transactions. Société Guernésiaise*, **18**, 167–77.

Girard, P. J. & Lovis, J. D. (1968). The rediscovery of ×*Asplenophyllitis microdon*, with a report on its cytogenetics. *British Fern Gazette*, **10**, 1–8.

Godwin, H. (1956). *The History of the British Flora*. Cambridge: Cambridge University Press.

Godwin, H. (1975). *The History of the British Flora. A Factual Basis for Phytogeography*, 2nd edn. Cambridge: Cambridge University Press.

Grime, J. P. (1985). Factors limiting the contribution of pteridophytes to a local flora. *Proceedings of the Royal Society of Edinburgh*, **86B**, 403–21.

Grose, D. (1957). *Flora of Wiltshire*. Devizes: Wiltshire Archaeological and Natural History Society.

Groves, H. & Groves, J. (1891). Is *Lycopodium complanatum* a British plant? *Journal of Botany, British and Foreign*, **29**, 178–9.

Hackney, P. (1977). *Polypodium australe* Fée in the north of Ireland. *Irish Naturalists' Journal*, **19**, 104–7.

Hackney, P. (1981). *Equisetum variegatum* Web & Mohr – a note on its occurrence in Co. Down and a comparison of some of its sites in Ireland. *Irish Naturalists' Journal*, **20**, 180–4.

Haufler, C. H. (1992). An introduction to fern genetics and breeding systems. In *Fern Horticulture: Past, Present and Future Perspectives*, ed. J. M. Ide, A. C. Jermy & A. M. Paul, pp. 145–55. Andover: Intercept Publishing.

Headley, A. D. & Callaghan, T. V. (1990). Modular growth of *Huperzia selago* (Lycopodiaceae: Pteridophyta). *Fern Gazette*, **13**, 361–72.

Heath, F. G. (1876). *The Fern Paradise*. London: Hodder & Stoughton.

Heath, J. & Scott, P. (1977). *Biological Records Centre. Instructions for Recorders*. London: Institute of Terrestrial Ecology, Natural Environment Research Council.

Hegi, G. (ed.) (1984). *Illustrierte Flora von MittelEuropa*, 3rd edn, Teil I, Band I, *Pteridophyta*, ed. K. Kramer. Berlin: Parey.

Hibberd, D. (1875). *The Fern Garden*, 5th edn. London: Groombridge.

Holland, S. (1968). On *Pteris vittata. Journal North Gloucester Naturalists' Society*, **19**, 318–21.

Holub, J. (1975). *Diphasiastrum*, a new genus in Lycopodiaceae. *Preslia*, **47**, 97–110.

Horwood, A. R. & Noel, C. W. F. (1933). *The Flora of Leicestershire and Rutland*. Oxford: Oxford University Press.

Hughes, W. W. (1969). The distribution of *Polypodium vulgare* L. subspecies *serrulatum* Archangeli in North Wales. *Nature in Wales*, **11**, 194–8.

Hulten, E. (1958). The Amphi-Atlantic Plants. *Kungliga Svenska Vetenskaps-Akademiens Handlingar*, Ser 4, 7(1), 1–340.

Hulten, E. (1964). The Circumpolar Plants. *Kungliga Svenska Vetenskaps-Akademiens Handlingar*, Ser 4, 7(5), 1–280.

Hutchinson, G. & Thomas, B. A. (1992). Distribution of Pteridophyta in Wales. *Watsonia*, **19**, 1–19.

Hutchinson, G. & Thomas, B. A. (1996). *Welsh Ferns*, 7th edn. Cardiff: National Museums & Galleries of Wales.

Hyde, H. A., Wade, A. E. & Harrison, S. G. (1978). *Welsh Ferns, Clubmosses, Quillworts and Horsetails*, 6th edn. Cardiff: National Museum of Wales.

Ide, J. M., Jermy, A. C. & Paul, A. M. (eds.) (1992). *Fern Horticulture: Past, Present and Future Perspectives*. Andover: Intercept Publishing.

Jalas, J. & Suominen, J. (eds.) (1972). *Atlas Florae Europaea (Distribution of Vascular Plants in Europe)*. 1. *Pteridophyta (Psilotaceae to Azollaceae)*. Helsinki: Societas Biologica Fennica & Committee for Mapping the Flora of Europe.

Jarrett, F. M. (1968). Plant notes (*Pteris vittata* L.). *Proceedings of the Botanical Society of the British Isles*, **7**, 387.

Jee, N. (1966). Botanical Report, 1965. *Report and Transactions. Société Guernésiaise*, **17**, 697–9.

Jee, N. (1994). The Guernsey Fern ×*Asplenophyllitis microdon*. *Report and Transactions, La Société Guernésiaise*, **23**, 724–49.

Jermy, A. C. (1968). Two new hybrids involving *Dryopteris aemula. British Fern Gazette*, **10**, 9–12.

Jermy, A. C. (1975). *Cystopteris*. In *Hybridisation and the Flora of the British Isles*, ed. C. A. Stace, p. 112. London: Academic Press.

Jermy, A. C. (1989). The history of *Diphasiastrum issleri* (Lycopodiaceae) in Britain and a review of its taxonomic status. *Fern Gazette*, **13**, 257–65.

Jermy, A. C., Arnold, H. R., Farrell, L. & Perring, F. H. (1978). *Atlas of Ferns of the British Isles*. London: The Botanical Society of the British Isles and the British Pteridological Society.

Jermy, A. C. & Camus, J. (1991). *The Illustrated Field Guide to Ferns and Allied Plants of the British Isles*. London: Natural History Museum.

Jermy, A. C. & Crabbe, J. A. (eds.) (1978). *The Island of Mull: a Survey of Its Flora and Environment*. London: British Museum (Natural History).

Jermy, A. C. & Harper, L. (1971). Spore morphology of the *Cystopteris fragilis* complex. *British Fern Gazette*, **10**, 211–13.

Jermy, A. C. & Page, C. N. (1980). Additional field characters separating the subspecies of *Asplenium trichomanes* in Britain. *Fern Gazette*, **12**, 112–13.

Jermy, A. C. & Walker, S. (1975). *Dryopteris*. In *Hybridisation and the Flora of the British Isles*, ed. C. A. Stace, pp. 113–18. London: Academic Press.

Jubrael, J. M. S., Sheffield, E. & Moore, D. (1986). Polymorphisms in the DNA of *Pteridium aquilinum*. In *Bracken. Ecology, Land Use and Control Technology*, ed. R. T. Smith & J. A. Taylor, pp. 309–13. Carnforth: Parthenon Press.

Kay, G. O. N. (1974). Diploid *Isoetes echinospora* in Britain. *Fern Gazette*, **11**, 56.

Kay, R. (1968). *Hardy Ferns*. London: Faber & Faber.

Keeble-Martin, W. & Fraser, G. T. (eds.) (1939). *Flora of Devon*. Arbroath: T. Buncle.

Kendall, A., Page, C. N. & Taylor, J. A. (1994). Linkages between bracken sporulation rates and weather and climate in Britain. In *Bracken: An Environmental Issue*, ed. R. T. Smith & J. A. Taylor, pp. 77–81. Aberystwyth: International Bracken Group.

Kent, D. N. (1963). Plant notes (*Pteris cretica* L.). *Proceedings of the Botanical Society of the British Isles*, **5**, 121.

Knox, E. M. (1951). Spore morphology in British ferns. *Transactions and Proceedings of the Botanical Society of Edinburgh*, **35**, 437–48.

Kukkonen, I. (1967). Studies in the variability of *Diphasium (Lycopodium) complanatum*. *Annales Botanici Societatis Zoologicae-Botanicae Fennicae Vanamo*, **4**, 441–70.

Labatut, A., Prelli, R. & Schneller, J. (1984). *Asplenium obovatum* in Britanny, N.W. France. *Fern Gazette*, **12**, 331–3.

Lamb, H. H. (1972). *The Changing Climate*. London: Methuen.

Lawton, J. H. (1976). The structure of the arthropod community on Bracken. *Botanical Journal of the Linnean Society*, **23**, 187–216.

Lawton, J. H. & MacGarvin, M. (1985). Interaction between bracken and its insect herbivores. *Proceedings of the Royal Society of Edinburgh*, **86B**, 125–31.

Le Sueur, F. (1984). *Flora of Jersey*. Jersey: Société Jersiaise.

Leach, W. (1930). A preliminary account of the vegetation of some non-calcareous British screes. *Journal of Ecology*, **27**, 321–32.

Lee, J. R. (1933). *Flora of the Clyde Area*. Glasgow: John Smith & Son.

Lees, F. A. (1988). *The Flora of West Yorkshire*. London: Reeve.

Lellinger, D. B. (1985). *A Field Manual of the Ferns and Fern-Allies of the United States and Canada*. Washington, DC: Smithsonian Institution Press.

Lindsay, S. & Dyer, A. F. (1997). Investigating the phenology of gametophyte development: an experimental approach. In *Pteridology in Perspective*, ed. J. Camus, M. Gibby & R. J. Johns, pp. 633–50. London: Royal Botanic Gardens, Kew.

Lindsay, S., Shefield, E. & Dyer, A. F. (1992). Soil spore banks, fern conservation and isozyme analysis. In *Fern Horticulture, Past, Present and Future Perspectives*, ed. J. M. Ide, A. C. Jermy & A. M. Paul, pp. 279–83. Andover: Intercept Publishing.

Lindsay, S. Sheffield, E. & Dyer, A. F. (1995) Dark germination as a factor limiting the formation of soil spore banks by bracken. In *Bracken: An Environmental Issue*, ed. R. T. Smith & J. A. Taylor, pp. 47–51. Leeds: International Bracken Group.

Lloyd, R. M. (1974). Reproductive biology and evolution in the Pteridophyta. *Annals of the Missouri Botanical Gardens*, **61**, 318–31.

Lousley, J. E. (1944). The flora of the bomb sites in the City of London in 1944. *Report of the Botanical Society and Exchange Club of the British Isles*, **12**, 875–83.

Lousley, J. E. (1964). Plant notes (*Pteris vittata*). *Proceedings of the Botanical Society of the British Isles*, **3**, 336–7.

Lousley, J. E. (1971). *Flora of the Isles of Scilly*. Newton Abbot: David & Charles.

Löve, A. & Kapoor, B. M. (1966). An alloploid *Ophioglossum*. *Nucleus*, **9**, 132–8.

Löve, A. & Kapoor, B. M. (1967). The highest plant chromosome number in Europe. *Svensk Botanisk Tidsskrift*, **61**, 29–32.

Lovis, J. D. (1955). The problem of *Asplenium trichomanes*. In *Species Studies in the British Flora*, ed. J. E. Lousley, pp. 99–103. London: Botanical Society of the British Isles.

Lovis, J. D. (1964). Autopolyploidy in *Asplenium*. *Nature*, **203**, 324–5.

Lovis, J. D. (1968a). Artificial reconstruction of a species fern, *Asplenium adulterinum*. *Nature*, **217**, 1163–5.

Lovis, J. D. (1968b). Fern hybridists and fern hybridising. II. Fern hybridising at the University of Leeds. *British Fern Gazette*, **10**, 13–20.

Lovis, J. D. (1975a). *Asplenium* L.× *Phyllitis* Hill = ×*Asplenophyllitis* Alston. In *Hybridisation and the Flora of the British Isles*, ed. C. A. Stace, pp. 104–6. London: Academic Press.

Lovis, J. D. (1975b). *Asplenium*. In *Hybridisation and the Flora of the British Isles*, ed. C. A. Stace, pp. 106–11. London: Academic Press.

Lovis, J. D. (1977). Evolutionary patterns and processes in ferns. In *Advances in Botanical Research*, ed. R. D. Preston & H. W. Woodhouse, vol. 4, pp. 229–415.

Lovis, J. D. (1981). Hybrids in European Aspleniaceae (Pteridophyta). *Botanica Helvetica*, **91**, 89–139,

Lovis, J. D. & Reichstein, T. (1964). A diploid form of *Asplenium ruta-muraria*. *British Fern Gazette*, **9**, 141–6.

Lovis, J. D. & Shivas, M. G. (1954). The synthesis of *Asplenium* × *breynii*. *Proceedings of the Botanical Society of the British Isles*, **1**, 97.

Lovis, J. D. & Vida, G. (1969). The resynthesis and cytogenetic investigation of ×*Asplenophyllitis microdon* and ×*A. jacksonii*. *British Fern Gazette*, **10**, 53–67.

Lowe, E. J. (1845). *Fern Growing*. London: Nimmo.

Manton, I. (1938). Hybrid *Dryopteris (Lastrea)* in Britain. *British Fern Gazette*, **7**, 165–7.

Manton, I. (1947). Polyploidy in *Polypodium vulgare*. *Nature*, **159**, 136.

Manton, I. (1950). *Problems of Cytology and Evolution in Pteridophyta*. Cambridge: Cambridge University Press.

Manton, I. (1955). The importance of ferns to an understanding of the British Flora. In *Species in the British Flora*, ed. J. E. Lousley, pp. 90–8. London: Botanical Society of the British Isles.

Manton, I., Lovis, J. D., Vida, G. & Gibby, M.

(1986). Cytology of the fern flora of Madeira. *Bulletin of the British Museum of Natural History (Botany)*, **15**, 123–61.

Manton, I. & Reichstein, T. (1961). Zu Cytologie von Polystichum brauni (Spenner) Fée und seiner Hybriden. *Bericht der Schweizerischen Botanischen Gesellschaft*, **71**, 370–83.

Marquand, E. B. (1901). *Flora of Guernsey and the Lesser Channel Islands*. London: Dulau & Co.

Marquand, E. D. (1951). *Flora of Guernsey*. London: Collins.

Marren, P. (1984). The history of Dickie's Fern in Kincardineshire. *Pteridologist*, **1**, 27–32.

McClintock, D. (1961). The ferns of the Channel Islands. *British Fern Gazette*, **9**, 34–7.

McClintock, D. (1964). Plant notes (*Pteris vittata* L.). *Proceedings of the Botanical Society of the British Isles*, **5**, 337–8.

McClintock, D. (1968). Plant notes: *Asplenium billotii* × *Phyllitis scolopendrium* = ×*Asplenophyllitis microdon* (T. Moore) Alston. *Proceedings of the Botanical Society of the British Isles*, **7**, 387–9.

McClintock, D. (1975). *The Wild Flowers of Guernsey*. London: Collins.

McClintock, D. (1987). Supplement to the Wild Flowers of Guernsey. St Peter Port: La Société Guernésiaise.

McVean, D. N. (1964). Moss and fern meadows. In *The Vegetation of Scotland*, ed. J. H. Burnett, chapter 13, pp. 514–21. Edinburgh: Oliver & Boyd.

Melville, R. (1938). *Isoetes hystrix* at the Lizard. *Journal of Botany, British and Foreign*, **76**, 17–19.

Merryweather, J. (1991). *The Fern Guide*. Shrewsbury: Field Studies Council.

Miller, D. R., Morrice, J. G. & Whitworth, P. L. (1989). *The Bracken Problem in Scotland*. Aberdeen: Macaulay Land Use Research Institute.

Milne-Redhead, E. & Trist, P. J. O. (1975). A remarkable population of *Ophioglossum vulgatum* in Suffolk. *Watsonia*, **20**(4), 415–16.

Milton, J. N. B. & Duckett, J. G. (1985). Potential allelopathy in *Equisetum*. *Proceedings of the Royal Society of Edinburgh*, **86B**, 468–9.

Mitchell, J. (1979). *A Report on the Past and Present Status of* Woodsia ilvensis *in the Moffat Hills (with Ecological notes)*. Balloch: Nature Conservancy Council, South West Region, Scotland.

Moore, T. (1860). *Lastrea remota*, a new British fern. *Phytologist*, **4**, 82–3.

More, D. (1878). On a new species of *Isoetes* from Ireland. *Journal of Botany*, **16**, 353–5.

More, A. G. *et al.* (1898). *Cybele Hibernica*, 2nd edn. Dublin: Dublin University Press.

Murray, C. W. (1981). *Equisetum* ×*trachyodon* in Skye, Western Scotland. *Fern Gazette*, **12**, 179–80.

Murray, C. W., Birks, H. J. B. & Murray, R. M. (1980), *The Botanist in Skye. A Guide to the Flowering Plants and Ferns*, 2nd edn. London: Botanical Society of the British Isles.

Naylor, F. (1866). On *Asplenium petrarchae*, D. C., as an Irish Plant. *Transactions and Proceedings of the Botanical Society of Edinburgh*, **9**, 365–6.

Nelson, E. C. (1992). Ferns in Ireland, cultivated and wild, through the ages. In *Fern Horticulture: Past, Present and Future Perspective*, ed. J. M. Ide, A. C. Jermy & A. M. Paul, pp. 57–86. Andover: Intercept Publishing.

Newman, E. (1853). *A History of British Ferns*. London: Van Voorst.

Newman, E. (1854). *Ophioglossum lusitanicum*. *Phytologist*, **5**, 80–1.

Ojika, M., Wakamatsu, K. Niwa, H. & Yamada, K. (1987). Ptaquiloside, a potent carcinogen isolated from bracken fern *Pteridium aquilinum* var. *latiusculum*: structure, elucidation based on chemical and spectral evidence, and reactions with amino acids, nucleosides and nucleotides. *Tetrahedron*, **43**, 5261–74.

Øllgaard, B. (1985). Observations on the ecology of hybridisation in the clubmosses (Lycopodiaceae). *Proceedings of the Royal Society of Edinburgh*, **86B**, 245–51.

O'Malley, D. J. S. (1979). *Asplenium cuneifolium* new to West Mayo. *Irish Naturalists' Journal*, **19**, 315.

Orth, R. (1938). Zur Morphologie der Primärblätter einheimischer Ferne. *Flora (Jena)*, **33**, 1–55.

Pacyna, A. (1972). Biometrics and taxonomy of the Polish species of the genus *Diphasium* Presl. *Fragmenta Floristica et Geobotanica*, **18**, 255–97.

Page, C. N. (1963). A hybrid horsetail from the Hebrides. *British Fern Gazette*, **9**, 117–19.

Page, C. N. (1967). Sporelings of *Equisetum arvense* in the wild. *British Fern Gazette*, **9**, 335–8.

Page, C. N. (1968). Spiral shoots in the Great Horsetail, *Equisetum telmateia* Ehrk. *Proceedings of the Botanical Society of the British Isles*, **7**, 173–6.

Page, C. N. (1971). Three pteridophytes new to the Canary Islands. *British Fern Gazette*, **10**, 205–8.

Page, C. N. (1972a). An assessment of inter-specific relationships in *Equisetum* subgenus *Equisetum*. *New Phytologist*, 355–69.

Page, C. N. (1972b). An interpretation of the morphology and evolution of the cone and shoot of *Equisetum*. *Journal of the Linnean Society (Botany)*, **65**, 359–97.

Page, C. N. (1973a). Two hybrids in *Equisetum* new to the British Flora. *Watsonia*, **9**, 229–37.

Page, C. N. (1973b). Ferns, polyploids, and their bearing on the evolution of the Canary Islands flora. *Monographiae Biologicae Canariense*, **4**, 83–8.

Page, C. N. (1974). *Equisetum* subgenus *Equisetum* in the Sino-Himalayan region – a taxonomic and evolutionary appraisal. *Fern Gazette*, **11**, 25–47.

Page, C. N. (1975). Some British Ferns and their cultivation. *Journal of the Scottish Rock Garden Club*, **14**, 263–76.

Page, C. N. (1976). The taxonomy and phytography of Bracken – a review. *Journal of the Linnean Society (Botany)*, **73**, 1–34.

Page, C. N. (1977). An ecological survey of the ferns of the Canary Islands. *Fern Gazette*, **11**, 297–312.

Page, C. N. (1978). Ferns as taxonomic tools and the future of Pteridology. *Transactions and Proceedings of the Botanical Society of Edinburgh*, **42**, Supplement, 37–41.

Page, C. N. (1979a). The diversity of ferns. An

ecological perspective. In *The Experimental Biology of Ferns*, ed. A. Dyer, pp. 10–56. London: Academic Press.

Page, C. N. (1979*b*). Experimental aspects of fern ecology. In *The Experimental Biology of Ferns*, ed. A. Dyer, pp. 551–9. London: Academic Press.

Page, C. N. (1979*c*). *Equisetum × trachyodon* in western Scotland. *Fern Gazette*, **12**, 57–9.

Page, C. N. (1979*d*). Macaronesian Heathlands. In *Heathlands and Related Shrublands of the World*, ed. R. L. Specht, pp. 117–23. The Hague: W. Junk.

Page, C. N. (1981). A new name for a hybrid horsetail in Scotland. *Fern Gazette*, **12**, 178–9.

Page, C. N. (1982*a*). *The Ferns of Britain and Ireland*, 1st edn. Cambridge: Cambridge University Press.

Page, C. N. (1982*b*). Field observations on the nectaries of Bracken, *Pteridium aquilinum*, in Britain. *Fern Gazette*, **12**, 233–40.

Page, C. N. (1982*c*). The history and spread of bracken in Britain. *Proceedings of the Royal Society of Edinburgh*, **81B**, 3–10.

Page, C. N. (1985). Pteridophyte biology: the biology of the amphibians of the plant world. *Proceedings of the Royal Society of Edinburgh*, **86B**, 439–42.

Page, C. N. (1986). The strategies of Bracken as a permanent ecological opportunist. In *Bracken. Ecology, Land Use and Control Technology*, ed. R. T. Smith & J. A. Taylor, pp. 173–81. Carnforth: Parthenon Press.

Page, C. N. (1988*a*). Two hybrids of *Equisetum sylvaticum* new to the British flora. *Watsonia*, **17**, 273–7.

Page, C. N. (1988*b*). *Ferns. Their Habitats in the Landscape of Britain and Ireland*. London & Glasgow: Collins New Naturalist.

Page, C. N. (1989*a*). Three subspecies of Bracken, *Pteridium aquilinum* (L.) Kuhn, in Britain. *Watsonia*, **17**, 429–34.

Page, C. N. (1989*b*). Compression and slingshot megaspore ejection in *Selaginella selaginoides* – a new phenomenon in pteridophytes. *Fern Gazette*, **13**, 267–75.

Page, C. N. (1990*a*). Herbarium sheets of the rare fern *×Asplenophyllitis microdon* (T. Moore) Alston in Edinburgh. *Watsonia*, **18**, 319.

Page, C. N. (1990*b*). Hybrids in the genus *Equisetum* in Europe: an updated annotation. In *Taxonomia, Biogeografia y Conservation de Pteridofitos*, ed. J. Rita, pp. 181–6. IME: Palma de Mallorca.

Page, C. N. (1990*c*). Taxonomic evaluation of the fern genus *Pteridium* and its active evolutionary state. In *Bracken Biology and Management*, ed. J. A. Thomson & R. T. Smith, pp. 23–4. Canberra: Australian Institute of Agricultural Science.

Page, C. N. (1995*a*). *Equisetum ×willmotii* C. N. Page – a new hybrid horsetail from County Cavan, Ireland. *Glasra*, **2**, 135–8.

Page, C. N. (1995*b*). Structural variation in western European Bracken – an updated taxonomic perspective. In *Bracken: An Environmental Issue*, ed. R. T. Smith & J. A. Taylor, pp. 13–15. Aberystwyth: International Bracken Group.

Page, C. N. & Barker, M. A. (1985). Ecology and geography of hybridisation in British and Irish horsetails. *Proceedings of the Royal Society of Edinburgh*, **86B**, 265–72.

Page, C. N. & Bennell, F. M. (1979). Preliminary investigation of two south-west England populations of the *Asplenium adiantum-nigrum* aggregate and the addition *A. cuneifolium* to the English flora. *Fern Gazette*, **12**, 5–8.

Page, C. N. & Bennell, F. (1986). Pteridophyta. In *The European Garden Flora*, ed. S. M. Walters *et al.*, pp. 1–67. Cambridge: Cambridge University Press.

Page, C. N. & Busby A. R. (1985). *Equisetum ×font-queri* in Shropshire. *Pteridologist*, **1**, 72.

Page, C. N., Dyer, A. F., Lindsay, S. & Mann, D. G. (1992). Conservation of pteridophytes; the *ex-situ* approach. *Fern Horticulture, Past, Present and Future Perspectives*, ed. J. M. Ide, A. C. Jermy & A. M. Paul, pp. 269–78. Andover: Intercept Publishing.

Page, C. N. & Gardner, M. F. (1994). Conservation of rare temperate rainforest tree species: a fast growing role for arboreta in Britain and Ireland. In *The Common Ground of Wild and Cultivated Plants*, ed. A. R. Perry & R. G. Ellis, pp. 119–44. Cardiff: National Museum of Wales.

Page, C. N. & McHaffie, H. S. (1991). Pteridophytes as indicators of landscape changes in the British Isles in the last hundred years. In *The History of British Pteridology*, ed. J. M. Camus, pp. 25–40. London: British Pteridological Society.

Page, C. N. & Mill, R. (1994). Scottish Bracken (*Pteridium*): new taxa and a new combination. *Botanical Journal of Scotland*, **47**, 139–40.

Page, C. N. & Mill, R. (1995). The taxa of Scottish Bracken in a European perspective. *Botanical Journal of Scotland*, **47**, 229–47.

Pajaron, S., Prada, C. Herrero, A. & Pangua, E. (1997). Isozymical study of genetic variability in *Asplenium foreziense* and related taxa. Preliminary results. In *Pteridology in Perspective*, ed. J. Camus, M. Gibby & R. J. Johns, pp. 307–11. London: Royal Botanic Gardens, Kew.

Pakeman, R. J. & Marrs, R. H. (1992). The conservation value of Bracken *Pteridium aquilinum* (L.) Kuhn dominated communities in the U.K., and an assessment of the ecological impact of bracken expansion or its removal. *Biological Conservation*, **62**, 101–14.

Pakeman, R. J., Marrs, R. H. & Jacob, P. J. (1994). A model of bracken (*Pteridium aquilinum*) growth and the effects of control strategies and changing climate. *Journal of Applied Ecology*, **31**, 145–54.

Panigraphi, G. (1965). Preliminary studies in the cytotaxonomy of the *Dryopteris villarii* (Bell) Woynar complex in Europe. *American Fern Journal*, **55**, 1–8.

Patton, D. (1924). The vegetation of Beinn Laoigh (Ben Lui). *Report of the Botanical Society and Exchange Club of the British Isles*, **7**, 268–319.

Paul, A. M. (1987). The status of *Ophioglossum azoricum* (Ophioglossaceae: Pteridophyta) in the British Isles. *Fern Gazette*, **13**, 173–87.

Payne, L. G. (1939). The Crested Buckler Fern. *London Naturalist*, 29–31.

Perring, F. H. & Walters, S. M. (1976). *Atlas of the*

British Flora, 2nd edn. London: E. P. Publishing for the Botanical Society of the British Isles.

Petch, C. P. (1980). *Lycopodiella inundata* (L.) Holub in West Norfolk. *Watsonia*, **13**, 128.

Petch, C. P. & Swann, E. L. (1968). *Flora of Norfolk*. Norwich: Jarrold.

Peterken, G. F. (1962). Plants notes: *Polypodium vulgare* L. *sensu lato*. *Proceedings of the Botanical Society of the British Isles*, **4**(14), 413.

Peterken, G. F. (1974). A method of assessing woodland flora for conservation using indicator species. *Biological Conservation*, **6**, 239–85.

Petersen, R. L. (1985*a*). Use of fern spores and gametophytes in toxicity assessments. *Proceedings of the Royal Society of Edinburgh*, **86B**, 453.

Petersen, R. L. (1985*b*). Towards an appreciation of fern edaphic niche requirements. *Proceedings of the Royal Society of Edinburgh*, **86B**, 93–103.

Pichi-Sermolli, R. E. G. (1970). A provisional catalogue of the family names of living pteridophytes. *Webbia*, **25**, 219–97.

Pichi-Sermolli, R. E. G. (1979). A survey of the pteridological flora of the Mediterranean region. *Webbia*, **34**, 175–242.

Pichi-Sermolli, R. E. G. (1983). Notes on Adanson's fern genera *Ceterach* and *Scolopendrium*. *Webbia*, **47**, 121–43.

Pichi-Sermolli, R. E. G. (1993). New studies on some family names of Pteridophyta. *Webbia*, **47**, 121–43.

Pickering, D. A. & Wigston, D. L. (1990). *Lycopodiella inundata* (Lycopodiaceae: Pteridophyta) on china-clay at Lee Moor, south Devon. *Fern Gazette*, **13**, 373–80.

Pinter, I. (1995). Progeny studies in the fern hybrid *Polystichum* ×*bicknellii* (Dryopteridaceae: Pteridophyta). *Fern Gazette*, **15**, 25–40.

Polunin, O. (1959). Circumpolar Arctic Flora. Oxford: Clarendon Press.

Pope, C. R. (1983). An aberrant form of *Equisetum telmateia* from the Isle of Wight. *Fern Gazette*, **12**, 303–4.

Porter, J. C. (1994). A study of the ecology of aspleniums in limestone grykes. *Fern Gazette*, **14**, 245–54.

Potter, D. M. & Pitman, R. M. (1994). The extraction and characterisation of the carcinogens from Bracken and the effect of composting. In *Bracken. Ecology, Land Use and Control Technology*, ed. R. T. Smith & J. A. Taylor, pp. 110–15. Carnforth: Parthenon Press.

Praeger, R. L. (1909). *Lastrea remota* in Ireland. *Irish Naturalist*, **28**, 13–19.

Praeger, R. L. (1934). *The Botanist in Ireland*. Dublin: Dublin University Press.

Praeger, R. L. (1951). Hybrids in the Irish flora: a tentative list. *Proceedings of the Royal Irish Academy* B, **54**, 1–141.

Proctor, M. H. (1972). Ecological and historical factors in the distributions of the British *Helianthemum* species. *Journal of Ecology*, **46**, 349–71.

Pugh, P. J. (1953). The distribution of *Dryopteris borreri* Newm. in the British Isles. *Watsonia*, **3**, 57–65.

Rasbach, K., Rasbach, O. & Wilmanns, O. (1976). Die Farnpflänzen Zentraleuropas. Gestalt, Geschichte, Lebensraum. Stuttgart: Fischer.

Ratcliffe, D. A. (1960) The mountain flora of Lakeland. *Proceedings of the Botanical Society of the British Isles*, **4**, 1–25.

Ratcliffe, D. A. (1968). An ecological account of Atlantic bryophytes in the British Isles. *New Phytologist*, **67**, 388.

Ratcliffe, D. A. (1977). *Highland Flora*. Inverness: Highlands and Islands Development Board.

Ratcliffe, D. A., Birks, H. J. B. & Birks, H. H. (1993). The ecology and conservation of the Killarney Fern *Trichomanes speciosum* Willd. In Britain and Ireland. *Biological Conservation*, **66**, 231–47.

Raven, J. A. (1985). Physiology and biochemistry of pteridophytes. *Proceedings of the Royal Society of Edinburgh*, **86B**, 37–44.

Reichstein, T. (1965). The ferns in Flora Europaea. *British Fern Gazette*, **9**, 230–33.

Reichstein, T. (1981). Hybrids in European Aspleniaceae (Pteridophyta). *Botanica Helvetica*, **91**, 89–139.

Richards, P. M. (1983). Phenolic chemistry distinguishes *Asplenium adiantum-nigrum* from *A. cuneifolium* Viv. *Watsonia*, **14**, 414–15.

Richards, P. W. & Evans, G. B. (1972). Biological Flora of the British Isles: *Hymenophyllum*. *Journal of Ecology*, **60**, 245–68.

Richardson, P. M. & Lorenz-Liburnau, E. (1982). C-glycosylxanthones in the *Asplenium adiantum-nigrum* complex. *American Fern Journal*, **72**, 103–6.

Rickard, M. H. (1972*a*). The rarest fern in Britain. *British Pteridological Society Newsletter*, **10**, 24–6.

Rickard, M. H. (1972*b*). The distribution of *Woodsia ilvensis* and *W. alpina* in Britain. *British Fern Gazette*, **10**, 269–80.

Rickard, M. H. (1989). Two spleenworts new to Britain – *Asplenium trichomanes* subsp. *pachyrachis* and *Asplenium trichomanes* nothosubsp. *staufferi*. *Pteridologist*, **1**, 244–8.

Riddelsdell, H. G., Hedley, G. W. & Price, W. R. (1948). *Flora of Gloucestershire*. Cheltenham: Cottswold Naturalists Field Club.

Ridley, H. N. (1936). Bracken sporelings in London. *Journal of Botany, British and Foreign*, **77**, 219.

Roberts, R. H. (1965). *Dryopteris assimilis* S. Walker in Snowdonia. *Nature in Wales*, **9**, 163–4.

Roberts, R. H. (1967). *Dryopteris abbreviata* (DC). Newm. widespread in North Wales. *Report of the Botanical Society and Exchange Club of the British Isles*, **7**, 82.

Roberts, R. H. (1969). Dwarf Male Fern. *Nature in Wales*, **11**, 141.

Roberts, R. H. (1970). A revision of some of the taxonomic characters of *Polypodium australe* Fée. *Watsonia*, **8**, 121–34.

Roberts, R. H. (1979). The Killarney Fern, *Trichomanes speciosum*, in Wales. *Fern Gazette*, **12**, 1–4.

Roberts, R. H. (1980). *Polypodium macaronesicum*

and *P. australe*: a morphological comparison. *Fern Gazette*, **12**, 69–74.

Roberts, R. H. & Page, C. N. (1979). A second British record for *Equisetum × font-queri*, and its addition to the English flora. *Fern Gazette*, **12**, 61–2.

Roberts, R. H. & Scannell, M. J. P. (1977). *Asplenium × ticinense* D. E. Meyer: a hybrid fern new to the British Isles. *Irish Naturalists' Journal*, **19**, 75–7.

Roberts, R. H. & Stirling, A. McG. (1974). *Asplenium cuneifolium* Viv. in Scotland. *Fern Gazette*, **11**, 7–14.

Roberts, R. H. & Synnott, D. M. (1972). *Polypodium australe* Fée in Scotland and north-east Ireland. *Watsonia*, **9**, 39–41.

Robinson, F. (1919). *Isoetes hystrix* Durieu in Cornwall. *Journal of Botany, British and Foreign*, **57**, 322.

Roger, J. G. (1954). The flora of Caern Lochan. *Transactions and Proceedings of the Botanical Society of Edinburgh*, **36**, 159.

Roger, J. G. (1958). Local plants of south-west Aberdeenshire. *Transactions and Proceedings of the Botanical Society of Edinburgh*, **37**, 215–16.

Roth, R. (1938). Zur Morphologie der Primärblätter einheimischer Farne. *Flora* (Jena), **33**, 1–55.

Rothmaler, W. (1944). Pteridophyten – Studien I. *Feddes Repertorium*, **54**, 55–82.

Rowlands, S. P. (1929). *Cystopteris fragilis* Bernh. var. *Dickieana*. *British Fern Gazette*, **6**, 18–19.

Rumsey, F. J., Headley, A. D., Farrar, D. R. & Sheffield, E. (1991). The Killarney Fern (*Trichomanes speciosum*) in Yorkshire. *Naturalist*, **116**, 41–3.

Rumsey, F. J. & Sheffield, E. (1997). Inter-generational ecological niche separation and the 'independent gametophyte' phenomenon. In *Pteridology in Perspective*, ed. J. Camus, M. Gibby & R. J. Johns, pp. 563–70. London: Royal Botanic Gardens, Kew.

Rumsey, F. J., Sheffield, E. & Farrar, D. R. (1990). British Filmy-fern gametophytes. *Pteridologist*, **2**, 40–2.

Rumsey, F. J., Thompson, P. & Sheffield, E. (1993). Triploid *Isoetes echinospora* (Isoetaceae: Pteridophyta) in northern England. *Fern Gazette*, **14**, 215–21.

Rush, B. J. (1983). The rediscovery of *Asplenium × confluens*. *Fern Gazette*, **12**, 301–2.

Rutherford, A. & Stirling, A. McG. (1972). *Polypodium australe* Fée and the tetraploid hybrid in Scotland. *British Fern Gazette*, **10**, 233–5.

Rutherford, A. & Stirling, A. McG. (1973). Observations on *Polypodium australe* Fée in Scotland. *Watsonia*, **9**, 357–61.

Ryan, P. (1990). The Land Quillwort in the Channel Islands. *Pteridologist*, **2**, 28–30.

Rymer, L. (1976). The history and ethnobotany of Bracken. *Journal of the Linnean Society (Botany)*, **73**, 151–76.

Saito, K., Nagao, T., Takatsuki, S., Koyama, K. & Natori, S. (1990). The sesquiterpenoid carcinogen from bracken fern and some analogues from the Pteridaceae. *Phytochemistry*, **29**, 1475–9.

Saito, K., Takayuki, N., Mataba, M., Kogama, K., Natori, S., Murakami, T. & Saiki, Y. (1989). Chemical array of ptaguiloside, the carcinogen of *Pteridium aguilinum*, and the distribution of related compounds in the Pteridaceae. *Phytochemistry*, **28**, 1605–11.

Salter, J. A. (1928). The altitude ranges of flowering plants and ferns in mid-Wales. *North Western Naturalist*, September & December.

Sarvela, J. (1978). A synopsis of the fern genus *Gymnocarpium*. *Annales Botanici Societatis Zoologicae-Botanicae Fennicae Vanamo*, **15**, 73–9.

Scannell, M. J. P. (1972). *Pilularia globulifera* L. new to East Mayo (H. 26) and a second record for West Donegal (H. 35). *Irish Naturalists' Journal*, **17**, 280.

Scannell, M. J. P. (1977a). *Polystichum × lochitiforme* (Haláesy) Becherer in Ireland. *Irish Naturalists' Journal*, **19**, 79.

Scannell, M. J. P. (1977b). *Equisetum variegatum* in Donegal East. *Irish Naturalists' Journal*, **19**, 53–4.

Scannell, M. J. P. (1977). *Asplenium cuneifolium* Viv. in West Galway, Ireland. *Irish Naturalists' Journal*, **19**, 245.

Scannell, M. J. P. & Synnott, D. M. (1972). *Census Catalogue of the Flora of Ireland*. Dublin: Department of Agriculture and Forestries.

Schneller, J. J. (1979). Biosystematic investigations in the Lady Fern (*Athyrium filix-femina*). *Plant Systematics and Evolution, Wien*, **132**, 255–77.

Schneller, J. J. & Holderegger, R. (1997a). Soil spore bank and genetic demography of populations of *Athyrium filix-femina*. In *Pteridology in Perspective*, ed. J. Camus, M. Gibby & R. J. Johns, pp. 663–5. London: Royal Botanic Gardens, Kew.

Schneller, J. J. & Holderegger, R. (1997b). Colonisation events and genetic variability within populations of *Asplenium ruta-muraria* L. In *Pteridology in Perspective*, ed. J. Camus, M. Gibby & R. J. Johns, pp. 571–80. London: Royal Botanic Gardens, Kew.

Scott, D. A. & Hill, T. G. (1900). The structure of *Isoetes hystrix*. *Annals of Botany*, **14**, 413–34.

Seddon, B. (1965). Occurrence of *Isoetes echinospora* in eutrophic lakes in Wales. *Ecology*, **46**, 447–8.

Sheffield, E. (1994). Alternation of generations in ferns: mechanisms and significance. *Biological Reviews*, **69**, 101–12.

Sheffield, E. (1997). From pteridophyte spore to sporophyte in the natural environment. In *Pteridology in Perspective*, ed. J. Camus, M. Gibby & R. J. Johns, pp. 541–9. London: Royal Botanic Gardens, Kew.

Sheffield, E. & Bell, P. R. (1987). Current studies of the pteridophyte life-cycle. *Botanical Review*, **53**, 442–90.

Sheffield, E., Wolf, P. G. & Haufler, C. H. (1989). How big is a bracken plant? *Weed Research*, **29**, 455–60.

Sheffield, E., Wolf, P. G., Haufler, C. H., Ranker, T. & Jermy, A. C. (1989). A re-evaluation of plants referred to as *Pteridium herediae*. *Botanical Journal of the Linnean Society*, **99**, 377–86.

Sheffield, E., Wolf, P. G., Rumsey, F. J., Robson, D. J., Ranker, T. A. & Challinor, S. M. (1993). Spatial distribution and reproductive behaviour of a triploid bracken (*Pteridium aquilinum*) clone in Britain. *Annals of Botany*, **72**, 231–7.

Shivas, M. G. (1955). The two subspecies of *Asplenium adiantum-nigrum* L. in Britain. In *Species Studies in the British Flora*, ed. J. E. Lousley, pp. 104–6. London: Botanical Society of the British Isles.

Shivas, M. G. (1961*a*). Contributions to the cytology and taxonomy of species of *Polypodium* in Europe and America. I. Cytology. *Journal of the Linnean Society (Botany)*, **58**, 27–38.

Shivas, M. G. (1961*b*). Contributions to the cytology and taxonomy of species of *Polypodium* in Europe and America. II. Taxonomy. *Journal of the Linnean Society (Botany)*, **58**, 39.

Shivas, M. G. (1962). The *Polypodium vulgare* complex. *British Fern Gazette*, **9**, 65–70.

Shivas, M. G. (1969). A cytotaxonomic survey of the *Asplenium adiantum-nigrum* complex. *British Fern Gazette*, **9**, 68–80.

Shivas, M. G. (1970). Names of hybrids in the *Polypodium vulgare* complex. *British Fern Gazette*, **10**, 152.

Sleep, A. (1971*a*). *Polystichum* hybrids in Britain. *British Fern Gazette*, **10**, 208–9.

Sleep, A. (1971*b*). A new hybrid fern from the Channel Islands. *British Fern Gazette*, **10**, 209–11.

Sleep, A. (1975). *Polystichum*. In *Hybridisation and the Flora of the British Isles*, ed. C. A. Stace, pp. 118–20. London: Academic Press.

Sleep, A. (1976). *Polystichum × lonchitiforme*, a fern hybrid new to the British Isles. *Watsonia*, **11**, 182.

Sleep, A. (1980). On the reported occurrence of *Asplenium cuneifolium* and *A. adiantum-nigrum* in the British Isles. *Fern Gazette*, **12**, 103–7.

Sleep, A. (1985). Speciation in relation to edaphic factors in the *Asplenium adiantum-nigrum* group. *Proceedings of the Royal Society of Edinburgh*, **86B**, 325–34.

Sleep, A. & Reichstein, T. (1967). Der Farnbastard *Polystichum × meyeri* hybr. nov. = *Polystichum braunii* (Spenner) Fée × *P. lonchitis* (L.) Roth und seine Cytologie. *Bauhinia*, **3**, 299–374.

Sleep, A., Roberts, R. H., Souter, J. I. & Stirling, A. McG. (1978). Further investigations on *Asplenium cuneifolium* in the British Isles. *Fern Gazette*, **77**, 345–8.

Sleep, A. & Ryan, P. (1972). The Guernsey Spleenwort – a new fern hybrid. *Report and Transactions. Société Guernésiaise*, **19**, 212–24.

Sleep, A. & Synnott, D. M. (1972). *Polystichum × illyricum* in a hybrid new to the British Isles. *British Fern Gazette*, **10**, 281–2.

Smith, A. G. (1970). The influence of mesolithic and neolithic man on British vegetation: a discussion. In *Studies on the Vegetational History of the British Isles*, ed. D. Walker & R. G. West, pp. 81–90. Cambridge: Cambridge University Press.

Smith, R. T. & Taylor, J. A. (eds.) (1994). *Bracken: An Environmental Issue*. Aberystwyth: International Bracken Group.

Soltis, D. E., Haufler, C. H., Darrow, D. C. & Gastony, G. J. (1983). Starch gel electrophoresis of ferns: a compilation of grinding buffers, gel and electrode buffers and staining schedules. *American Fern Journal*, **73**, 9–27.

Sommerville, A. (1977). *A Guide to Biological Recording in Scotland*. Edinburgh: Biological Recording in Scotland Committee, Scottish Wildlife Trust.

Southwood, T. R. E. (1977). Habitat, the template for ecological strategies? *Journal of Animal Ecology*, **46**, 337–65.

Sowerby, J. E. (1866). *English Botany*, vol. **12**, *Cryptogamia*. London.

Sporne, K. R. (1962). *The Morphology of Pteridophytes*. London: Hutchinson University Library.

Stace, C. A. (1975). *Hybridisation and the Flora of the British Isles*. London: Academic Press.

Stansfield, F. W. (1916). *Lastrea dilatata* and its allies. *British Fern Gazette*, **3**, 104–11.

Stansfield, F. W. (1923). *Asplenium lanceolatum microdon*. *British Fern Gazette*, **5**, 14–17, 32–5.

Stansfield, F. W. (1924). *Asplenium lanceolatum microdon* and *A. adiantum-nigrum microdon*. *British Fern Gazette*, **5**, 77–9.

Stansfield, F. W. (1927). *Asplenium trichomanes* and its varieties. *British Fern Gazette*, **5**, 174–7.

Stansfield, F. W. (1931). Experiments in the propagation from spores of hybrid ferns. *British Fern Gazette*, **6**, 86–8.

Stansfield, F. W. (1934). '*Lastrea dilatata* Boydii' (F. W. S.) and *Nephrodium subalpinum* Borbasio. *British Fern Gazette*, **6**, 281–3.

Stark, G. (1991). Variation in *Equisetum variegatum* (Equisetaceae: Pteridophyta). *Fern Gazette*, **14**, 1–4.

Stirling, A. McG. (1972). *Polypodium australe* Fée in the Edinburgh area. *Transactions and Proceedings of the Botanical Society of Edinburgh*, **41**, 549–51.

Stirling, A. McG. (1974). A fern new to Scotland – *Polystichum × illyricum* in West Sutherland. *Watsonia*, **10**, 231.

Stokoe, R. (1978). *Isoetes echinospora* Durieu new to Northern England. *Watsonia*, **12**, 51–2.

Synnott, D. M. (1970). Evidence of *Polypodium australe* Fée in Scotland. *Transactions and Proceedings of the Botanical Society of Edinburgh*, **40**, 623–4.

Synnott, D. & Baird, H. (1980). *Ferns of Ireland*. Dublin: Irish Environment Library Series, No. 68.

Tate, G. R. (1853). *Asplenium germanicum* etc. at Kyloe, Northumberland. *Phytologist*, **4**, 909–10.

Taylor, J. A. (1985). The relationship between land-use change and variations in bracken encroachment rates in Britain. In *The Biogeographical Impact of Land Use Changes*, ed. R. T. Smith, pp. 19–28. Norwich: BSB/Geo Books.

Taylor, J. A. (1986). The Bracken problem: a local hazard and a global issue. In *Bracken. Ecology, Land Use and Control Technology*, ed. R. T. Smith & J. A. Taylor, pp. 21–42. Carnforth: Parthenon Press.

Taylor, J. A. (1990). The bracken problem: a global perspective. In *Bracken Biology and Management*, ed. J. A. Thomson & R. T. Smith, pp. 3–19. Canberra: Australian Institute of Agricultural Science.

Taylor, J. A. (ed.) (1991). *Bracken Toxicity and Carcinogenicity as Related to Animal and Human Health*. Aberystwyth: International Bracken Group.

Taylor, J. A. (1994). Coming to terms with the Bracken problem. In *Bracken: An Environmental Issue*, ed. R. T. Smith & J. A. Taylor, pp. 1–11. Aberystwyth: International Bracken Group.

Tennant, D. J. (1966). *Cystopteris dickieana* R. Sim in the central and eastern Scottish Highlands. *Watsonia*, **21**, 135–9.

Tennant, D. J. (1995). *Cystopteris fragilis* var. *alpina* Hook. in Britain. *The Naturalist*, **120**, 45–50.

Thomas, B. A. & Spicer, R. A. (1987). *The Evolution and Palaeobiology of Land Plants*. London: Croom Helm.

Thomas, D. W. (1975). Wild gametophytes of *Diphasium alpinus* (L.) Rothm. in North Wales. *Watsonia*, **10**, 277–9.

Thompson, H. S. (1939). Brake Fern on Bristol walls. *Journal of Botany, British and Foreign*, **77**, 218–19.

Thomson, J. A. & Smith, R. T. (eds.) (1990). *Bracken Biology and Management*. Canberra: Australian Institute of Agricultural Science.

Thomson, J. A., Tan, M. K. & Weston, P. H. (1995). Principal lineages amongst the bracken ferns worldwide: a molecular analysis. In *Bracken: An Environmental Issue*, ed. R. T. Smith & J. A. Taylor, pp. 21–8. Aberystwyth: International Bracken Group.

Thomson, J. A., Willoughby, C. & Shearer, C. M. (1986). Factors affecting the distribution, abundance and economic status of bracken (*Pteridium esculutum*) in New South Wales. In *Bracken. Ecology, Land use and Control Technology*, ed. R. T. Smith & J. A. Taylor, pp. 109–19. Leeds: Parthenon Publishing.

Townsend, F. (1864). Contributions to a flora of the Scilly Isles. *Journal of Botany, British and Foreign*, **2**, 102–20.

Tryon, A. F. (1980). Classifications, spores and nomenclature of the Marsh Fern *Rhodora*, **82**, 461–74.

Tryon, R. M. & Tryon, A. F. (1982). *Ferns and Allied Plants, with Special Reference to Tropical America*. New York, Heidelberg, Berlin, Springer-Verlag.

Turner, J. (1964). The anthropogenic factor in vegetational history. I. Tregaron and Whixall Mosses. *New Phytologist*, **63**, 73–90.

Turner, J. (1965). A contribution to the history of forest clearance. *Proceedings of the Royal Society*, **B161**, 343–54.

Turner, J. (1970). Post-neolithic disturbance of British vegetation. In *Studies in the Vegetational History of the British Isles*, ed. D. Walker & R. G. West, pp. 98–116. Cambridge: Cambridge University Press.

Turner, J. S. & Watt, A. S. (1939). The oakwoods (*Quercetum sessiliflorae*) of Killarney, Ireland. *Journal of Ecology*, **27**, 202–33.

Vida, G. (1969). Tetraploid *Dryopteris villarii*. *Botanikai Közlemények*, **56**, 11–15.

Vida, G. (1970). The nature of polyploidy in *Asplenium ruta-muraria* L. and *A. lepidium* C. Presl. *Caryologia*, **23**, 525–47.

Vida, G. (1972). Cytotaxonomy and genome analysis of the European ferns. In *Evolution in Plants*, pp. 51–60. Budapest: Akademiai Kiado.

Vida, G. (1974). Genome analysis of the European *Cystopteris fragilis* complex. *Acta Biologica Academiae Scientiarum Hungaricae*, **20**, 181–92.

Vida, G. & Reichstein, T. (1975). Taxonomic problems in the fern genus *Polystichum* by hybridisation. In *European Floristic and Taxonomic Studies*, ed. S. M. Walker & C. J. King, pp. 126–35. London: The Botanical Society of the British Isles.

Villalobos-Salazar, J., Meneses, A. & Salas, J. (1990). Carcinogenic effects in mice of milk from cows fed on bracken fern *Pteridium aquilinum*. In *Bracken Biology and Management*, ed. J. A. Thomson & R. T. Smith, pp. 247–51. Canberra: Australian Institute of Agricultural Science.

Villalobos-Salazar, J., Mora, J., Meneses, A. & Pashov, B. (1994). The carcinogenic effects of Bracken spores. In *Bracken: An Environmental Issue*, ed. R. T. Smith & J. A. Taylor, pp. 102–3. Aberystwyth: International Bracken Group.

Vogel, J. C. (1997). Conservation status and distribution of two serpentine restricted *Asplenium* species in central Europe. In *Pteridology in Perspective*, ed. J. Camus, M. Gibby & R. J. Johns, pp. 187–8. London: Royal Botanic Gardens, Kew.

Vogel, J. C., Russell, S. J., Barrett, J. A. & Gibby, M. (1997). A non-coding region of chloroplast DNA as a tool to investigate reticulate evolution in European *Asplenium*. In *Pteridology in Perspective*, ed. J. Camus, M. Gibby & R. J. Johns, pp. 313–27. London: Royal Botanic Gardens, Kew.

Von Travel, F. (1937). *Dryopteris borreri* Newm. und ihr Formenkreis. *Verhandlungen der Schweizerischen Naturforschenden Gesellschaft*, 153–4.

Wagner, W. H. (1954). Reticulate evolution in the Appalachian aspleniums. *Evolution*, **8**, 103–8.

Wagner, W. H., Jr (1969). The role and taxonomic treatment of hybrids. *BioScience*, **19**, 785–9.

Wagner, W. H., Jr, Wagner, F. S. & Beitel, J. M. (1985). Evidence for interspecific hybridisation in pteridophytes with subterranean mycoparasitic gametophyes. *Proceedings of the Royal Society of Edinburgh*, **86B** 273–81.

Walters, M. (1984). The relation between British and European floras. In *The Flora and Vegetation of Britain. Origins and Changes – The Facts and their Interpretation*, ed. J. L. Harley & D. H. Lewis, pp. 3–13. London: Academic Press.

Walker, S. (1909). Identification of a diploid ancestral genome in the *Dryopteris spinulosa* complex. *British Fern Gazette*, **10**, 97–9.

Walker, S. (1961). *Dryopteris assimilis*. *American Journal of Botany*, **48**, 607.

Walker, S. (1955). Cytogenetic studies in the *Dryopteris spinulosa* complex. I. *Watsonia*, **3**, 193–209.

Walker, S. (1961). Cytogenetic studies in the *Dryopteris spinulosa* complex. II. *American Journal of Botany*, **48**, 607–14.

Watt, A. S. (1947). Pattern and process in the plant community. *Journal of Ecology*, **43**, 490–506.

Webb, D. A. (1983). The flora of Ireland in its European context. *Journal of Life Science of the Royal Dublin Society*, **4**, 143–60.

Webb, D. A. & Scannell, M. J. P. (1983). *Flora of Connemara and the Burren*. Cambridge: Cambridge University Press.

Wells, A. J. & McNally, R. (1994). An appraisal of the spatial association of Bracken and cancer in England and Wales. In *Bracken: An Environmental Issue*, ed. R. T. Smith & J. A. Taylor, pp. 104–9. Aberystwyth: International Bracken Group.

Westwood, M. R. I. (1989). An aberrant form of *Equisetum telmateia* (Pteridophyta) from the west of Ireland. *Fern Gazette*, **13**, 277–81.

Wheldon, J. A. & Wilson, A. (1907). *Flora of West Lancashire*. Liverpool: Henry Young & Sons.

White, D. J. B. (1961). Some observations on the vegetation of Blakeney Point, Norfolk, following the disappearance of the rabbits in 1954. *Journal of Ecology*, **49**, 113–18.

White, D. J. B., White, M. F. & Peterken, G. F. (1970). *Polypodium* on Blakeney Point, Norfolk. *Transactions of the Norfolk and Norwich Naturalists' Society*, **21**, 373–7.

Whyte, J. H. (1930). The spread of bracken by spores. *Transactions and Proceedings of the Botanical Society of Edinburgh*, **35**, 209–11.

Wigston, D. J. B., Pickering, D. & Jones, S. (1981). *Lycopodiella inundata* (L.) Holub at Smallhanger, South Devon. *Watsonia*, **13**, 325–6.

Wigston, S. L. (1929). *Lycopodiella inundata* (L.) Holub at Fox Tor Mires, South Devon. *Watsonia*, **2**, 343–4.

Wilce, J. H. (1965). Section *Complanata* of the genus *Lycopodium*. *Beiheft zur Nova Hedwigia, Weinheim*, **19**, 1–233.

Willmot, A. (1977). A pteridophyte flora of the Derbyshire Dales National Nature Reserve. *Fern Gazette*, **11**, 279–84.

Willmot, A. (1979). An ecological survey of the ferns of The Burren, Co. Clare, Eire. *Fern Gazette*, **12**, 9–28.

Willmot, A. J. (1981). An ecological survey of the ferns of Berwickshire, Scotland. *Fern Gazette*, **12**, 133–54.

Willmot, A. J. (1983). An ecological survey of the ferns of the Killarney district, Co. Kerry, Ireland. *Fern Gazette*, **12**, 249–365.

Willmot, A. J. (1985). Population dynamics of *Dryopteris* in Britain. *Proceedings of the Royal Society of Edinburgh*, **86B**, 307–13.

Willmot, A. J. (1989). The phenology of leaf life spans in woodland and populations of the ferns *Dryopteris filix-mas* (L.) Schott and *D. dilatata* (Hoffm.) A. Gray in Derbyshire. *Biological Journal of the Linnean Society*, **99**, 387–95.

Wilson, A. (1956). *The Altitudinal Range of British Plants*, 2nd edn. Arbroath: Buncle.

Wolf, P. G., Haufler, C. H. & Sheffield, E. (1987). Electrophoretic evidence for genetic diploidy in the Bracken Fern (*Pteridium aquilinum*). *Science*, **236**, 947–9.

Wolf, P. G., Haufler, C. H. & Sheffield, E. (1988a). Maintenance of genetic variation in the clonal weed *Pteridium aquilinum* (Bracken). *Bulletin of Biochemistry and Biotechnology*, **1**, 46–50.

Wolf, P. G., Haufler, C. H. & Sheffield, E. (1988b). Electrophoretic variation and mating system of the clonal weed *Pteridium aquilinum* (L. Kuhn) (Bracken). *Evolution*, **42**, 1350–5.

Wolf, P. G., Sheffield, E., Thomson, J. A. & Sinclair, R. B. (1995). Bracken taxa in Britain: a molecular analysis. In *Bracken: An Environmental Issue*, ed. R. T. Smith & J. A. Taylor, pp. 16–20. Aberystwyth: International Bracken Group.

Wolf, P. G., Sheffield, E. & Haufler, C. H. (1991). Estimates of gene flow, genetic substructure and population heterogeneity in bracken (*Pteridium aquilinum*). *Biological Journal of the Linnean Society*, **42**, 407–23.

Wollaston, G. B. (1875). Three species of *Lastrea filix-mas*. *British Fern Gazette*, **3**, 20–4.

Wollenweber, E. (1985). Flavenoids as biochemical markers in fern taxonomy. *Proceedings of the Royal Society of Edinburgh*, **86B**, 469–70.

Young, J. E. (1985). Some effects of temperature on germination and protonemal growth in *Asplenium ruta-muraria* and *A. trichomanes*. *Proceedings of the Royal Society of Edinburgh*, **86B**, 454–5.

Zu, L. R. (1992). Bracken poisoning and enzootic haematuria in cattle in China. *Research in Veterinary Science*, **53**, 116–21.

Index

Note on synonymous botanical names: For convenience of usage, all commonly used (and synonyms of) botanical names of pteridophytes not used or listed in the body of this work are nevertheless included within this index and refer to the main page entries of the plant to which they relate.